Philosophy of Mathematics

Textbooks in Mathematics

Series editors:
Al Boggess, Kenneth H. Rosen

https://www.routledge.com/Textbooks-in-Mathematics/book-series/CANDHTEX-BOOMTH

Philosophy of Mathematics

Classic and Contemporary Studies

Ahmet Çevik

CRC Press
Taylor & Francis Group
Boca Raton London New York

CRC Press is an imprint of the
Taylor & Francis Group, an **informa** business

A CHAPMAN & HALL BOOK

First edition published 2022
by CRC Press
6000 Broken Sound Parkway NW, Suite 300, Boca Raton, FL 33487-2742

and by CRC Press
2 Park Square, Milton Park, Abingdon, Oxon, OX14 4RN

ISBN: 9781032121284 (hbk)
ISBN: 9781032022680 (pbk)
ISBN: 9781003223191 (ebk)

DOI: 10.1201/9781003223191

Typeset in CMR
by KnowledgeWorks Global Ltd.

Contents

Preface

Philosophy of mathematics is an exciting subject studied by a small number of philosophers today and even less by mathematicians. It is strongly related to logic and foundations of mathematics. The Golden Age of the foundations of mathematics began with the foundational crisis, which is usually considered to be the period between the late 19th century and mid-20th century. Unfortunately, the foundations and philosophy of mathematics is not receiving the attention it deserves by the mathematical community. I wrote this book with the hope of bringing *back* this intriguing subject to the attention of mathematical community to rekindle an interest in philosophical subjects surrounding the foundations of mathematics and introduce various philosophical positions ranging from the classic views to more contemporary ones, including those which are more mathematical oriented. Ideally, as an outcome, I am hoping to engage all philosophically minded mathematicians in philosophical debates and foundational discussions. Another purpose in writing this book is to contribute to the philosophical literature from the perspective of a mathematician and encourage like-minded scholars to make similar contributions. I hope that this book motivates mathematicians to argue about the foundations by getting involved in the trending philosophical discussions and to collaborate with philosophers, as this was happening in the Golden Age of the foundations of mathematics.

Intended Audience

This book is primarily for upper division undergraduate and graduate students in mathematics or philosophy. Students in theoretical computer science can also benefit from this material. I should emphasise, however, that it is particularly aimed for philosophically minded mathematicians due to the selected content and the presentation style. I would like to encourage young mathematicians into thinking about the philosophical issues behind fundamental concepts in mathematics and about different views one can have regarding mathematical objects and mathematical knowledge. It is important to know how an adopted philosophical view may dramatically affect the mathematical practice. So a course in philosophy of mathematics may help the reader realise how the methodology for mathematical practice changes in accordance with the supported view. It also provides the philosophical background of many basic notions used in mathematics, such as the concepts of set, number, space, proof, computation, and so on. Philosophers, *ipso facto*, question all kinds of philosophical problems about mathematics more often than mathematicians do. For this reason, I believe that philosophers know more about the nature of mathematics—particularly its relation to other branches of philosophy—whereas mathematicians know more about the practice. So reading about the philosophy of mathematics might do a *red pill* effect—reference to the movie *The Matrix*—on mathematicians about the nature of mathematics.

Philosophy and mathematics have entirely different methodologies. Traditionally, if two people were to argue about the truth or falsity of an ordinary mathematical statement, only one of them would have to be correct. This is not, however, usually the case in philosophy. I have given lectures on philosophy of mathematics to both philosophy and mathematics students, and as a result, I came to the conclusion that mathematicians tend to demand a definite description of the philosophical terms being used, such as truth, existence, universe, knowledge, intuition, etc. They want to know outright the definition of these metaphysical concepts whenever I get to mention them in the lectures, where, in fact, defining what they are *is* the main goal of the entire course. I found their urge perfectly normal as they have been taught to take statements in the binary "must be either this or that" sense. Yet philosophical problems are not always supposed to be solved, settled, or finalised with a definite solution. One of the big philosophical questions is whether or not mathematical objects exist. This question may not and does not need to have a single answer, but it may have possibly many answers based on different views. It is essential to keep in mind that the aim of philosophy of mathematics *is* to determine and argue about the criteria for existence, truth, views of attaining mathematical knowledge, and many others. The reader will notice when reading this material that, in fact, there are many different viewpoints for these concepts, each of which establishes a distinct philosophical position. Mathematicians need to understand, before starting to read this material, that none of these philosophical views is regarded as the absolute *correct* view, for there may be no correct view and that there may be no universal definition of "correctness" or the concepts mentioned above. So, I hope to guide young mathematicians in looking at a philosophical problem of mathematics from different perspectives and let them decide or even define their own philosophy of mathematics, and practice mathematics in accordance with the philosophical position they support.

How to Read This Book

The core material of this volume is from my book [73] published in Turkish, which is based on my lecture notes for the philosophy of mathematics course that I taught at the Middle East Technical University in Turkey for many years. However, I later added many new chapters, subsections, new figures, and many references. In the end, it became twice as large as my earlier work. This material, thus, can be used as a textbook for a one-semester or even one-year course on philosophy of mathematics.

I always suggest the reader pay attention to the footnotes, as many of them explain the concepts used in the text in more detail. This will help the reader understand the subject matter in a much better way. The book contains 18 chapters throughout which I occasionally give my personal views on the related issue. All references are numbered uniquely in the bibliography where articles are specified in italic, books in bold. When a reference is cited for the first time in the text or in the footnote, the reference number is also given. All figures are drawn in the eps vector format by myself with GeoGebra 5. Sections with a more sophisticated mathematical content are marked with asterisk symbol * in the Table of Contents. Philosophy students may feel free to skip these sections, or at least omit the proofs of the theorems. I strongly recommend the reader to start with Chapter 1 for the fact that it overviews the subject matter of philosophy of mathematics. It will also assist the reader to select specific chapters for more careful review, depending on their level of interest.

I assume that the reader is familiar with basic axiomatic set theory. For completeness though, in Chapter 2, Mathematical Preliminaries, I give a summary of propositional and predicate logic, methods of proof, and the necessary background needed for set theory. As the chapter title suggests, Chapter 2 can be considered as the mathematical preliminaries for the book. Readers who are comfortable with predicate logic and set theory may skip this chapter. Subjects covered in Chapter 2 are actually taught as a one-semester course to undergraduate students in the mathematics departments in most universities. Readers who would like to refresh their knowledge in predicate logic and set theory may refer to Chapter 2 when necessary.

I first cover the four classical views in Chapters 3 through 6. In Chapter 3, I explain the philosophical position called *Platonism*, originated from Plato's ideas, as the reader may have guessed. The term Platonism, in some texts, refers to Plato's ancient philosophy in the broader sense. I use the general term *realism* to refer to the common view for modern versions of Plato's philosophy regarding existence and reality. In the same chapter, I also reflect on Aristotle, who planted the seeds of what we will later call *empiricism*.

Chapter 4 concerns another classical view called *intuitionism*. The origins of intuitionism go back to Immanuel Kant. I will explain Kant's classification of knowledge, and then look at Brouwer's *constructivism*.

Chapter 5 investigates *logicism*, a view which was predominantly defended by the great logician Gottlob Frege and led by philosophers such as Bertrand Russell, Alfred North Whitehead, and Rudolf Carnap. Logicism, broadly speaking, claims that mathematics can be reduced to pure logic, and that all mathematical entities can be described solely by using logical concepts and the logical language. In the same chapter, I also look at Russell's type theory for solving the foundational problems appeared in the beginning of the 20th century.

Chapter 6 concerns *formalism* and Hilbert's *formalisation programme*, a project initiated by David Hilbert in the early 20th century. Formalism is a view in the philosophy of mathematics which abandons semantics. In fact, it argues that there is no meaning in mathematics, that mathematical activity merely consists of symbol manipulation rules, and that mathematics is just a symbolic game with no obligation to assign any meaning to the symbols.

Chapter 7 studies a seminal result in mathematical logic and foundations of mathematics. I state and prove Gödel's Incompleteness Theorems. Proving the theorem will first require us to define the notion of primitive recursive functions and the concept of recursive enumerability. In the same chapter, I also refer to mathematical concepts given in Chapter 2. Gödel's theorems tell us that Hilbert's formalisation programme could not be entirely achieved for consistent formal systems which satisfy a desired amount of expressive power.[1] I then discuss the reasons behind the incompleteness phenomenon from the viewpoint of algorithmic information theory.

In Chapter 8, I argue about the definition of algorithmic computability. In modern mathematics, the definition of computability is relied on a philosophical statement called the *Church-Turing Thesis*. Chapter 8 is reserved for the study of this thesis and its alternatives. I may, in fact, consider Chapters 6 through 8 as interconnected topics.

Chapter 9 gives a historical account of the philosophy of infinity. I begin from the ancient Greek period, move on with the problems that appeared in the Middle Ages, and then conclude the discussion with Cantor's set theory and his discovery of different sizes of infinity.

In Chapter 10, I look at *supertasks* (sometimes known as *hypercomputation* or *transfinite computation*) from philosophical and mathematical point of views. The reader may want to review, before starting to read this chapter, some set theoretical concepts such as ordinal and cardinal numbers, whose summary is given in Chapter 2. It is important to be familiar with the notions introduced in Chapter 7. Although the theorems given in Chapter 10 may look rather technical to some readers, it would be crucial to review ordinal number arithmetic beforehand.

Chapter 11 is another section which is particularly of special interest to mathematicians. I examine the notion of models, truth in a model, and prove some fundamental theorems of mathematical logic such as the compactness theorem, Gödel's Completeness Theorem, and the Löwenheim-Skolem Theorem. I will also discuss Skolem's paradox, an interesting antinomy of the Löwenheim-Skolem Theorem.

Chapter 12 studies one of the formerly controversial axioms of set theory, at least among the constructivists, called the *Axiom of Choice*. The axiom is widely used today, mostly indirectly, in the proof of many useful theorems of mathematics. I look at the functionality of the Axiom of Choice and try to explain with different scenarios what this axiom really does. I argue that based on what grounds this axiom should be accepted (or rejected). For this, I will try to answer the question that under which conditions a mathematical axiom could be accepted as an "intrinsically plausible" statement. At the end of the chapter, I give the Axiom of Determinacy, a well-known alternative

---

[1] I will explain what I mean by these terms.

to the Axiom of Choice. Axiom of Determinacy states that for every two-player infinite game, there exists a winning strategy for some player. I will give the proof of the fact that the Axiom of Determinacy contradicts the Axiom of Choice. We will make use of ordinals, upper, and lower bounds in this subsection.

In Chapter 13, I review Quine and Maddy's *naturalism*. Naturalism abandons first philosophy and allows mathematical practice to act within itself without relying on extra-mathematical assumptions. The aim of Chapter 13 is also to study Quine's, Gödel's, and Maddy's versions of realism, and discuss the ontological and epistemological relationship between natural sciences and mathematics. I also discuss in the same chapter that on what naturalistic ground the axioms of set theory are true.

Chapter 14 is about the view called *structuralism*, defended by Benacerraf, Shapiro, Resnik, *et al.* A natural number may have more than one set-theoretical representation. But then which one of these representations is the "real" set-theoretic description of numbers? This is not an easy question to answer. Consequently, structuralism argues for that we should instead consider the abstract structure of natural numbers even though there may be many exemplifications. According to structuralism, mathematics is the study of the relationship between structures. Mainly, none of the objects of a structure can individually exist independent of other objects in the same structure.

Chapter 15 concerns a paradox which does not use self-reference or circularity in the traditional sense. The antinomy is called *Yablo's paradox* in the logical literature. Any statement of the kind "This sentence is false" admits self-reference. A typical property of many of the logical paradoxes is that they rely on the notion of circularity and self-reference. However, it was claimed that Yablo's paradox was shown to be immune to this property. I also argue that Yablo's paradox is rather "circular" in the limit sense. For this, I will again use the notions provided in Chapter 2.

Chapter 16 covers a relatively new philosophical position called *pluralism*. Mathematical pluralism is the view which argues for the plurality of mathematical universes, each of which can be considered as "legitimate" on its own right. Here I did not hide my personal view on pluralism, where I criticised its radical forms and discussed about a possible solution in favour of reconciling pluralism and realism.

In Chapter 17, I discuss the debate on whether or not mathematics needs new axioms. The question is related to the problem of understanding the set-theoretical universe and of proving independent statements not known to be solvable within the current system of set theory. I discuss the attempt of set theorists to enlarge the constructible universe so as to obtain an ultimate expansion of the constructible hierarchy in which all large cardinal axioms hold.

Chapter 18 is devoted to mathematical nominalism, a rival to realism and structuralism. On nominalism, abstract objects, in particular mathematical entities, do not exist. Although there are different forms of nominalism, I largely focus on a position called *fictionalism*, advocated by Hartry Field and revisited by many philosophers. We look at Balaguer's account on fictionalism. We also summarise the critique by Burgess and Rosen, and then give a few more contemporary positions in nominalism.

Acknowledgements. I would like to thank Mark Balaguer, Mark Colyvan, Hartry Field, Michèle Friend, Joel David Hamkins, Penelope Maddy, Antonio Montalbán, and Ken Ross for their comments and many useful suggestions. Special thanks to Bob Ross, senior editor at CRC Press, for his patience and cooperation. I would like to thank the Gendarmerie and Coast Guard Academy of Turkey for allowing me to offer this course elsewhere, which led to the preparation of this book. I also owe a debt of gratitude to Ali Nesin for his kind permission to include some of the core material that appeared in my earlier book. Last, but certainly not least, I am grateful to my parents who have always supported and encouraged me in my life and made me become the person I am today.

Ahmet Çevik
Ankara

1

Introduction

The word *mathematics* originates from the ancient Greek word μάθημα (*máthēma*) which means "that to be learned", "that to be studied", and "knowledge". Despite that, the term mathematics has been used with different meanings throughout history, contrary to the meaning used in ancient Greece; it is often said that for John von Neumann, as stated in Zukav [311], mathematics is not something to be learned or understood but rather something to get used to.[1] The more we get used to it, the more we understand it and the more we recognise the mathematical truth. For Bertrand Russell [247], mathematics may be defined as 'the subject in which we never know what we are talking about, nor whether what we are saying is true'.[2]

Mathematics is the name of the discipline which studies quantity, change, structure, and space. In the modern day, there is a separation made between *pure mathematics* and *applied mathematics*. Pure mathematics concerns the study of mathematical concepts independent of any real-life applications. For Russell, pure mathematical studies conducted before the 19th century were not properly distinguished from applied mathematics, and so he claims that pure mathematics was rather discovered by George Boole in his *Laws of Thought* [38].[3] Applied mathematics, on the other hand, studies mathematical models of physical phenomena and applications of mathematical methods in science, engineering, and the industry. Subjects such as analysis, abstract algebra, topology, number theory, and logic are some of the examples of the study matter of pure mathematics. Probability theory, cryptology, financial mathematics, game theory, social choice theory, and similar areas are the branches of applied mathematics. For our purpose, in this book, we study the philosophy of pure mathematics rather than problems of applied mathematics. Apart from the philosophical discussions, we will also give some fundamental theorems of mathematical logic which are relevant to the philosophy of mathematics. Mathematical logic, to put simply, is the mathematical study of logic. Mathematical logic and foundations is divided into four pillars: Set theory, model theory, computability theory (recursion theory), and proof theory.[4]

---

[1] Zukav, p. 208, 1979.

[2] Russell, p. 84, 1901.

[3] ibid, p. 83.

[4] Computability theory is traditionally known as recursion theory. Since the late 20th century, especially in the European continent, the term "computable" has, in fact, been used more often than the term "recursive".

DOI: 10.1201/9781003223191-1

According to Russell [248], *pure mathematics* consists of statements of the form "If p, then q" for two propositions p and q.[5] Modern mathematics relies on what we call the *axiomatic method*.[6] The usage of the axiomatic method goes as far back as to Euclid's *Elements* in the ancient Greek period. In the *Elements*, Euclid proves geometrical propositions based on the undefined intuitive notions of "point", "line", and on other postulates which we call the *axioms*. This systematic discourse of establishing the truth or falsity of a claimed proposition based on a given set of assumptions determines the methodology of modern mathematics, namely the axiomatic method.

Mathematics is ultimately grounded on definitions and axioms. Stephen Cole Kleene [164] mentions two kinds of axiomatics: material axiomatics and formal axiomatics.[7] According to Kleene, axiomatised systems wherein the objects of the system known prior to the axioms are distinguished as *material axiomatics*. Axiomatised systems in which the mathematical entities are suppose to model or denote some physical phenomena are all instances of material axiomatics. Euclid's geometry is an example of material axiomatics, since the physical space is supposed to be modelled by a three-dimensional coordinate system such that each point refers to a "position" in the physical space. In this case, the notion of "point" in the coordinate system has a corresponding meaning in the physical system. Formal axiomatics, on the other hand, is an axiomatised system wherein the axioms are known prior to any specification of the system of objects which the axioms are about.[8] A *definition* is a clear

---

[5] Russell, p. 3, 1903.

[6] The name *modern mathematics* is also known as *abstract mathematics* and it refers to the foundational system of mathematics which was axiomatised in the beginning of the 20th century. We shall use these two names interchangably.

[7] Kleene, p. 28, 1950.

[8] An interesting question that can be raised about the relationship between axioms and definitions concerns their order of priority in the sense that which comes before the other. My personal view is that, to a certain degree, we should beforehand reach to an agreement on the "meaning" of the non-logical symbols or concepts contained in an axiom. This meaning is acquired directly by means of empirical evidence in Kleene's material axiomatics. For instance, the meaning of the notion of "point" in a three-dimensional Euclidean space is derived from perceiving the physical environment as a three-dimensional space consisting of sufficiently dense particles so that each particle is represented by a "point" in the Euclidean space. In formal axiomatics, the mathematical entities are constructed from semantically undefined objects, e.g., axioms of a formal symbolic system. For instance, the concept of *membership* in set theory is mathematically undefined within the theory. Mathematicians of course agree on what "membership" intuitively means. The agreement is simply made by common sense relying on the empirical phenomenon of *containment*. Personally, I believe it is hard to neglect the need for a sufficient amount of semantics even within the bold versions of formal axiomatics for the simple reason that, regarding set theory for instance, we neither interpret the "membership" $\in$ symbol arbitrarily nor do we write arbitrary formal axioms. We want the axioms in the first place to tell something about sets, in the way we perceive them. We state the axioms relying on the "standard interpretation" of the objects of the system. For example, consider the axiom "$x + 0 = x$" that we know from elementary arithmetic. Even treated formally, the axiom is grounded on the standard interpretation of the natural number 0 and its identity property under addition. My personal view is in favour of the existence of a "standard interpretation". Studying the consequences of the existence, or otherwise, of a standard interpretation is not the aim of this chapter; we

and unambiguous description of a mathematical entity, usually based on previously defined objects.[9] Suppes [276] gives the conditions for a definition to be traditionally considered as "legitimate" as follows[10]:

1. A definition must give the essence of that which is to be defined.

2. A definition must not be circular.

3. A definition must not be in the negative when it can be in the positive.[11]

4. A definition must not be expressed in figurative or obscure language.

Suppes gives a well-defined theory of definitions, but unfortunately studying this subject in detail is beyond the scope of our book.[12]

We will refer to the initial assumptions of a mathematical system as *axioms*. In (pure) mathematical disciplines, we derive conclusions from the axioms using the rules of inference. On the basis of the axioms, we claim certain statements that we wish to prove or disprove. But then what is a *proof*? Roughly speaking, a *proof* is a sequence of statements $S_0, S_1, S_2, \ldots, S_k$, where i and k are natural numbers such that $i \leq k$ and that either every S_i is an axiom or, for every $j < i$, each S_i is derived logically from at least one S_j. The final statement of a proof is called a *theorem*. Sometimes we may need to use a previously established statement in the proof of the theorem as an auxiliary statement. Such a statement is called *lemma*. The immediate consequence of a theorem is called a *corollary*. In some cases, we claim a statement and we believe that it is most likely to be true without presenting any proof. These kinds of statements are called *conjectures*. So a conjecture is a statement which we *believe* to be true, yet we provide no formal evidence for it.[13]

Mathematicians demand proof. Every proof ultimately rests on axioms and definitions. However, in disciplines which use the axiomatic method, it is not possible to define everything. For instance, concepts such as "point" and "line" are accepted as undefined terms in Euclidean geometry, just as concepts like "set" and "membership" are undefined in set theory. The meaning of these concepts is acquired through common sense and intuition that all or at least the majority of humans have. Hence, mathematics begins with undefinable

---

shall leave the discussion for Chapter 16. For these reasons, I defend the motto "semantics prior to axioms", meaning that the axioms are determined by how we interpret (under the standard interpretation) the objects in the system. The definitions can be seen as acronyms or synonyms. We encourage the reader to argue and think about the order between the definitions and axioms.

[9]cf. Russell [248] (1903), p. 429. Also cf. Quine [229] (1936), p. 329. Page reference of Quine (1936) to the reprint version in Benacerraf and Putnam (1983).

[10]Suppes, p. 151, 1957.

[11]In other words, a definition must not describe what the term is *not*, but must describe what it *is*.

[12]For a detailed account, we refer the reader to Suppes [276], §8. See also Brown [46] (2008), §7, for further discussion.

[13]By formal evidence we mean a mathematical proof.

terms. We tend to agree on the intuitive meaning of these terms. Common sense tells us that the concept of set membership is interpreted as *containment* in a collection. Since mathematics is built on the undefinables, it is *atomic* in the sense that it should be based on objects that cannot be further defined other than what they are. We cannot do mathematics if there is no consensus on the meaning of undefinable terms. In fact, any discipline which uses the axiomatic method, not just mathematics, relies on this principle. The nature of the axioms and definitions is usually articulated by philosophers and equally by mathematicians who are interested in the foundations of mathematics.

It is often said that mathematics is "beautiful". But what do we mean when we say a mathematical work or a proof is beautiful? How do we assess the quality of a mathematical work in the first place? This usually depends on various factors. Broadly speaking, some of the criteria for writing "good" mathematics can be listed as follows:

1. Internal *consistency* of the definitions, logical arguments, and most importantly *rigour*. The work should contain no ambiguities and it should leave both the author and the reader with no doubt in mind. A mathematical paper should be organised systematically. Definitions are expected to be given in *order* from the simplest to the most complex.

2. Consistency in the usage of the variable symbols. For example, if we denote the natural numbers by lowercase Latin letters such as i, j, k, l, m, n, then we do not want to denote the same objects, unless explicitly mentioned, by a different type of symbols like A, B, C or α, β, γ.

3. A good mathematical work is usually expected to be *self-contained*. That is, it should contain all the background and information that is needed for the understanding of the work by fellow mathematicians who are not familiar with that particular subject. Of course, on many occasions it may not be possible to give everything at once. But the less the reader feels the need to refer to other texts, the better it is.

4. The claims are expected to have "strong" implications. This could be achieved by proving logically "strong" theorems that cover all the general cases of a problem, or by presenting results that reduce a complex idea to a simpler result. When we say a statement is logically strong, this is not solely related to the formal structure of the statement. We call a statement *universal* if it starts with the universal quantifier "for all". A statement is called *existential* if it starts with the existential quantifier "there exists". For an existential statement of the form "there exists some x with property P" to be logically strong, it must be as specific as possible about the property P. The more information it gives about the property of the object x, the stronger the statement becomes. For universal statements, however, we usually want to have as less assumptions as possible in the hypothesis.

For example, saying that "every algebraist thinks analytically" is weaker than saying "every mathematician thinks analytically". Similarly, saying that "every philosopher is a wise person" is weaker than saying "every moral philosopher is a wise person". So the more information a statement gives, the stronger it is. The reason why we want to prefer less assumptions in universal statements is due to *Occam's Razor* which can be stated as follows.

Occam's Razor: Plurality is not to be posited without necessity.

Occam's Razor says that if there are two competing explanations of the same phenomenon, under the same circumstances, then the one with less assumptions is preferred over the one with more assumptions. In existential statements, saying that "there exists a set" is weaker than saying "there exists a non-empty closed Cantor set whose members satisfy the property P". Clearly the latter implies the former but not vice versa. So the latter is logically stronger. We often encounter similar situations in real life. In a social scenario, say, describing an unknown person as "one with curly hair" is less informative than describing the same person as "one with blonde curly hair and blue eyes". The latter description clearly extends the former. Hence, the latter description is stronger. In existential statements, we want to give as much information as possible about the asserted object. In universal statements, we want to have as less assumptions as possible in the hypothesis. So in mathematics, one criterion which determines the impact of the work is to look at the claims and see if they are logically strong. Nevertheless, one may also give logical equivalency results by showing that a previously shown theorem is logically equivalent to a seemingly more intuitive or an interesting statement.[14]

5. The proofs need to be perfect. A good mathematical paper would look "beautiful" in the sense that reading them would be as riveting as reading an exciting novel. Using neat language in proving the theorems and the creativity factor are major elements in writing good mathematics.[15] A nice proof may be short and elegant, or it may provide a surprisingly simple solution to a complicated problem. But the first and foremost requirement is that all proofs need to be sound. The aesthetic features are not prior to soundness. We may demonstrate this so-called "mathematical beauty" with an example from geometry. It should not be hard to see the creativity and elegance behind the proof of the Pythagorean theorem.

Pythagorean Theorem: Given a right triangle, the square of the length of the hypotenuse is equal to the summation of the squares of the length of the other two sides (see Figure 1.1).

---

[14]Logical equivalency in propositional logic will be discussed in Chapter 2.

[15]The great mathematician Paul Erdös was fond of the idea that there is a sacred volume which he referred to as *The Book* that he believed God kept inside the most elegant proofs of every mathematical theorem.

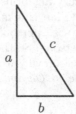

FIGURE 1.1
Pythagorean theorem tells us that $a^2 + b^2 = c^2$.

A geometric proof of the Pythagorean theorem can be given as follows:

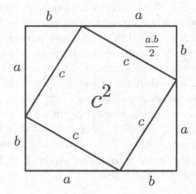

FIGURE 1.2
A geometric proof of the Pythagorean theorem.

As it can be seen from Figure 1.2, the summation of the areas of all four triangles and the area of the inner square with side c is equal to the area of the outer square with side $a + b$. Simplifying the equality, it follows that $c^2 = a^2 + b^2$.

The question that what constitutes mathematical rigour has been contemplated for a long time. As a matter of fact, Aristotle's *Posterior Analytics* settles this problem to some extent for scientific demonstrations. According to John Burgess [52], mathematical demonstration, however, is a different concept. Burgess concludes that mathematical rigour should comply with the following requirements.[16]

(i) Every new proposition must be deduced from previously established

---

[16]See Burgess (2015), pp. 6–8, for more details.

propositions. Similarly, every new notion must be defined from previously explained notions.

(ii) There must be a set of postulates from which our deductions begin. Likewise, there must exist some unexplained primitive notions.

(iii) The meaning of the primitives and the truth of the postulates must be evident.

The level of rigour, the elegance of the given proofs, unambiguity and clarity of the definitions, and the fact that the theorems are eternally and undisputably (logically) true in the mathematical realm are delightful to the mathematician.[17] The aim of the mathematician is to find *why* the given claim is true rather than finding *what* is true. The philosopher can argue about the definitions and axioms regardless of reaching to a conclusion. However, the mathematician is usually interested in what follows from the axioms. Another way of doing mathematics is to find what kinds of axioms one should accept in order to reach to a certain conclusion. This approach is called *reverse mathematics*.[18]

Modern mathematics is based on the notion of "sets". Everything in mathematics, except *category theory* and *Neumann-Gödel-Bernays set theory*, can be expressed in terms of sets. For this reason, we accept today the axioms of set theory as our mathematical basis.

On the one hand, we want to be able to use, at least in our scientific theories, mathematical objects like sets, numbers, functions, relations, geometrical objects, etc. On the other hand, philosophers and philosophically minded mathematicians may ask whether or not these objects exist in the first place. But talking about mathematical objects is one thing, their own reality is another thing. Do mathematical objects really exist? If so, then where do they exist? What does mathematical knowledge depend on? How do we acquire mathematical knowledge? Is mathematics discovered or created? Is mathematics just logic? How do we prove the existence of an object? Can we prove its existence from the impossibility of its non-existence? What is infinity? Is there an absolute infinite? What is so special about the Axiom of Choice in set theory? What is the relationship between mathematical "truth" and "provability"? Can mathematical "truth" be completely captured in a formal axiomatic system? Does every mathematical statement have a truth value (either true or false)? Or do there exist undecidable statements? What does it mean for a statement to be decidable? Is it possible to perform an infinite number of tasks in a finite interval of time? On what basis do we accept

---

[17]Mathematicians tend to be quite meticulous. Apparently, one way for certain that mathematicians can contribute to the society is to verify legal texts and see if they are logically consistent. The great logician Kurt Gödel found a fatal flaw, called *Gödel's loophole*, in the U.S. constitutional law that could have allowed one to transform the existing constitutional democracy into a dictatorship.

[18]See Stillwell [275] (2018) or Simpson [266] (2009).

axiom candidates? These kinds of questions are related to the subject matter of philosophy of mathematics.

Philosophy of mathematics studies the following philosophical issues concerning mathematics. The first is regarding the *metaphysics*, or in other words *ontology* of mathematics.[19] Ontology is a branch of philosophy which studies the nature of *being* and *existence*. Ontological questions in philosophy of mathematics concern problems about the origin, essence and the existence of mathematical objects. There are three main views on the ontology of mathematical objects: *Realism, idealism,* and *nominalism.* Realism is traditionally called *Platonism,* although the term realism refers to contemporary positions, for the fact that it originates from Plato's philosophy. Mathematical realism is the view that mathematical objects exist independent of the physical reality as non-spatiotemporal and non-mental abstract entities. According to this view, even if the physical world did not exist, mathematical concepts would still exist in their own Platonic world. Mathematical idealism is the philosophical position which claims that mathematical objects are the direct product of the mind. The view goes back to Immanuel Kant. According to idealism, mathematical objects exist only in the mind, and that if it was not for our capability of cognition, mathematical entities would not come into existence. Finally, nominalism is the philosophical view which adheres for that abstract objects do not exist, not even in our minds. A nominalist does not believe in the existence of metaphysical or abstract concepts and they have no relation whatsoever to the physical world.

Another subject matter of the philosophy of mathematics concerns *epistemology,* i.e., study of knowledge. Epistemological questions concern problems about mathematical knowledge, its properties and how we acquire it. The three philosophical positions we mentioned above are, in fact, not limited to the ontology of mathematics. Mathematical realism can as well be interpreted in the domain of epistemology. The philosophical position that the truth value of every mathematical statement is independent of us will be later referred to as *epistemological realism.* According to this view, mathematical statements have an absolute truth value and that they are either true or false independent of the mind, language, and sensory experience.

Is mathematical knowledge rational or empirical? The conflict between rationalism and empiricism led to controversies and debates about the answers to this question. According to rationalism, for instance, "true knowledge" is acquired by reasoning. Originating from Plato, rationalism influenced many 17th- and 18th-century philosophers including Descartes, Spinoza, and Leibniz. The competing view, namely *empiricism,* asserts that "true knowledge" is acquired by sensory experience. On empiricism, the only way to obtain true knowledge is by sensory experience. The idea goes as far back as to Aristotle and it especially influenced the 18th-century British philosophers such as John Locke, George Berkeley, David Hume, and John Stuart Mill, which, in

---

[19]The word *metaphysics* literally means "after physics" or "beyond physics".

turn, established the school of British empiricism. Empiricism also influenced the 20th century logical positivism which was formed by the *Vienna Circle*, a well-known group of philosophers and scientists, led by Moritz Schlick, Rudolf Carnap, visited by Alfred J. Ayer, Ludwig Wittgenstein, and many other great thinkers.

Questions regarding the methodology of mathematics are also included in the subject matter of the philosophy of mathematics. What is the mathematical method? What is a proof? Another major concern is about the semantics. What are mathematical propositions all about? What do they really tell us? For instance, according to formalism, mathematical statements and concepts have no meaning at all. On formalism, mathematics is just a game played with symbols, following predetermined symbol manipulation rules. The meaning of the symbols, in this regard, is neglected or has no importance. We may ask many other questions similar to what we have put so far. In the next chapters, we will try to find possible answers to these questions by introducing different philosophical views. I shall emphasise that, however, we may not always be able to draw mathematical conclusions from philosophical answers. In fact, when trying to answer these questions, we often find ourselves in a position of choosing one of the conflicting claims. Many ancient philosophical problems are still being discussed today as a common practice. So the reader should not expect a definite single answer, particularly a mathematical answer, to these philosophical questions. If a mathematical statement is true, it remains true *forever* unless we change the axioms and definitions. However, this may not always the case to be in philosophy as it is more subjective, whereas mathematics is supposed to be objective. We may give reasonable answers in philosophy, but not to the degree that it is irrefutable. It is quite common in philosophical practice to oppose a presented view. Even the ideas of greatest philosophers still receive many objections. "Do numbers exist?", "What is the condition for an object to *exist*?". Such questions do not have a definite answer. There are established philosophical views in the literature which reject the existence of numbers, as well as views which claim the opposite. Our purpose in this book is to introduce a large spectrum of philosophical positions to readers so that they can decide in the end which philosophy suits their beliefs and ideas.

Discussion Questions

1. Discuss about the order of priority between definitions and axioms. Which one should come first, if any? Why?

2. Try to describe the criterion for a proof to be credited as "beautiful".

3. Discuss the consequences of having a mathematical system with no basis/axiom. Is it even possible? If so, what kind of mathematics would it lead to?

4. Godfrey Harold Hardy, an English mathematician, denigrated applied mathematics in his *Mathematician's Apology* [135] as being *trivial* and *ugly*. Discuss what could have motivated him for his views on this matter.

2

Mathematical Preliminaries

In this chapter, I shall give the notation and some mathematical concepts used in the book. Readers may skip this chapter, without loss of generality, if they are familiar with basic propositional and predicate logic, methods of proof, and axiomatic set theory.

2.1 Summary of Propositional and Predicate Logic

Mathematics is the language of science, logic is the language of mathematics. Roughly defined, *logic* is the study of valid rules of reasoning and laws of thought. In the following subsections, we will summarise *propositional* and *predicate* logic.

2.1.1 Propositional logic

A *proposition* is a statement which can be true or false. The following statements are some of the examples of propositions we may encounter in daily life.

(i) Water boils at 100 degrees Celsius.

(ii) It is raining outside.

(iii) All crows are black.

(iv) $2 + 2 = 5$.

(v) If you drop an object, it will fall.

Some examples of statements which are not regarded as a proposition may be given as follows.

(i) Could you please pass the salt? (Question)

(ii) Wake up early tomorrow. (Imperative)

(iii) $x + 2 = 2x$. (Truth value depends on an unknown variable x)

DOI: 10.1201/9781003223191-2

(iv) Mondays are boring. (Contains subjective term)

We also have statements which are neither true nor false, for instance, "This statement is false". Such a statement is called a *paradox*. That is, the sentence "This statement is false" is true if and only if it is false. Paradoxes should not be a part of any "consistent" system.

We will use the letters A, B, C or p, q, r to denote propositional variables. Given any proposition A, we denote the *negation* of A by $\neg A$. In words, the negation of the proposition "All crows are black" is "Not all crows are black". The negation of "Water boils at 100 degrees celsius" is the proposition "Water does not boil at 100 degrees celsius".

A proposition may also be in a *compound* form. For example, the statement "Water boils at 100 degrees celsius or all crows are black" can be seen as a compound statement of two propositions. A compound statement in propositional logic uses the following operators: *negation* ($\neg$), *and* ($\wedge$), *or* ($\vee$), *if-then* ($\rightarrow$), *if and only if* ($\leftrightarrow$).[1] So if A and B are propositions, then so are $\neg A$, $A \vee B$, $A \wedge B$, $A \rightarrow B$, and $A \leftrightarrow B$.

Propositions of the form $A \rightarrow B$ are called *conditional statements*. We read $A \rightarrow B$ as "If A, then B". Here, A is called the *hypothesis* (or *assumption*), and B is called the *consequent*. The truth value of a compound proposition is determined by the *truth tables* given below. Let us denote "true" and "false" by T and F, respectively. Each propositional variable either can get T or F as a valuation. Each row on the truth table denotes a possible valuation of the propositional variables and the corresponding truth value of the compound proposition.

A	$\neg A$
T	F
F	T

A	B	$A \vee B$
T	T	T
T	F	T
F	T	T
F	F	F

A	B	$A \wedge B$
T	T	T
T	F	F
F	T	F
F	F	F

---

[1] The symbols given in the parentheses denote the symbolic representation of the operators.

A	B	$A \to B$
T	T	T
T	F	F
F	T	T
F	F	T

Given a conditional statement $A \to B$, A is said to be a *sufficient condition* for B, and B is said to be a *necessary condition* for A. This means that, given that $A \to B$ is true, in order B to be true it is sufficient that A holds, and in order for A to be true it is necessary that B holds. The conditional statement $A \to B$ becomes *vacuously true* whenever A is false. If B is true, then $A \to B$ is *trivially true*. Let us consider the following statements.

(i) If snow is black, then all apples are blue.

(ii) If snow is black, then some apples are green.

These two propositions are of the form $A \to B$. Now, let us say straight away that both statements are true. Consider the first proposition: "If snow is black, then all apples are blue". In this statement, the assumption is false and so by the truth table of the conditional statement, the truth value of the consequent is irrelevant. That is, given a statement of the form $A \to B$, if A is false, then no matter what B is, $A \to B$ is true. If, however, we had "If snow is white, then all apples are blue", then by the truth table, this would be a false statement. To explain in detail, let us analyse statement (i). Why do we actually define this statement to be true? Let us for a moment determine its truth value from the negative. The statement "If snow is black, then all apples are blue" would be false if snow was white. So the statement "If snow is black, then all apples are blue" is *not false*. But in classical logic, *not false* means *true*. Of course, since the assumption of (i) is false, the statement is just vacuously true. In (ii), the assumption is false but the consequent is true. Since it is really the case that some apples are green, this very fact does not depend on a condition. That is, if the consequent of a conditional statement is true, even if the assumption is false, the statement is always true. Regardless of whether snow is black or otherwise, this does not change the fact that some apples are green. So the statement (ii) is trivially true (and by definition, it is also vacuously true since the assumption is false).

A statement of the form $A \leftrightarrow B$ is called *biconditional* and it is defined as $(A \to B) \wedge (B \to A)$. The truth table of $A \leftrightarrow B$ is then defined as below.

A	B	$A \leftrightarrow B$
T	T	T
T	F	F
F	T	F
F	F	T

If a compound proposition contains n many propositional variables, then the propositional variables can take in total of 2^n different valuations in the truth table. So the number of rows in the truth table of a statement with n propositional variables is basically 2^n. For example, the statement $(A \rightarrow B) \wedge (C \vee \neg A)$ contains three propositional variables, hence its truth table contains $2^3 = 8$ rows.

A proposition which is true under every valuation is called a *tautology*.[2] An example of a tautology is $A \vee \neg A$. A proposition which is false under every valuation is called a *contradiction*. For example, the proposition $A \wedge \neg A$ is a contradiction. If a proposition is neither a tautology nor a contradiction, then it is called *contingent*.

Given a conditional statement $A \rightarrow B$, we define its *converse* as $B \rightarrow A$. We define its *contrapositive* as $\neg A \rightarrow \neg B$ and define the *inverse* as $\neg A \rightarrow \neg B$. We will use some of these forms later in this chapter.

Definition 2.1 Let A and B be two propositions. If $A \leftrightarrow B$ is a tautology, then A and B are called *logically equivalent*, and this is denoted by $A \equiv B$.

As it can be verified using truth tables, the logical equivalent of $A \rightarrow B$ is $\neg A \vee B$. Furthermore, $A \rightarrow B$ is also logically equivalent to its contrapositive form $\neg B \rightarrow \neg A$. The equivalencies given below are called *De Morgan's Laws* which the reader can easily verify using truth tables.

$$\neg(A \wedge B) \equiv \neg A \vee \neg B,$$
$$\neg(A \vee B) \equiv \neg A \wedge \neg B.$$

Other equivalencies include but not limited to:

(i) $\neg(\neg A) \equiv A.$

(ii) $A \vee B \equiv B \vee A,$

$A \wedge B \equiv B \wedge A.$

(iii) $(A \vee B) \vee C \equiv A \vee (B \vee C),$

$(A \wedge B) \wedge C \equiv A \wedge (B \wedge C).$

(iv) $A \vee (B \wedge C) \equiv (A \vee B) \wedge (A \vee C),$

$A \wedge (B \vee C) \equiv (A \wedge B) \vee (A \wedge C).$

(v) $A \vee (A \wedge B) \equiv A,$

$A \wedge (A \vee B) \equiv A.$

What we have given regarding propositional logic will be sufficient for our purpose. The reader may refer to [127] or [165] for a detailed account on this matter.

---

[2] We use the word tautology to mean logically valid.

2.1.2   Predicate logic

We often encounter sentences like "$x > 3$", "$x = y + 5$", and so forth. These sentences can neither be claimed as true nor false since the truth value depends on the unknown variables x and y. For this reason, these types of sentences are often called *open sentences*. In general, open sentences may be true for some values of the unknown variable, false for other values. In some cases, it may be true (or false) for all values that the variable can take. An open sentence is formed by two components: The *subject* and the *predicate* of the sentence. For example, in the sentence "$x > 3$", the subject is x and the predicate is "greater than 3". If we denote this predicate by P, we can let $P(x)$ denote the entire sentence.[3] That is, $P(x)$ means "$x > 3$". Now, $P(x)$ itself does not have a truth value unless we fix x. For instance, $P(x)$ is true for $x = 4$, false for $x = 2$.

As discussed above, we imagine that the predicate takes objects from a fixed domain to get saturated so as to obtain a truth value. This domain is called the *domain of discourse* or simply the *universe*. The universe of $P(x)$, as given earlier, may be taken as the set of integers. So the universe is usually determined from the context of the predicate. Obviously, the domain of discourse for "x is a lawyer" cannot consist of cats, cars, or numbers. The domain in this case should simply consist of people. For if our predicate were, say, "x has blue eyes", then our domain would consist of objects that are more appropriate for the predicate such as living beings, in particular, animals and humans. In predicate logic, given a predicate $P(x)$, we *quantify* over the objects in the universe. If the quantification is carried over the *elements* of the universe, then this is simply *first-order logic*. If we quantify over *subsets* of the universe, then it is second-order. Quantification over the *subsets of subsets* of the universe gives us third-order logic, and so on. In our study, we will only consider first-order logic. We use two types of quantification in predicate logic: "For all" and "there exists". We denote them, respectively, by $\forall$ and $\exists$.

It is easier to understand how we use quantifications with an example. Consider a property $P(x)$. The meaning of $\forall x\, P(x)$ is "for all x, $P(x)$ holds". On the other hand, $\exists x\, P(x)$ means "there exists some x such that $P(x)$ holds".

Suppose now that we are given two predicates, namely $Q(x)$ and $R(x)$. Let $Q(x)$ mean "x is a rational number" and let $R(x)$ mean "x is a real number". The following examples given below will help the reader understand the usage of propositional connectives in predicate logic and how we interpret them.

Every rational number is a real number	$\forall x\, (Q(x) \rightarrow R(x))$.
No rational number is a real number	$\forall x\, (Q(x) \rightarrow \neg R(x))$.
Some rational numbers are real	$\exists x\, (Q(x) \wedge R(x))$.
Some rational numbers are not real	$\exists x\, (Q(x) \wedge \neg R(x))$.

---

[3] We may also call $P(x)$ a *property* or *relation*. Relations are not restricted to one-place predicates. We may also have two-place or three-place predicates such as $P(x, y)$ or $P(x, y, z)$.

At this point, the reader may ask why the first sentence is expressed as $\forall x\,(Q(x) \to R(x))$. Suppose that our universe consists of numbers. As a matter of fact, the sentence "Every rational number is a real number" is equivalent to saying "If a number is rational, then it is a real number". This is why we need to use the conditional implication operator. Had we wrote $\forall x\,(Q(x) \wedge R(x))$, this would mean "Every number is rational and real".

So then when does the sentence $\forall x\,P(x)$ become true and when does it become false? If $P(x)$ holds for every x in the universe, then $\forall x\,P(x)$ is true. If there exists some x such that $\neg P(x)$ holds, then $\forall x\,P(x)$ is false. What about the truth value of $\exists x\,P(x)$? If there exists some x such that $P(x)$ holds, then $\exists x\,P(x)$ is true. On the other hand, if $P(x)$ holds for no x, in other words if $\neg P(x)$ holds for every x, then $\exists x\,P(x)$ is false. It is then important to know the following equivalencies:

$$\neg \forall x\,P(x) \equiv \exists x\,\neg P(x), \ \neg \exists x\,P(x) \equiv \forall x\,\neg P(x).$$

Note that $\exists x\,P(x)$ being true does not mean $P(x)$ is true for *arbitrary* x, but it rather means the sentence is true for some *specific* value(s) of x. However, if $\forall x\,P(x)$ holds, then $P(x)$ holds for arbitrary values of x. The distinction between the notion of *arbitrary* and *specific* is crucial. The following sentences are valid in predicate logic.[4]

If $\forall x\,P(x)$, then $P(c)$ is true for arbitrary c.
If $P(c)$ is true for arbitrary c, then $\forall x\,P(x)$.
If $\exists x\,P(x)$, then $P(c)$ is true for some c.
If $P(c)$ holds for some c, then $\exists x\,P(x)$.

Notice how the order of the negation ($\neg$) operator changes the meaning of the sentence. The sentence $\neg \forall x\,P(x)$ is not equivalent to $\forall x\,\neg P(x)$. The former says it is not the case that $P(x)$ holds for every x, i.e., there exists some x for which $P(x)$ is false, while the latter says $P(x)$ is false for every x.

In predicate logic, a variable appearing in the sentence may not be restricted by every quantification operator used in the sentence. Consider the sentence,

$$\forall x\,P(y).$$

The variable y is not quantified by $\forall x$ for the obvious fact that the operator merely quantifies x. Here, y is not bound to any quantification. The effective range of a quantification symbol determines its *scope*. Consider the following examples, where the scope of the quantifier $\forall x$ is underlined.

---

[4]Although validity in predicate logic will be discussed in Chapter 11 and is defined significantly different than that in propositional logic, for now we use the term validity to mean that the sentence is true in every interpretation of the predicate.

(i) $\forall x \, \underline{(P(x))} \wedge Q(x)$

(ii) $\exists y \, \forall x \, \underline{(P(x,y) \rightarrow \forall z \, Q(z))}$

(iii) $\forall x \, \forall y \, \underline{(P(x,y) \wedge Q(y,z))} \wedge \exists x \, P(x,y)$

(iv) $\forall x \, \underline{(P(x) \wedge \exists x \, Q(x,z) \rightarrow \exists y \, R(x,y))} \vee Q(x,y)$

If a variable x is in the scope of a quantifier $\forall x$ or $\exists x$, then x is called a *bound* variable. Otherwise, it is called a *free variable*. For example, consider the sentence

$$\forall x \, P(x,y).$$

The variable x appearing in $P(x,y)$ is a bound variable, whereas y is a free variable. In (i) given above, x appearing in $P(x)$ is a bound variable, yet the one appearing in $Q(x)$ is a free variable since it is not in the scope of a quantifier related to x. In (ii), the variable x appearing in $P(x,y)$ is a bound variable. In (iii), the x appearing in the first $P(x,y)$ is quantified by $\forall x$, whereas the one appearing in the second $P(x,y)$ is quantified by $\exists x$. Hence, both instances of x are bound. In (iv), x is a bound variable in $P(x)$, $Q(x,z)$ and $R(x,y)$, free in $Q(x,y)$.

Let us repeat some of the essential points about predicate logic. If $\forall x \, P(x)$ holds, then so does $P(c)$ for any c. That is, $\forall x \, P(x) \rightarrow P(c)$ is true. To see why this is the case, consider the following argument. In order for $\forall x \, P(x) \rightarrow P(c)$ to be false, by the truth table of conditional statements $P(c)$ must be false. But then $\forall x \, P(x)$ would have to be false. So $\forall x \, P(x) \rightarrow P(c)$ is true no matter what c is. Let us also argue that $P(c) \rightarrow \exists x \, P(x)$ holds. If $P(c)$ is true for some constant c, then so is $\exists x \, P(x)$ since the x that makes $P(x)$ is true is actually the specific object that c refers to. We should also look at the case when $P(c)$ is false. If $P(c)$ is false, without taking the truth value of $\exists x \, P(x)$ into account, by the truth table of conditional statements

$$P(c) \rightarrow \exists x \, P(x)$$

will be true.

We may also use nested quantification in predicate logic. Consider the following two sentences:

(i) $\forall x \, \exists y \, Q(x,y)$
(ii) $\exists x \, \forall y \, Q(x,y).$

Now (i) says that "for every x, there exists some y such that $Q(x,y)$ holds". On the other hand, (ii) says "there exists some x such that $Q(x,y)$ holds for every y." Notice the difference between two sentences and how the order of the quantifier changes the meaning. Finally, consider the sentence

(iii) $\forall x \, \exists y \, \forall z \, R(x,y,z).$

The sentence (iii) means "for every x there exists some y such that $R(x, y, z)$ holds for every z".

2.2　Methods of Proof

In this section, we review methods of proof. Generally speaking, a proof consists of a sequence of claims that is used to justify a hypothesis. The method of the proof of a given statement p may vary. The reader will be able to determine, with enough practice, which type of proof would suit in various cases.

In mathematics, a claim cannot be deducted from nothing. To make any conclusion and deduce a mathematical statement, we need assumptions. In fact, we would like to deduce new statements from these assumptions. To do this, we need deduction rules, i.e., rules of inference. A deduction rule is a schema that allows us to infer a new statement from a given set of assumptions. Let us consider the example given below.

> If you know your password, you can sign in to your account.
> You know your password.
> Therefore, you can sign in to your account.

Let us denote the proposition "You know your password" by p and denote "You can sign in to your account" by q. Then, the inference given in the example can be written symbolically as follows:[5]

$$p \to q$$
$$p$$
$$\overline{}$$
$$\therefore q$$

This inference rule is known as *modus ponens*. In fact, we may write many inference rules, however, any rule can be written in the implicational form of $p \to q$ for two propositions p and q.[6] For this reason, modus ponens can be

---

[5] When writing an inference in the symbolic form, we usually use $\therefore$ to mean "therefore".

[6] An important point we should note here is that the proposition $p \to q$ means two things depending on the context. First, it means "If p is true, then so is q". Second, we will see that it also means "p proves q". In fact, two interpretations (truth vs. provability) are the same for propositional and first-order predicate logic. We will later refer to this as the *completeness theorem*. Not every logic system is guaranteed to have the completeness property. Now $p \to q$ is a statement written in the object language of propositional logic. On the other hand, "If p is true, then so is q" is merely a conditional statement in meta-language. The distinction between object language and meta-language is critical particularly in logical philosophy. For a detailed account, we refer the reader to the 3rd and 7th chapters of Quine's book [238].

considered as the unique inference rule. A proof in the tableux form can be, in general, written in the following manner:

$$\text{Assumption } 1$$
$$\text{Assumption } 2$$
$$\vdots$$
$$\text{Assumption } n$$
$$\overline{\phantom{\text{Assumption } n}}$$
$$\therefore \text{Conclusion}$$

Before introducing methods of proof, let us state the following propositions as tautologies which can be easily verified by the reader using the truth table method.

(i) $p \to (p \vee q)$

(ii) $(p \wedge q) \to p$

(iii) $p \to (q \to (p \wedge q))$

(iv) $[p \wedge (p \to q)] \to q$

(v) $[\neg q \wedge (p \to q)] \to \neg p$

(vi) $[(p \to q) \wedge (q \to r)] \to (p \to r)$

(vii) $[(p \vee q) \wedge \neg p] \to q$

(viii) $[(p \vee q) \wedge (\neg p \vee r)] \to (q \vee r)$

(ix) $(\neg q \to \neg p) \to (p \to q)$

It is worth explaining some of these tautologies. The proposition (i) basically states the $\vee$-introduction rule. It states that the disjunction of a tautology with any proposition is still a tautology. For example, consider the tautology $p \vee \neg p$. Then for any q, the proposition $(p \vee \neg p) \vee q$ is also a tautology. Proposition (ii) is the $\wedge$-elimination rule, stating that if a conjunctive proposition is true, then so is the proposition obtained by eliminating any of the components of the conjunction. We can give an example of this in daily life. If the proposition "I ate salad and I had an orange juice" is true, then so is the proposition "I had an orange juice". Proposition (iii) states that if p and q are two true propositions, then their conjunction is a true proposition. Proposition (iv) is just modus ponens. The proposition stated in (v) is, in fact, the contrapositive of modus ponens, called *modus tollens*. Proposition (vi) states the transitivity property of propositional logic. For this, consider the statements "If I put ice in the glass, the water will spill over" and "if the water spills over, then the table gets wet". From these two statements, we can conclude that "If I put ice in the glass, the table will get wet". The explanation of the rest of the tautologies is left to the reader as an exercise.

Now we introduce the methods of proof. Roughly defined, a logical/ mathematical proof consists of a finite sequence of propositions. When proving any statement, we have to start our argument with a statement and so we need initial assumptions. These assumptions are called *axioms*. According to Oxford English Dictionary, the word axiom refers to a statement or proposition which is regarded as being established, accepted, or self-evidently true. If we can deduce in finitely many steps, using the rules of inference, the desired statement from a given collection of axioms, then this sequence of propositions constitutes a proof of the desired statement. We give the formal definition as follows.

Definition 2.2 Let S be a collection of propositions. If for any sequence of propositions $B_1, \ldots, B_n$ in S, $(B_1 \wedge \ldots \wedge B_n) \to A$ is a tautology, then we say that S *proves* A. In this case, the sequence $B_1, \ldots, B_n$ is a *proof* of A from S.

Although mathematical proofs in daily practice are given in natural language, they all, in principle, can be seen formally as a sequence of propositions. Proving a theorem formally can be very tedious. Mathematicians do not prove theorems in the formal manner as formal proofs are in most cases used by computer programmes or formal axiomatic systems for verification and automation purposes. Fortunately, we do not need to worry about formal proofs in this book.

2.2.1   Direct proof

A *direct proof* is a way of proving any statement of the form $p \to q$, using the rules of inference, starting from the assumption p to arrive at the conclusion q. We give the following example.

Proposition 2.1 For any integers m and n, if m and n are odd, then $m + n$ is even.

Proof. If m and n are odd numbers, then for some integers a and b, we have that $m = 2a + 1$ and $n = 2b + 1$. Therefore,

$$
\begin{aligned}
m + n &= (2a + 1) + (2b + 1) \\
&= 2a + 2b + 2 \\
&= 2(a + b + 1).
\end{aligned}
$$

Since $m + n$ is two times the number $a + b + 1$, we conclude that $m + n$ is an even number.   □

2.2.2   Proof by contrapositive

Another way of proving a statement of the type $p \to q$ is by proving its *contrapositive* form. It can be easily shown using the truth table method that

$p \rightarrow q$ is logically equivalent to $\neg q \rightarrow \neg p$. So proving the contrapositive form is same as proving the original statement.

Proposition 2.2 If n^2 is odd, then so is n.

Proof. The contrapositive form of the given statement can be wrtten as "If n is even, then so is n^2". We shall prove the contrapositive form. If n is even, then for some k, $n = 2k$. Then, $n^2 = (2k)^2 = 4k^2 = 2(2k^2)$. Therefore, since n^2 is two times the number $2k^2$, n^2 must be even. □

2.2.3 Proof by contradiction

In logic and mathematics, *proof by contradiction* is a powerful method to establish the truth or falsity of a statement. Suppose we want to prove a proposition p. To prove p by contradiction, we first assume that $\neg p$ holds. Then we derive a contradiction and conclude that our assumption $\neg p$ must be false, meaning that it cannot be the case that $\neg p$. That is, $\neg(\neg p)$ must hold. In classical logic, $\neg(\neg p)$ is logically equivalent to p. From this equivalence, we deduce p. This method is called *proof by contradiction*, which is widely used in many areas of mathematics and it is often a very convenient method. Let us first prove the classic result that $\sqrt{2}$ is irrational.

Proposition 2.3 $\sqrt{2}$ is irrational.

Proof. Suppose for a contradiction that $\sqrt{2}$ is rational. Then, for some natural numbers a and b

$$\sqrt{2} = \frac{a}{b}. \qquad (*)$$

Assume without loss of generality that $\frac{a}{b}$ is in the simplest form, that is a and b are *relatively prime*, meaning that their greatest common divisor is equal to 1. Taking the square of each side of the equation $(*)$, we get

$$a^2 = 2b^2.$$

Then, a^2 is an even number. Thus, a is even. Hence, $a = 2k$ for some natural number k. Substituting $a = 2k$ above, we get $(2k)^2 = 2b^2$. From this, we have $4k^2 = 2b^2$, and by simplifying both sides we get

$$2k^2 = b^2.$$

Hence, b^2 is an even number and so is b. Then, $b = 2l$ for some natural number l. Substituting $a = 2k$ and $b = 2l$ in the equation $(*)$, we get

$$\sqrt{2} = \frac{2k}{2l}.$$

This simplifies to

$$\sqrt{2} = \frac{k}{l}.$$

But $\dfrac{a}{b}$ was in the simplest form. A contradiction. Therefore, $\sqrt{2}$ is irrational.

$\square$

Let us now look at Euclid's proof of the infinitude of primes. It is worth noting that Euclid did not state this directly, as ancient Greeks in that period avoided using infinity as a completed totality. The fact that there are infinitely many prime numbers was rather implied by the fact that for every prime number, there exists a larger prime number. This is what Euclid proved. But let us phrase it in the usual way that there are infinitely many primes.[7]

Proposition 2.4 There are infinitely many prime numbers.

Proof. Assume for a contradiction that there are finitely many prime numbers. Let us list them by $a_1, \ldots, a_n$. We should derive a contradiction from this. Now consider the number

$$b = a_1 a_2 \cdots a_n + 1.$$

This number cannot be divided by any number in the list $a_1, \ldots, a_n$ since if it were divisible by some a_i, then the remainder would always be 1. Then, b must be a prime number. Yet b is not in the list. But then b cannot be a prime number, since we assumed that the list contained all prime numbers. A contradiction. Therefore, there must be infinitely many primes. $\square$

Let us now show how to prove a conditional statement of the form $p \to q$ using proof by contradiction. The idea is the same. We should assume that it is not the case that $p \to q$, i.e., $\neg(p \to q)$, derive a contradiction and conclude that $\neg\neg(p \to q)$, hence $p \to q$. So then what is the negation of $p \to q$? We shall use its logical equivalent. In fact, $p \to q$ is logically equivalent to $\neg p \vee q$. Then, using De Mogan's Law, $\neg(p \to q)$ is equivalent to $\neg(\neg p \vee q)$ which is equivalent to $p \wedge \neg q$. Indeed, the only time when $p \to q$ is false is if the assumption is true and the conclusion is false. So in order to prove $p \to q$ by contradiction, we should start by assuming that p is true and q is false. Then we should derive a contradiction and conclude that $\neg(p \wedge \neg q)$. Using De Morgan's Law, this is equivalent to $\neg p \vee q$ which is equivalent to $p \to q$. Consider the following example.

Proposition 2.5 If $n^3 + 5$ is odd, then n is even.

Proof. This proposition is a conditional statement of the form $p \to q$. Let p denote "$n^3 + 5$ is odd" and let q denote "n is even". We start by assuming that p is true and q is false. That is, we assume $n^3 + 5$ and n are odd. If n is

---

[7]Similar idea was also used in Euclidean geometry that a line segment could be extended indefinitely. This does not explicitly tell us that there are lines of infinite length, although it is implied by so, but it rather tells us that a line can potentially become infinite in length.

odd, then for some number k, $n = 2k + 1$. Therefore,

$$n^3 + 5 = (2k + 1)^3 + 5$$

$$= 8k^3 + 12k^2 + 6k + 6$$

$$= 2(4k^3 + 6k^2 + 3k + 3).$$

Then, $n^3 + 5$ must be an even number. A contradiction. Therefore, when $n^3 + 5$ is odd, n must be even. $\qquad\square$

2.2.4 Proof by induction

The last method we shall introduce is perhaps one of the most commonly used proof methods, particularly in situations when one needs to establish that a certain property holds for *all* natural numbers. *Proof by induction* is a proof technique which is usually used when we want to prove that a property holds for infinitely many consecutive objects.

Let us first discuss the nature of mathematical induction. An apparent difference between physics and mathematics is that physics is empirical, whereas mathematics is rational, keeping aside that it can be argued to what degree it really is. Mathematics is generally believed to be independent of the physical reality. It operates on an "idealised" abstract ground. Physics, due to our limited empirical observation capabilities, does not permit us to arrive at universal or general conclusions for infinitely many cases. As a matter of fact, positivism is opposed to the metaphysical nature of asserting universal judgements in science. Since empirical observation merely gives a fact about a particular instance and since we cannot make infinitely many observations, we cannot come to general conclusions through empirical observations. Otherwise, this would require us to refer to metaphysics by transcending all particular instances of facts. For example, the statement that "At every point of the universe, there is gravitational pull" is a conclusion we arrive after experimenting on finitely many points in space. But then how do we know there is no gravitational pull billions of lightyears away? The fact that we have not observed *every* point of space, does not give us the entitlement of arriving at a general conclusion that there is really gravitational pull at every point of the universe. "Gravitational pull exists" is a fact limited to spaces we have observed so far. In physics, we cannot exhaust infinitely many cases. What we can do is to make assumptions and say something like "if an arbitrary point x in space has the same properties as all the points we have observed so far, then gravitational pull exists at point x as well". The property of inductively deducing (we will later call this the principle of induction) a universal statement which quantifies all objects in an infinite domain is something exclusive to mathematics. In fact, this inductive property refers to metaphysics by simply covering an infinite totality as a result of establishing a successive relation between consecutively observed instances. Invoking the successor relation arbitrarily often leads to covering the domain as a totality. In this sense, mathematics uses some level of metaphysics. In fact, contrary to physics, mathematics must refer to metaphysics if we want to form abstractions and generalisations over infinite domains. The reason physics does not have this metaphysical feature is, perhaps, that no two physical entities are identical to each other. If we assume that any two points in space have the same properties, then the empirical observation about a *particular* point in space would, in fact, hold for *all* points.

Suppose now that we want to prove some property P about natural numbers or similar structures. For this, we should begin with showing that P holds for the first element of the set of natural numbers. That is, we first need to show that $P(0)$ holds. We call this the *basis step* of the proof. Next we do the following: we show that if P holds up to some natural number n, then it should as well hold for $n + 1$. That is, we assume $P(n)$ and then show that $P(n + 1)$ holds. This is called the *induction step*. In the induction step, the assumption $P(n)$ is called the *inductive hypothesis*. If we prove the basis step and the induction step, we conclude that $P(n)$ holds for all n. To see this explicitly, consider the following argument. Proving the basis step shows that P holds for 0. Proving the induction step will imply that since $P(0)$ holds, then so does $P(1)$. But then applying the induction step again, since $P(1)$ holds, then so does $P(2)$, ad infinitum. Then, we conclude that P holds for *all* natural numbers.[8] Let us now give an example of proof by induction. For any natural number n, let us denote the following property by $P(n)$:

$$0 + 1 + 2 + \cdots + n = \frac{n(n + 1)}{2}.$$

Proposition 2.6 For every natural number n, $P(n)$ holds.

Proof. First we show that $P(n)$ holds for 0. Showing that $P(0)$ holds is straightforward. Let us first calculate the left-hand side of the given equation. The sum of the numbers from 0 to 0 is simply 0. The right-hand side will be

$$\frac{0 \cdot (0 + 1)}{2}.$$

If we look at both sides of the equation, we see that

$$0 = 0.$$

---

[8]The arrived conclusion is due to the *Principle of Mathematical Induction*: $[P(0) \wedge \forall n(P(n) \to P(n + 1))] \to \forall n \, P(n)$. The reason inductive proofs are valid is that the generalisation made about the natural numbers in the Principle of Mathematical Induction is congruent with the definition of the set of natural numbers. As we will see in the next subsection, the set of natural numbers is defined as the smallest set that contains 0, and that if it contains any number n, it must also contain $n + 1$.

Since both sides of the equation are equal, $P(0)$ is true. We just proved the basis step. Now we need to show that if $P(n)$ is true, so is $P(n + 1)$. The inductive hypothesis assumes that $P(n)$ is true. From this assumption, we have to show that $P(n + 1)$ holds. We can write $P(n + 1)$ as follows:

$$(0 + 1 + \cdots + n) + (n + 1) = \frac{(n + 1)((n + 1) + 1)}{2}.$$

We know that, on the left-hand side of the equation, the sum of the numbers from 0 to n, by the inductive hypothesis, is equal to $\frac{n(n + 1)}{2}$. The left-hand side can be then written as

$$\frac{n(n + 1)}{2} + (n + 1).$$

If we can show that this is equal to the right hand side, then we are done. So let us show this.

$$\frac{n(n + 1)}{2} + (n + 1) = \frac{n(n + 1) + 2(n + 1)}{2}$$

$$= \frac{(n + 1)(n + 2)}{2}$$

$$= \frac{(n + 1)((n + 1) + 1)}{2}.$$

Then, $P(n + 1)$ must be true. By the Principle of Mathematical Induction, $P(n)$ is true for every natural number n. $\square$

It is also worth to discuss about *existence and uniqueness* proofs. Existence of an object can be proved in two ways. The first way is to construct the object and explicitly show its existence. The second way is to prove that its absence (non-existence) leads to a contradiction. Then, using the equivalence $\neg(\neg p) \equiv p$, we conclude its impossible that the object cannot exist, and so it must exist. So the second way proves the existence rather indirectly. Uniqueness proofs use a kind of proof by contradiction. First we show that an object x with the desired property exists. Then we assume as if there were another object y with the same property, and then we show that $x = y$. From this equality, we conclude that since x is no different than y, there must be a unique object satisfying the desired property. What about absence proofs? How can we prove the non-existence of an object? We do this simply by assuming first the existence of the object and then deriving a contradiction from this assumption. Then we conclude that the object cannot exist. We will study existence proofs in Chapter 4 in more detail.

Another important subject in predicate logic is the notion of "truth in structures". We will study this notion in Chapter 11. Things we have given so far will be sufficient for now. For a more detailed account on predicate logic, we refer the reader to Hamilton [127], Kleene [165], or Enderton [84].

2.2.5 Proof fallacies

Mathematicians present proofs to establish the truth value of a proposition. Naturally, writing proofs involve logical inferences. When doing so, students sometimes tend to "break" the rules of the mathematical game. We will now discuss about common mistakes done when writing proofs. The first fallacy is called the *fallacy of affirming the conclusion*. It would be best explained with an example.

Example 2.1 If you run, you will get tired.
You got tired.
Therefore, you ran. (Fallacy)

Just because you are tired, does not imply you ran. Recall that running is just a *sufficient* condition to get tired, not a necessary condition. So there may be other reasons that you got tired. In other words, $[(p \to q) \land q] \to p$ is not a tautology.

Another fallacy is called the *fallacy of denying the hypothesis*.

Example 2.2 If the weather is sunny, I will go to the beach.
The weather is not sunny.
Therefore, I did not go to the beach. (Fallacy)

In real life, this inference may not always look like a fallacy. For instance, when we say "all young men are required to serve in the military", this would mean "women are not required to serve in the military". But the fact that the inference given in the example is a fallacy in propositional logic comes from the truth table of conditional statements. That is, $[(p \to q) \land \neg p] \to \neg q$ is not a tautology in propositional logic.

Finally, there is the infamous *circular reasoning*. This is a fallacy where one attempts to prove a statement using another statement which is logically equivalent to the original statement that we want to prove.

Example 2.3 If n^2 is even, then so is n.

Fallacious proof. Suppose that n^2 is an even number. Then, for some number k, we have that $n^2 = 2k$. Let $n = 2l$ for some number l (fallacious step). Therefore, n is even. □

There is no justification behind letting $n = 2l$. This is a circular reasoning for the fact that $n = 2l$ is a statement logically equivalent to the statement we want to prove, which is the claim that n is even.

Circular reasoning is also sometimes mistakenly used in existence proofs. If $p \to q$ and $q \to p$ are both true, then it does not necessarily follow that p or q must be true. It just tells us that they have the same truth value, i.e., either both are true or false. Applying the same reasoning, if the existence of p implies the existence of q, and vice versa, this does not say that they exist. It rather says that either they both exist or neither of them exists.

We end this section with Oskar Perron's (1880–1975) illustration of how assuming false propositions might lead to false conclusions. The argument is called *Perron's paradox* and it runs as follows. The claim is that *if* there was to exist a largest natural number N, then $N = 1$. Suppose for a contradiction that $N \neq 1$. Then, $N^2 > N$. But this contradicts the assumption that N being the greatest number. So it's not the case that $N \neq 1$. Therefore, $N = 1$. The fallacy here is due to the usage of the *ex falso* rule that from contradiction follows anything. We can conclude literally anything about an object that does not exist. For instance, "Every four sided triangle is a circle" is vacuously true simply because of the fact that a four sided triangle is non-existent. In the same manner, we can say anything about the "greatest natural number".[9]

2.3   Basic Mathematical Notions

In several parts of the book, we will be using set theoretical notions. For completeness, we shall give these notions to the reader. Although we do not assume that the reader is familiar with the formal treatment of set theoretical concepts, some background in the intuitive meaning of these concepts is still desired. So the purpose of this section is to review some set theoretical concepts and the notions used in the book. This section should be considered rather as a short summary of axiomatic set theory for the reader to get them familiar with our notation. For a detailed account on axiomatic set theory, the reader may refer to Jech [155] or Suppes [277].

A formal axiomatic system, consists of a formal language, derivation rules, and a set of axioms. Anything in mathematics, except category theory and Neumann-Gödel-Bernays set theory (NGB), can be in principle expressed in terms of sets. The language of set theory consists of logical symbols $=, \neg, \vee, \wedge, \to, \leftrightarrow, (,), \exists, \forall$, and the *membership* symbol $\in$. It also contains variable symbols $x_1, x_2, x_3, \ldots$ to denote sets. If x is an element of y, we denote this relation by $x \in y$.

---

[9]See L. C. Young [308], §10, pp. 22–23 for further discussion.

Definition 2.3 Let x and y be sets. The *atomic formulas* of set theory are

$$x \in y \text{ and } x = y.$$

Every atomic formula is a *formula* of set theory. If φ and ψ are formulas, then so are the following sentences:

$$(\neg \varphi), \quad (\varphi \wedge \psi), \quad (\varphi \vee \psi), \quad (\varphi \rightarrow \psi), \quad (\varphi \leftrightarrow \psi), \quad (\forall x \, \varphi), \quad (\exists x \, \varphi).$$

For convenience, we write $x \notin y$ to mean $\neg(x \in y)$. Occasionally, we will ignore the outer parentheses. For example,

$$\forall x \; x \in y$$

is not actually a formula of set theory according to our defintion. However, instead of writing this sentence formally as

$$(\forall x(x \in y))$$

we may remove the outer parentheses and write $\forall x \; x \in y$. So we also accept that such sentences without parentheses are formulas of set theory. We use the parentheses just to avoid ambiguity.

2.3.1   Axioms of ZFC set theory

Zermelo-**F**raenkel Set Theory with the Axiom of **C**hoice (ZFC) set theory consists of eight axioms and two axiom schemas. We will give some of these axioms in the formal language of set theory for the reader to see how it is used in expressing the axioms. Set theoretical axioms alone are not sufficient to do mathematics however. Apart from the axioms of the theory, we also need the logical axioms. It is best to give some of these logical axioms, to be more systematic, before presenting the axioms of set theory. Every tautology is accepted as a logical axiom. Along with these tautologies, some of the logical axioms include sentences given as follows:

(i) $x = x$

(ii) $x = y \rightarrow y = x$

(iii) $(p \wedge q) \rightarrow p$

(iv) $p \vee \neg p$

(v) $(\varphi(x) \wedge x = y) \rightarrow \varphi(y)$

We will remind the reader these rules when necessary. Now we give the ZFC axioms.

Axiom of Empty Set. *There exists a set with no element.*

In the formal language of set theory, we can write this axiom as

$$\exists y \, \forall x \, (x \notin y).$$

This axioms asserts the existence of a set. Furthermore, this set contains no element. Axiom of Empty Set is an existential axiom. It assumes no background ontology. From now on, we will accept the existence of a set with no elements. The more skeptic readers may, of course, argue against this. Nevertheless, we need to have at least one set to be able to construct objects. The set whose existence is asserted in the Axiom of Empty Set does not rely on another set.[10]

Axiom of Extensionality. *Two sets are equal if they contain the same elements.*

In formal language, we express this axiom as

$$\forall a \, (a \in x \leftrightarrow a \in y) \to x = y.$$

The inverse direction, that is,

$$x = y \to \forall a \, (a \in x \leftrightarrow a \in y)$$

is a consequence of the logical axioms. Using the Axiom of Extensionality, we can prove that the set whose existence is asserted in the Axiom of Empty Set is in fact unique. For this, suppose that there are sets x and y satisfying the Axiom of Empty Set such that $x \neq y$. In this case, by the Axiom of Extensionality, not all elements of x and y must be the same. Then, one of them must contain an element that other does not have. But since neither of these sets contains any element, they cannot be distinct from each other. Therefore, it must be that $x = y$. So then, there must be a unique set with no element. Since it deserves a special name, we shall call it the *empty set* and denote it by $\emptyset$. From now on, we will use the symbol $\emptyset$ in our formulas to denote the empty set.

Let us introduce another abbreviation. Let A and B be two sets. If every element of A is also an element of B, then A is a *subset* of B. We denote this by $A \subseteq B$. That is,

$$A \subseteq B \leftrightarrow \forall x \, (x \in A \to x \in B).$$

---

[10]If we wanted to write a simpler version of the Axiom of Empty Set, we could assert the sentence $\exists x \, x = x$. This sentence tells us that there exists an object in the mathematical universe which is identical to itself, yet it tells nothing about its property. At least the Axiom of Empty Set asserts that the existing set has no element. Of course, since we are working with sets, our interpretation of the sentence $\exists x \, x = x$ is that the set whose existence is asserted is actually a set. In the set theoretic literature, this sentence is sometimes known as the *Set Existence Axiom*.

From now on, we will use the symbol $\subseteq$ as an abbreviation for the subset relation.

Axiom Schema of Specification. *Let x be a set and $\varphi(z)$ be a formula in the language of set theory. Then, the collection of elements z of x satisfying the formula $\varphi(z)$ is a set.*

Formally, we write this as

$$\forall x \, \exists y \, \forall z \, (z \in y \leftrightarrow (z \in x \wedge \varphi(z))).$$

If we denote this set by y, whose existence is asserted, then y is in fact a subset of x. Note that this is not a single axiom, but it expresses infinitely many instances. In other words, it is an *axiom schema*. We get a separate axiom for each formula $\varphi(z)$.

Another important point is that the Axiom Schema of Specification is conditioned on the existence of other sets. That is, *if x is a set*, then any subcollection of x is also a set. Notice its difference from the Axiom of Empty Set. In ZFC, some axioms are existential and some are conditional. That is to say, some axioms "create" sets relying on no ontological basis and some "construct" new sets using other sets.

Axiom of Union. *If x is a set, then there exists a set y whose elements are precisely the elements of the elements of x. We denote this set by $\bigcup x$.*[11]

We may write this axiom in the language of set theory as

$$\forall x \, \exists y \, \forall a \, (a \in y \leftrightarrow \exists z \, (a \in z \wedge z \in x)).$$

Just like the Axiom Schema of Specification, Axiom of Union is a statement that asserts the existence of a set based on the existence of others.

Axiom of Pairing. *Given any two sets, there exists a set whose members are exactly the two given sets.*

Formally, we write this as

$$\forall x \, \forall y \, \exists z \, \forall a \, (a \in z \leftrightarrow (a = x \vee a = y)).$$

The set whose existence is asserted is denoted by $\{x, y\}$. Moreover, this set is unique. If x and y are the same (the axiom does not say they should be distinct), then instead of writing $\{x, y\}$, we simply write $\{x\}$. The elements

---

[11]The *union* of two sets x and y is denoted by $x \cup y$ and defined as $z \in (x \cup y) \leftrightarrow (z \in x \vee z \in y)$. The *intersection* of two sets x and y is denoted by $x \cap y$ and defined as $z \in (x \cap y) \leftrightarrow (z \in x \wedge z \in y)$. The fact that the intersection of two sets is a set is followed from the Axiom Schema of Specification.

of the pair $\{x, y\}$ is not ordered. Hence, we may call it an *unordered set*. An *ordered set* is denoted by (x, y) and is defined as the set $\{\{x\}, \{x, y\}\}$ which satisfy the following characteristic property:

$$(x, y) = (z, t) \leftrightarrow ((x = z) \wedge (y = t)).$$

In fact, the way (x, y) is defined is irrelevant as long as it satisfies that $(x = z) \wedge (y = t)$ whenever $(x, y) = (z, t)$. As we shall investigate in Chapter 14, according to the structuralist philosophy of mathematics, mathematical objects are defined by their characteristic and structural properties rather than their "own" intrinsic features.

Axiom of Power Set. *If x is a set, then there exists a set whose elements consists of all subsets of x.*

In the language of set theory, we write this formally as

$$\forall x \, \exists y \, \forall z \, (z \in y \leftrightarrow z \subseteq x).$$

The set of all subsets of x is called the *power set* of x and it is denoted by $\mathcal{P}(x)$.[12]

The reader may ask if our system, so far, is consistent. One can indeed prove that the axioms presented up to now are consistent with each other. Another problem is how much mathematics we can actually express with these axioms given so far. The answer to this question is, unfortunately, not much. Given what we have now, we are only able to prove the existence of "finite" sets.[13] That is, our set theoretic universe at this stage contains all finite sets. For any natural number, we can prove the existence of a set containing n elements. But how many finite sets are there? Obviously, according to our intuition, there should be infinitely many. Applying the power set axiom, we may obtain sets with more elements as much as we want. If a set x has n many elements, then the power set of x must contain 2^n many elements. We give this as a theorem.

Theorem 2.1 Let S be a finite set containing n elements. Then, $\mathcal{P}(S)$ contains 2^n elements.

---

[12]There is a serious ontological commitment behind the Axiom of Power Set. The commitment is, more specifically, to assume the existence of *all* subsets of x as a completed totality. This is not a major problem for finite sets. However, speaking about the totality of *all* subsets of an infinite set is controvertial for some philosophical views, e.g., constructivism. According to these views, the collection of *all* subsets of an infinite set may not even be a definable object.

[13]If we were to define natural numbers within this system, how many numbers would it be possible to define? We can define as many as we want.

Proof. We prove by induction on n. Clearly, for the basis case, for $n = 0$, the empty set has only $2^0 = 1$ subset and that is the empty set itself. Suppose now that the theorem holds for some natural number n and that S is a set containing n elements, and that $\mathcal{P}(S)$ contains 2^n elements. Let T be a set containing $n + 1$ elements. We must show that T has 2^{n+1} subsets. So there exists some $a \in T$ such that $T - \{a\} = S$. In other words, $S \cup \{a\} = T$. By the induction hypothesis S has 2^n subsets which does not contain a. But similarly, S has 2^n subsets which contain a. In total, there are $2^n + 2^n = 2^{n+1}$ subsets of T, which is what we wanted to prove. $\qquad\square$

We can intuitively see that the set theoretical universe contains infinitely many sets. But can we prove this fact within our system? Unfortunately not yet. We do not know if there are infinite sets in the system. For this reason, we will now introduce a new axiom which is another existential statement. This axiom will assert the existence of an infinite set. As a matter of fact, in some sense, this set will have the "structure" of the naturals. First and foremost, we should require this set to contain the element 0. Furthermore, if it contains an element x, it should also contain the "successor" of x, which is defined as the set $x \cup \{x\}$ (see Section 2.3.2).

Axiom of Infinity. *There exists a set I such that $\emptyset \in I$, and such that whenever any x is in I, so is $x \cup \{x\}$.*

In the formal language of set theory, we write this axiom as follows:

$$\exists I \, (\emptyset \in I \wedge \forall x \, (x \in I \to x \cup \{x\} \in I)).$$

A set that satisfies the Axiom of Infinity is called an *inductive set*. Although it is not straightforward, the intersection of all inductive sets is a set. Furthermore, this intersection itself gives an inductive set. The *set of natural numbers* is defined as the intersection of all inductive sets and it is denoted by $\mathbb{N}$. The elements of $\mathbb{N}$ are called *natural numbers*.

Axiom Schema of Replacement. *Let X be a set and let $\varphi(x, y)$ be a formula. If for all $x \in X$ there exists a unique y such that $\varphi(x, y)$ holds, then there exists a set whose members satisfy the formula*

$$\exists x \, (x \in X \wedge \varphi(x, y)).$$

That is,

$$\{y : \exists x \, (x \in X \wedge \varphi(x, y))\}$$

is a set.

Axiom of Regularity. *If x is a non-empty set, then there exists some $y \in x$ such that $x \cap y = \emptyset$.*

More formally,

$$x \neq \emptyset \to \exists y \, (y \in x \land x \cap y = \emptyset).$$

In fact, the Axiom of Regularity is rather an artificially invented axiom which avoids circular membership relations like $x \in x$ or $x \in y \in x$. The axiom was first introduced by John von Neumann [204] for conventional purposes and to avoid Russell's paradox (see Section 5.2). It also prevents membership related infinite regressions and infinite decreasing chains such as

$$x_1 \ni x_2 \ni x_3 \ni \dots.$$

By this way, every set can be built up from $\emptyset$ and can have a "rank" in the set theoretical hierarchy. It is also worth noting that, Axiom of Regularity does not really expand and enrich the set theoretical universe. On the contrary, it puts restriction on certain kinds of sets to exist. Sets that satisfy the Axiom of Regularity are called *well-founded*, otherwise we call them *ill-founded*. Although there is a rich literature on ill-founded sets, we are not going to study them in this book. Readers who are interested in ill-founded set theory may refer to Quine [230] or Aczel [1].

The set of axioms we have given so far is called ZF (Zermelo-Fraenkel Set Theory) axioms. The last axiom we shall introduce is the Axiom of Choice. All axioms together will form the axioms of ZFC. Before writing the Axiom of Choice, let us give some definitions.

Relations, Functions, and Orders.

If A and B are two sets, the *Cartesian product* of A and B is defined as

$$\{(a, b) : a \in A \land b \in B\}$$

and denoted by $A \times B$. For any given sets A and B, $A \times B$ is a set.[14]

Definition 2.4 Let Y and Z be two sets and let $R \subseteq Y \times Z$. If for all $x \in R$, there exist $y \in Y$ and $z \in Z$ such that $x = (y, z)$, then R is called a *relation*. The *domain* of a relation R is defined as

$$\mathrm{dom} R = \{x : \exists y \, (x, y) \in R\}.$$

The *range* of a relation R is defined as

$$\mathrm{ran} R = \{y : \exists x \, (x, y) \in R\}.$$

---

[14] We leave the proof of this fact to the reader. Hint: Note that for $x \in A$ and $y \in B$, since the ordered pair (x, y) is defined as $\{\{x\}, \{x, y\}\}$, (x, y) is an element of the set $\mathcal{P}(\mathcal{P}((A \cup B)))$.

We leave the reader the proof of the fact that these two collections are in fact sets.

Definition 2.5 Let X be a set, $R \subseteq X \times X$ be a relation, and let $x, y, z \in X$.

(i) If $(x, x) \in R$, then R is called *reflexive*.

(ii) If $(y, x) \in R$ whenever $(x, y) \in R$, then R is called *symmetric*.

(iii) If $(x, z) \in R$ holds whenever $(x, y) \in R$ and $(y, z) \in R$ hold, then R is called *transitive*.

(iv) If $x = y$ whenever $(x, y) \in R$ and $(y, x) \in R$, then R is called *anti-symmetric*.

Consider the usual $<$ relation defined on the set of natural numbers $\mathbb{N} = \{0, 1, 2, \ldots\}$. The relation $<$ is only a transitive, relation. The relation $\leq$ defined on the same set, however, is reflexive, transitive, and anti-symmetric. If a relation R satisfies the first three properties, then R is called an *equivalence relation*. If $R \subseteq X \times X$ is an equivalence relation, then for any $x \in X$, the set

$$\{y : (x, y) \in R\}$$

is called the *equivalence class* of x.

Definition 2.6 Let X and Y be two sets and let $f \subseteq X \times Y$ be a relation. If $\mathrm{dom} f = X$ and for every $x \in X$ there exists a unique $y \in Y$ such that $(x, y) \in f$, then f is called a *function* from X to Y. In this case, we write $f(x) = y$ instead of writing $(x, y) \in f$. If $a \in X$, we call a the *argument* of f. We denote a function f from X to Y by $f : X \to Y$.

If $f : X \to Y$ is a function such that $\mathrm{ran} f = Y$, then f is called *onto* (or *surjection*). If $f(x) = f(y)$ implies $x = y$, then f is called *one-to-one* (or *injection*). If a function is one-to-one and onto, then it is called *one-to-one correspondence* (or simply *bijection*). The *graph* of a function $f : X \to Y$ is the set $\{(x, y) : f(x) = y\}$. Let $f : X \to Y$ be a function. The *image* of a set $A \subseteq X$ under a function f is the set $\{y : f(x) = y \text{ and } x \in A\}$. The *preimage* of a set $B \subseteq Y$ under f is the subset of X defined by

$$f^{-1}(B) = \{x \in X : f(x) \in B\}.$$

Now we can give the Axiom of Choice. Let X be a collection of non-empty sets. If there exists a function f such that for every $x \in X$, we have $f(x) \in x$, then f is called the *choice function* of X.

Axiom of Choice. *Every collection of non-empty sets has a choice function.*

Let I be a set and let $i \in I$. When we are given sets A_i, we should understand that I is used for indexing the sets A_i over I. Here, i is referred to as an *index*, and I is called the *index set*. The collection A_i is usually referred to as a *family*. A family can basically be thought of as a collection whose members are indexed by I. A family is usually denoted by $\{A_i\}_{i \in I}$ or shortly by A_i, depending on the context. If I is an index set and $i \in I$, then the Cartesian product of the family A_i is

$$\textstyle\prod_{i \in I} A_i = \{f : f \text{ is a function on } I \text{ and for every } i \in I,\ f(i) \in A_i\}.$$

One equivalent form of the Axiom of Choice is related to the Cartesian product of a family of sets. The statement is that the Cartesian product of a family of non-empty sets is non-empty. That is, Axiom of Choice is logically equivalent to the statement

$$\text{"For a family } X_i \text{ of non-empty sets, } \textstyle\prod X_i \neq \emptyset \text{"}.$$

There are many other equivalent forms, but we will particularly be interested in this statement for a later discussion. Chapter 12 will be devoted to the Axiom of Choice where we will look at other equivalent forms and discuss some of the philosophical issues surrounding this axiom.

Definition 2.7 Let X be a set and let $R \subseteq X \times X$ be a relation.

(i) If R is reflexive, anti-symmetric and transitive, then R is called a *partial order*.

   In addition to this condition, if for every $x, y \in X$, either $(x, y) \in R$ or $(y, x) \in R$ holds, then R is a *total order* (or *linear order*).

(ii) Let X be a set and let $R \subseteq X \times X$ be a partial order on X. Then we call (R, X) a *partially ordered set*. In addition, if R is a total order, then (R, X) is called a *totally ordered set*.

Instead of denoting partially ordered or totally ordered sets by (R, X) as a pair, we simply denote them by the set X on which the relation is defined. It is almost standard to denote partial orders by the symbol $\leq$. We also write $x < y$ to mean $(x \leq y \wedge x \neq y)$.

Definition 2.8 Let P be a set, let $\leq$ be a partial order defined on P and let $a \in P$.

(i) If $a < b$ for no $b \in P$, then a is called a *maximal element* of P.

(ii) If $b \leq a$ for every $b \in P$, then a is called the *greatest element* of P.

(iii) If $b < a$ for no $b \in P$, then a is called a *minimal element* of P.

(iv) If $a \leq b$ for every $b \in P$, then a is called the *least element* of P.

Let P be a totally ordered set and let $X \subseteq P$. If $x \leq a$ for all $x \in X$, then a is called an *upper bound* of X. If $a \leq x$ for every $x \in X$, then a is called a *lower bound* of X. If a is an upper bound of X and $a \leq b$ for every upper bound b of X, then a is called the *least upper bound* (or *supremum*) of X. If a is a lower bound of X and $b \leq a$ for every lower bound b of X, then a is called the *greatest lower bound* (or *infimum*) of X. We shall denote the supremum of X by $\sup X$ and denote the infimum by $\inf X$. Note that the supremum and the infimum must be unique since if a and b were both supremum elements, say, then $a \leq b$ and $b \leq a$ would both hold and it would follow that $a = b$.

2.3.2   Ordinal and cardinal numbers

In ZFC, every object is a set. Even the most fundamental objects of mathematics, such as natural numbers or functions, are defined in terms of sets. For example, if we have to define the number 5 as a set, it is "natural" to describe it as a set containing five elements. We may define the natural number 5 as the set of natural numbers "smaller" than 5.

Definition 2.9 Let S be a totally ordered set. If every subset of S has a least element, then S is called a *well-ordered* set.

In terms of sets, we define natural numbers by well-ordered sets as

$$0 = \emptyset,$$
$$1 = \{0\} = \{\emptyset\},$$
$$2 = \{0, 1\} = \{\emptyset, \{\emptyset\}\},$$
$$3 = \{0, 1, 2\} = \{\emptyset, \{\emptyset\}, \{\emptyset, \{\emptyset\}\}\},$$
$$\vdots$$

We define the *successor* of a natural number n to be

$$n \cup \{n\}.$$

Although there are other notations for denoting the numbers in the set theoretic language, the notation given above originally goes back to John von Neumann, and we will adopt this notation from now on.

For now let us denote the successor of a natural number n by Sn. According to this definition, the number 4, for instance, is defined as

$$4 = \{0, 1, 2, 3\} = \{0, 1, 2, \} \cup \{3\} = 3 \cup \{3\} = S3.$$

We will later define the addition operation $+$ so as to satisfy $Sn = n + 1$.

Axiom of Infinity asserts the existence of at least one inductive set. Considering the fact that it posits the existence of sets containing "infinitely many" elements, accepting the existence of the set of natural numbers requires a serious metaphysical commitment. Let alone accepting an infinite totality, claiming that this totality is actually a *set* is a big metaphysical faith. We shall discuss this issue in Section 4.2 of Chapter 4 and partially in Section 9.1 of Chapter 9. We should now leave this and introduce now a special type of well-ordered sets.

Definition 2.10 Let A be a set. If $x \in A$ implies $x \subseteq A$, then A is called a *transitive set*. If a set is transitive and well-ordered by the $\in$ relation, then it is called an *ordinal*. For any ordinals α and β, we define

$$\alpha < \beta \text{ if and only if } \alpha \in \beta.$$

Moreover, $\alpha = \{\beta : \beta \in \alpha\}$.

Every natural number is an ordinal. The least "transfinite" ordinal which is greater than every natural number is denoted by ω and it is defined as the set of natural numbers $\mathbb{N}$. That is, $\omega = \mathbb{N}$. For every ordinal α, we define $\alpha + 1 = \alpha \cup \{\alpha\}$.

By the Axiom of Choice, we know that every set can be well-ordered. Let $(P, \leq)$ and $(Q, \leq)$ be partially ordered sets. We call a function $f : P \to Q$ *order-preserving* if $f(x) \leq f(y)$ whenever $x \leq y$ for $x \in P$ and $y \in Q$. Let $f : P \to Q$ be a bijection. If f is order-preserving, then f is an *isomorphism* between P and Q. In this case, we say that P and Q are *isomorphic*. Using the Axiom Schema of Replacement, one can show that every well-ordered set is isomorphic to an ordinal.

Definition 2.11 Let α and β be ordinals. If there exists some β such that $\alpha = \beta + 1$, then α is called a *successor ordinal*. Otherwise, α is called a *limit ordinal*.[15]

If there exists a one-to-one correspondence between a set A and a natural number n, then A is a "finite" set. According to this definition, 0 is a limit ordinal. However, it is still a natural number and so it is finite. The smallest non-zero limit ordinal is ω.

By the Axiom of Choice and the Axiom Schema of Replacement, we said that every set corresponds to an ordinal number. Furthermore, every set has also a "cardinal" number. We will define cardinal numbers from ordinals. However, we should be more careful in choosing which ordinals deserve to be called cardinals. Cardinals will be ordinal numbers which satisfy a specific property.

Before giving the definition of cardinal numbers, let us first define equipollency. Let X and Y be two sets. If $f : X \to Y$ is a bijection, then X has the

---

[15]The reader may ask, at this point, what kind of ordinals the set of limit ordinals really contains. In order to define an object in terms of the complement of another object, we need to know beforehand exactly which set we are taking the complement of. If an ordinal is not a limit ordinal, then it is a successor ordinal. So the set of successor ordinals is defined as the complement of the set of limit ordinals. But then what exactly is the complement of the set of limit ordinals? We naturally think that we are taking the complement of limit ordinals *with respect to* the class of *all* ordinals. But then which objects are included in the set of non-successor ordinals? Does it solely contain limit ordinals? If the answer is yes, then it follows that we pretty much know everything about the class of *all* ordinals and that the complement of the set of successor ordinals, with respect to the class of *all* ordinals, only contains limit ordinals and nothing more. If not, what else is there other than limit ordinals and successive ordinals?

same *cardinality* as Y, and we denote this by $|X| = |Y|$. Here, $|X|$ denotes the cardinality of X. If two sets have the same cardinality, we say that they are *equipollent*. Cantor's Theorem (see Theorem 9.1) tells us that for any set X, we have that $|X| < |\mathcal{P}(X)|$.

Definition 2.12 Let α be an ordinal. If $|\alpha| \neq |\beta|$ for every $\beta < \alpha$, then α is called a *cardinal*.

So a cardinal number is an ordinal which is not equipollent to any smaller ordinal. Every set is equipollent to a unique cardinal.

Definition 2.13 Let X be a set. If $|X| \leq |\mathbb{N}|$, then X is called a *countable* set. If $|X| = |\mathbb{N}|$, then X is called *countably infinite*. If $|\mathbb{N}| < |X|$, then X is *uncountable*. If $|X| < |\mathbb{N}|$, then X is called a *finite* set.

Every natural number is a finite cardinal number. The smallest infinite cardinal is ω. When we want to emphasise that we are working in the cardinal number context, we will denote it by $\aleph_0$.

Recall that we defined the successor of a natural number n as $Sn = n \cup \{n\} = n + 1$. We shall use the same definition to define the successor of an ordinal. We define the *successor* of an ordinal α to be $S\alpha = \alpha \cup \{\alpha\}$. In general, if α and β are ordinals, then their sum is defined inductively as follows:

$$\alpha + 0 = \alpha.$$
$$\alpha + S\beta = S(\alpha + \beta).$$
If γ is a limit ordinal, then $\alpha + \gamma = \bigcup_{\beta < \gamma}(\alpha + \beta).$

Ordinal multiplication is defined as follows:

$$\alpha 0 = 0.$$
$$\alpha S(\beta) = \alpha\beta + \alpha.$$
If γ is a limit ordinal, then $\alpha\gamma = \bigcup_{\beta < \gamma} \alpha\beta.$

What remains now is to define ordinal exponentiation. We define exponentiation similarly as follows:

$$\alpha^0 = 1.$$
$$\alpha^{S(\beta)} = \alpha^\beta \alpha.$$
If γ is a limit ordinal, then $\alpha^\gamma = \bigcup_{\beta < \gamma} \alpha^\beta.$

If α and β are ordinals, then $\alpha + \beta$, $\alpha\beta$, and α^β are also ordinals. Note that ordinal addition and multiplication are not commutative. For instance, for any natural numbers $n \geq 1$ and $m > 1$

$$n + \omega = \omega \neq \omega + n$$
$$m \cdot \omega = \omega \neq \omega \cdot m = \underbrace{\omega + \cdots + \omega}_{m \text{ times}}.$$

The Principle of Mathematical Induction can be generalised to transfinite ordinals. This yields the *transfinite induction* theorem. The Principle of Mathematical Induction merely consists of the basis and induction steps. Extending this principle to transfinite ordinals, we now have an induction step for limit ordinals.

Theorem 2.2 (Transfinite Induction) Let C be a class of ordinals and assume that:

(i) $0 \in C$;

(ii) if $\alpha \in C$, then $\alpha + 1 \in C$;

(iii) if α is a non-zero limit ordinal and $\beta \in C$ for all $\beta < \alpha$, then $\alpha \in C$.

Then, C is the class of all ordinals.

Alephs. We can talk about two types of numbers: ordinal numbers and cardinal numbers. Now, the ordinals

$$\omega + 1, \omega + 2, \omega + 3, \dots, \omega + \omega, \dots, \omega^2, \omega^2 + 1, \omega^2 + 2, \dots, \omega^\omega, \dots, \omega^{\omega^\omega}, \dots$$

are actually equipollent to ω. So they are all countable ordinals. Sometimes the cardinal ω is denoted by ω_0, referring to the least infinite cardinal. Using Cantor's Theorem, we can show that for any cardinal, there is a larger cardinal. In fact, the smallest cardinal number which is greater than ω_0 is defined as the ordinal number

$$\omega_1 = \bigcup_{|\alpha| \leq \omega_0} \alpha$$

where α is an ordinal. If β is a successor ordinal, then $\beta = \alpha + 1$ for some α, and in this case, ω_β is defined to be the smallest cardinal which is greater than ω_α. If γ is a limit ordinal, then, for $\alpha < \gamma$, we define

$$\omega_\gamma = \bigcup_{\alpha < \gamma} \omega_\alpha.$$

We sometimes may wish to refer to the same object in different contexts using different symbols. For every ordinal α, ω_α is at the same time an infinite cardinal number. Instead of using the ω (omega) notation, we denote infinite cardinal numbers by the $\aleph$ (aleph) notation. That is, for every ordinal α

$$\aleph_\alpha = \omega_\alpha.$$

Using the Axiom of Choice, we can define the successor of a cardinal κ as

$$\kappa^+ = |\inf\{\lambda : \lambda \text{ is an ordinal and } \kappa < |\lambda|\}|.$$

Despite that cardinals are well-defined mathematical objects, it is not easy to understand the relationship between them. In the proceeding chapters, we will often remind the reader that ZFC is not strong enough to determine the relationship between an infinite cardinal and the next smallest cardinal that proceeds it.

Cofinality. We now give an order-theoretic notion concerning infinite sequences. Let $\alpha > 0$ be a limit ordinal. For a limit ordinal β, we say that a β-sequence (sequence of length β) $\langle \alpha_\xi : \xi < \beta \rangle$ is *cofinal* in α if $\lim_{\xi \to \beta} \alpha_\xi = \alpha$. Similarly, $A \subseteq \alpha$ is cofinal in α if $\sup A = \alpha$. If α is an infinite limit ordinal, the *cofinality* of α is denoted by $\mathrm{cf}(\alpha)$ and it is defined as the least limit ordinal β such that there exists an increasing β-sequence $\langle \alpha_\xi : \xi < \beta \rangle$ such that $\lim_{\xi \to \beta} \alpha_\xi = \alpha$.

Definition 2.14 An infinite cardinal $\aleph_\alpha$ is called *regular* if $\mathrm{cf}(\omega_\alpha) = \omega_\alpha$. It is *singular* if $\mathrm{cf}(\omega_\alpha) < \omega_\alpha$.

By the Axiom of Choice, we know that every infinite successor cardinal is a regular cardinal. Moreover, the following statements hold.

(i) If λ is an uncountable cardinal, then $\mathrm{cf}(\lambda)$ is always a regular cardinal.

(ii) If λ is regular, then $\mathrm{cf}(\lambda) = \lambda$.

(iii) If λ is singular, then $\mathrm{cf}(\lambda) < \lambda$.

The reader should feel free to refer to this chapter, or other texts, back and forth, whenever we use these mathematical notions in the proceeding chapters. Though we will occasionally remind the reader necessary definitions in the related chapters.

Review and Discussion Questions

1. Prove De Morgan's Law using the truth table method.

2. Write the truth table of the following propositions.

    (a) $\neg p \to (q \vee \neg r)$.

    (b) $(p \wedge \neg q) \leftrightarrow (\neg q \to r)$.

    (c) $\neg((p \to q) \to (\neg p \to \neg r))$.

3. Consider the proposition p : "The ball bounces whenever it hits the ground".

    (a) Write the contrapositive of p.

    (b) Write the negation of p.

    (c) Write the inverse of p.

4. Write the negation of the proposition "If a number is Brouwerian, then it is neither positive nor negative".

5. A statement of propositional logic is *consistent* if there is a valuation of the propositional variables which makes the proposition true. Is $p \rightarrow \neg p$ consistent?

6. Let $A(x)$ mean "x is an artist" and let $C(x)$ mean "x is creative". Translate the following statements to English.

   (a) $\forall x(\neg A(x) \lor C(x))$.
   (b) $\forall x(A(x) \rightarrow \neg C(x))$.
   (c) $\neg \exists x(A(x) \land \neg C(x))$.
   (d) $\exists x(A(x) \lor C(x))$.

7. Let $A(x)$ and $C(x)$ be defined as previously. Write in English the negation of the following statements.

   (a) $\exists x \neg C(x)$.
   (b) $\forall x(A(x) \rightarrow \neg C(x))$.

8. Consider the statement "Either every artist is creative or there is a non-artist who is not creative whenever there is an artist who is creative". What is the negation of this statement?

9. Let $S(x, y)$ mean "student x has class y". Translate the following statements of predicate logic to English.

   (a) $\forall x \exists y S(x, y)$.
   (b) $\forall x \neg \exists y S(x, y)$.
   (c) $\neg \forall x \exists y S(x, y)$.
   (d) $\exists x \forall y S(x, y)$.
   (e) $\exists x \neg \exists y S(x, y)$.

10. Prove by using direct proof that every even integer can be written as the sum of two odd integers.

11. Prove the statement above using proof by contrapositive.

12. Let a, b, and c be natural numbers. Prove that a divides c whenever a divides b and b divides c.

13. (Pigeonhole Principle) Prove that if $n + 1$ pigeons are placed in n pigeonholes, then one pigeonhole must contain at least two pigeons. Use proof by contradiction.

14. Prove the Pigeonhole Principle using proof by induction.

15. Suppose that x and y are real numbers. Prove that if xy is irrational, then either x or y is irrational.

16. Prove or disprove the following statements.

   (a) For every non-negative number s, there exists a non-negative number t such that $s \geq t$.

   (b) There exists a non-negative number t such that for all non-negative numbers s, it holds that $s \geq t$.

   (c) For every non-negative number t, there exists a non-negative number s such that $s \geq t$.

   (d) There exists a non-negative number s such that for all non-negative numbers t, it holds that $s \geq t$.

17. Consider the following arguments and determine what kind of fallacy occurs in each.

   (a) If Romeo loves Juliet, then Juliet loves Romeo. And if Juliet loves Romeo, then Romeo loves Juliet. Therefore, Romeo loves Juliet.

   (b) If there are Martians on Earth, then there are humans on Mars. There are no Martians on Earth. Therefore, there are no humans on Mars.

   (c) If John wears black suit, then Jack wears Grey suit. Jack wears Grey suit. Therefore, John wears black suit.

   (d) Numbers exist if and only if sets exist. Therefore, both numbers and sets exist.

   (e) Numbers exist if and only if sets exist. Suppose numbers exist if and only if functions exist. Therefore, sets exist.

18. Using solely the first five axioms of ZFC, describe how one could define, for example, negative integers in set-theoretical notation.

19. Let A be a set. The function $f : \emptyset \to A$ is called the *empty function*. Show that the empty function is unique.

20. Show that for any two given finite sets A and B, if $f : A \to B$ and $g : B \to A$ are one-to-one functions, then there exists a bijection between A and B.

21. Find which properties the following relations have.

   (a) The property that "line a is perpendicular to line b" in the two-dimensional plane.

   (b) The property that "a is a sibling of b" in the domain of all people.

   (c) The property that "a owes money to b" in the domain of all people.

   (d) The property that "a is divisible by b", for natural numbers a and b.

22. Let $\mathbb{N}$ and $\mathbb{Z}$ denote, respectively, the set of natural numbers and the set of integers.

(a) Define a one-to-one but not onto function $f : \mathbb{N} \to \mathbb{Z}$.

(b) Define an onto but not one-to-one function $f : \mathbb{N} \to \mathbb{Z}$.

(c) Define a bijection between $\mathbb{N}$ and $\mathbb{Z}$.

23. Let $P = \{2, 4, 6, 8, 10, 12\}$ be a set and let $a|b$ denote the divisibility relation on P, i.e., a divides b. Find the maximal, minimal, least, and greatest elements of the poset $(P, |)$, if any.

24. Define a poset in which there are infinitely many maximal and infinitely minimal elements.

25. Let A be a set, and let x be some element (not necessarily in A). Prove that $|A \times \{x\}| = |A|$.

26. Prove that for any set A, A is infinite if and only if it contains an infinite subset.

27. Show that the set of all words over the Latin alphabet is countably infinite.

3

Platonism

"To them, I said, the truth would be literally nothing but the shadows of the images." —Plato, The Republic, Book VII

In this chapter, we introduce one of the oldest and most fundamental views in philosophy of mathematics which goes back to Plato (427 B.C.–347 B.C.), called *Platonism*, often referred to as *realism*. In order to explain this philosophical position, we should first look at Plato's ontological and epistemological philosophy.

Plato, with no doubt, is regarded as one of the most influential figures in the history of philosophy. His ideas have influenced many philosophers for centuries and his works have survived for more than two millennia. Despite that Plato stayed loyal to the ideas of his mentor Socrates, he developed his own views in his late period. Plato's views ultimately rest on two doctrines: *Theory of Forms* and *immortality of the soul*. Much of his works are generally based on these two themes. In fact, as we shall mention later, immortality of the soul is intertwined with the theory of forms, as they are supposed to complement each other. These themes were studied thoroughly in one of the most important works of Plato, *The Republic* [216].

3.1 Theory of Forms

We will be particularly interested in Plato's philosophy of mathematics. However, to investigate this, one should first look at the *theory of forms* (or *theory of ideas*) which is really the backbone of Plato's ontological philosophy. Theory of forms was emerged from the quest for finding the *definition of virtue* in the Socratic period. The word *idea*, however, was first used in the *Euthypro* [212] dialogue so as to define "piety" (holiness), the universal property that all pious things have in common.[1] The eternal, unchanging, and unique property that all pious things have in common was referred to as *idea*.[2] All courageous people have a certain property in common, which is nothing but the *form of*

---

[1] *Euthypro*, 10A–11B. Also in Guthrie, p. 114, 1995.
[2] The word *idea* may also be interchanged with the term *form* or *universal*.

DOI: 10.1201/9781003223191-3

courage. Essentially, the form of an object is a definition which refers to the property such that all particular objects of that kind have in common.

According to Plato, every object and quality has a form. There are two worlds in Plato's ontology: the *world of becomings* (or the physical world) and the *world of beings* (or the world of forms).[3] Physical objects are subject to change, they can break down, they are dynamic, fragile, and imperfect, whereas forms are static, eternal, immortal, and perfect.

For Plato, reality should be sought within the world of forms. He explains that the reality is independent and separate from the physical existence by the famous *cave allegory* given in Book VII of *The Republic*. Imagine a number of prisoners, sitting in a cave, all of whom are chained to the ground and unable to move their body and head. Suppose that they only see the wall in front of them. On a platform above, just behind the prisoners, assume that there is a fireplace and imagine people passing right in front of the fire holding all sorts of puppets of human and animal shape. The prisoners only see the shadows of the puppets, perhaps thinking that they are real humans and real animals. The reality, for the prisoners, is nothing but the shadows and reflections on the wall. What is more real, however, is what they do not see. Once they are free, they will be able to see the fire behind them. They will later see entire cave and perhaps the outside. Eventually, they will be able to connect more with the reality by seeing the Sun. In this allegory, Plato associates the reflections on the wall with the objects of the physical world and associates the Sun with the forms. The physical world is merely an imperfect and mortal manifestation of the reality. The allegory explains that there is a higher reality beyond the physical world and our sensory experience. According to Plato, the reality should be sought within the world of forms. Since every object of the physical world imitates an imperfect instance of the form, and since physical objects exhibit a restricted type of reality, the realm of becoming does not tell us much about the absolute reality.

The world of forms is usually referred to, mostly by philosophers of mathematics, as the *Platonic universe*. According to Plato, forms exists independent of the human mind, language, senses, space/time, and physical world. They are non-physical, non-spatiotemporal, and non-mental *abstract* beings. Every object or quality in the realm of becoming bears some relation to its form. For example, each book we see around us is a *particular* object. However, there is only one *form* of book that grants all instances of books their physical existence. This form is a *universal* and it is independent of every physical object. Plato distinguishes *beings* from *becomings*. Beings consist of forms. Becomings are merely the physical manifestations of forms. All objects that are perceived by sensory experience are particulars, whereof the form of an object is universal and it is the property that all objects of that kind have in common.

---

[3]This assumption makes Plato a dualist philosopher. In the philosophy of mind, *dualism* is the theory that body and mind are distinct and separable.

The most fundamental assumption in Plato's ontological philosophy is then the following: forms *exist* as abstract beings and they are *independent* of us.

All forms are, in fact, placed and arranged in a hierarchy in the Platonic world. The highest form is the *Form of the Good,* since objects, according to Plato, aspire to be good.[4] For Plato, physical things are rooted in forms, but forms are rooted in the Form of the Good. Thus, the Form of the Good is a property that all other forms have in common. All objects imitate the Form of the Good and so "goodness" is a feature that every object exhibits. Then, the following observation would constitute the basis of Plato's ontological philosophy: *existence and the goodness are correlated.* That is, for Plato, the Form of the Good exists eternally, and since the source of all objects comes from the Form of the Good, any existing object imitates goodness and improves itself towards perfection. The strong connection between existence and goodness forms the groundwork of Plato's ontological philosophy. In this aspect then, all objects in the realm of becomings imitate their corresponding forms to complete themselves and evolve into perfection and into becoming "The Good".

3.2   Plato's Epistemological Philosophy

We want to know how the realm of forms and the realm of physical objects are linked to each other. Then, the question we shall ask is the following: *How can one reach to the forms and come in contact with the realm of beings?* As a natural consequence of asking this question, the *immortality of the soul* was emerged together with the theory of forms so as to find an answer. For Plato, as described in the dialogue Phaedo, the soul resembles what is divine and immortal. In fact, the soul is a thing that resembles the forms most.[5] He presents four different arguments to prove the fact that, although the body of the human perishes after death, the soul still exists and remains eternal. It is the soul through which we get to experience the forms.

Plato's epistemological philosophy is based on the idea of *recognising* and revealing knowledge that is already present in the soul. As described in the dialogue Meno, we do not learn knowledge, but we *recognise* it.[6] According to Plato, as it was argued in the Statesman dialogue, the sole method of acquiring true knowledge and establishing truth is via the *dialectical method,* also came to known as the *Socratic method.*[7] Dialectic is a type of discourse between two or more people holding different positions about a subject but wishing to establish the truth through reasoned arguments by arguing on both

---

[4]Plato, *The Republic,* 509b.
[5]Plato, *Phaedo,* 78c-80b.
[6]Plato, *Meno,* 85d.
[7]Plato, *Statesman,* 266d.

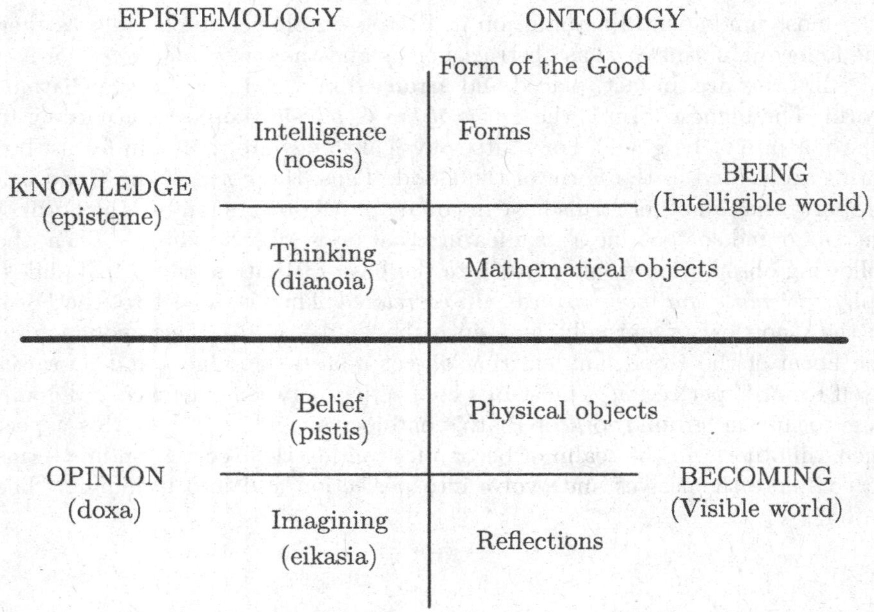

EPISTEMOLOGY   ONTOLOGY

Form of the Good

	Intelligence (noesis)	Forms	
KNOWLEDGE (episteme)			BEING (Intelligible world)
	Thinking (dianoia)	Mathematical objects	

	Belief (pistis)	Physical objects	
OPINION (doxa)			BECOMING (Visible world)
	Imagining (eikasia)	Reflections	

FIGURE 3.1
According to the divided line, mathematical entities exist as abstract objects in the realm of being.

sides, expounding truth, and exposing errors. Dialectic comprises three stages of development. First, a thesis or statement of an idea, which gives rise to a second step; a reaction or antithesis that contradicts or negates the thesis. Third, the synthesis, a statement through which the differences between the two points are resolved. At the base of the dialectical method, there exists reasoning and argumentation. For this very reason, the Socratic method gave birth to the epistemological view called *rationalism*, which is described as the view that true knowledge is acquired, for the most part, by reasoning.

Now we shall discuss how mathematics comes into play. Since one reaches the realm of forms through reasoning, mathematics must also take part in the process of the discovery of truth and reality. True knowledge, as eternally exists in the realm of beings, is attained through the forms. Plato considers mathematical objects as non-mental, non-physical, non-spatiotemporal, static, and eternal beings. Similarly, on Plato's epistemological philosophy, mathematical knowledge is assumed to bear true knowledge, if not exactly as perfect as the Form of the Good. In Book VI of *The Republic*, Plato gives the *divided line* to explain the difference between the realm of being and the realm of becoming (see Figure 3.1). It is clear to observe that mathematical objects in this hierarchy is placed above all physical objects.

As it can be seen in Figure 3.1, Plato separates beings, i.e., forms and mathematical entities, from becomings, i.e., physical entities and reflections. The realm of *becoming* consists of two levels of apparent realities where physical objects are assumed to have a higher sense of reality than that of reflections. Similarly, the realm of *being* contains two levels of realities where forms are located over mathematical objects. Notice that the *Form of the Good* is placed at the highest level of the realm of being. It can be inferred from the divided line that physical objects are merely thought of as imperfect representations of mathematical entities. A circle in the realm of becoming, for instance, can only exist as an imperfect physical manifestation of the "Form of the circle" as a "circular-like" object. That is to say, every object in the physical universe is an imperfect *instantiation* of its form. None represents the perfect, idealised, static, and eternal circle, as perfect things can only exist in the realm of being. Consider, as another example, a line drawn on a paper. We may even have used a ruler. Nevertheless, for Plato, the drawn line does not represent the perfect Form of the line, but it rather denotes a perishable and imperfect instance. To make more sense of this, it is worth examining the paragraph from *The Republic* we give below.

> This at least will not be disputed by those who have even a slight acquaintance with geometry, that this science is in direct contradiction with the language employed in it by its adepts. [...] Their language is most ludicrous, though they cannot help it, for they speak as if they were doing something (adding, extending auxiliary lines, etc.) and as if all their words were directed towards action. For all their talk is of squaring and applying and adding and the like, whereas in fact the real object of the entire study (of geometry) is pure knowledge [...] the knowledge of that which always is, and not of a something which at some time comes into being and passes away. [...] for geometry is the knowledge of the eternally existent.[8]

Assume for a moment that Plato is right about that the knowledge of geometry is pure, static, and eternal. Of course, in this case, geometry should not use such a dynamic language, and moreover, the language and the methodology should be congruent with the nature of geometry. The constructions in Euclidean geometry, if they really extend our knowledge and introduce new entities, do not mean much to a Platonist. Euclidean geometry is based on a non-temporal planar system on which we perform dynamical constructions like rearranging geometrical objects, adding or extending lines and so on.

Apart from the ontological questions, we may also ask questions about the epistemology of mathematics. Our primary concern in epistemology is the nature of mathematical knowledge and how it is acquired. Not only can realism

---

[8]Plato, *The Republic*, 527a. Parentheses added.

be interpreted in the domain of ontology, but also in epistemology of mathematics, particularly regarding the truth value of propositions. *Epistemological realism* argues for that every mathematical proposition has an absolute truth value independent of the human mind, language, physics, etc. According to Plato, mathematical truth is independent of the physical universe, hence it lies beyond our sensory experience.

We shall now look at Plato's interpretation of numbers and counting. His views on this matter were partly given in the *Sophist* [217] dialogue.[9] For Plato, the general platonic distinction between two realms remains true for numbers and their instantiations. On the one hand, there are quantities that make reference to physical entities. On the other hand, there are numbers in their pure abstract form, which exist in the realm of being, independent of any physical instantiations. Numbers that are used to quantify physical objects are perceived by sensory experience. The pure form of a number, however, is perceived through rational grounds. Plato, in his *Philebus* [215], mentions two types of arithmetics: *daily arithmetic* which is really used for ordinary purposes, and *philosopher's arithmetic* that is inevitably used for the purpose of contemplation and finding the truth.[10] In daily arithmetic, one works with mutually non-identical physical objects. That is to say, for ordinary people, the number "two" is really just limited to two books, two cups, two desks, etc. No two physical objects are thought to be identical to each other. For the philosopher, however, the situation is supposed to be slightly different. In philosopher's arithmetic, physical objects are irrelevant when talking about numbers since philosophers tend to consider the pure abstract form of the number "two", for example, that is independent of any of its physical instance. The quantified entity, in philosopher's arithmetic, is not thought to be a physical object, but thought to be its abstraction as a "unit" in which all the physical properties of the quantified object are omitted. So the quantification is merely over *abstract units* in an equivalent class. In either case, both in daily arithmetic and philosopher's arithmetic, the quantification is performed over a *thing* as we cannot think of using the number "two" without introducing a noun to be quantified. For daily arithmetic, that *thing* is considered to be a physical object, whereas it is considered as an abstract unit in philosopher's arithmetic.

When we explained Plato's epistemological philosophy, we said that knowledge was rather recognised. Attaining knowledge, thereof, is about discovering and recognising what has already been present eternally in the soul. The same idea holds for mathematical knowledge as well: Mathematical knowledge is recognised, not learned. Since Platonism argues for the independence of all sorts, both metaphysically and epistemologically, mathematics is not invented. In fact, according to realism, mathematics is discovered. Let us give a short argument as to why this is the case. Realism is based on the

---

[9] Plato, *Sophist*, 238a.
[10] Plato, *Philebus*, 56.

assumption that mathematical objects exist eternally (past and present) and that they are independent of the mind, language, and sensory experience. What eternally exists has no beginning. However, all invented objects must have a point in time prior to which they did not exist. Since the act of invention conflicts with the idea of past eternal existence, inventing mathematical objects and mathematical knowledge is impossible in realism. This, in turn, can determine as to whether mathematics is a science or an art form. For a realist, since mathematical activity is merely a discovery, mathematics can be classified as a science.

3.3 Aristotelian Realism

Aristotle (384 B.C.–322 B.C.), along with Plato, is one of the most important and most influential ancient philosophers. He is one of the few great thinkers who studied wide range of fields of philosophy, from logic to ethics, from rhetoric to politics, from physics to metaphysics. Although we started our investigation with Plato, it would not be a complete study without discussing Aristotle's view on philosophy of mathematics, as he criticises Plato's fundamental ideas. Aristotle's writings regarding the philosophy of mathematics can be found, for the most part, in Books M and N of his *Metaphysics* [9].

Rationalism, as we mentioned earlier, is a philosophy of knowledge which claims that true knowledge is only attained by reasoning. In other words, rationalism regards all knowledge acquired through sensory experience as *doxa*, i.e., common belief or opinion. Whereas, according to Plato, *episteme* (true knowledge) can only be attained through reason. In contrast with rationalism, *empiricism* is the view that true knowledge is acquired by sensory experience. In the scene of history of philosophy, there have been many debates and discussions revolving around rationalism and empiricism as two views are regarded as competing beliefs. There were times such as the ancient Greek period and the Age of Enlightenment when rationalism was the popular belief. There were also times after the 17th century when empiricism became more influential, particularly with the works of philosophers from the British empiricism school. Just as Plato is regarded as one of the founders of rationalism, we may regard Aristotle as the philosopher who planted the seeds of empiricism.

Aristotle does not object the idea that forms exist. He believes that forms exist as perfect ideal beings. Yet he opposes to the idea that forms are independent of the physical reality. This is the primary difference between Plato and his pupil. According to Aristotle, form is the *essence* of a physical substance that makes the object what it fundamentally is. Aristotle claims that forms are in fact dependent on physical objects, for if we destroyed all books, for example, there would be no form of book. In Plato's philosophy, forms are ontologically prior to physical objects. This is not the case to be in Aristotle, as forms and physical objects ontologically depend on each other. Forms require physical objects to actualise themselves; a physical object, on the other hand, remains chaotic without its form. The question that should be asked for an Aristotelian realist, thereof, is not whether or not mathematical objects exist but *in what* they exist, if they do so.

Aristotle claims that mathematical objects are not actually independent from the objects of the physical realm. This is supported in a paragraph in Book II of *Physics* as follows.

> The next point to consider is how the mathematician differs from the physicist. Obviously physical bodies contain surfaces and volumes, lines and points, and these are the subject-matter of mathematics. [...] Now the mathematician, though he too treats of these things, nevertheless does not treat of them as the limits of a physical body; nor does he consider the attributes indicated as the attributes of such bodies. [...] While geometry investigates physical lines but not qua physical, optics investigates mathematical lines, but qua physical, not qua mathematical.[11]

Physical objects embody mathematical concepts like surface, line segments, and points in the form of their physical instantiations. According to Aristotle, the geometer does not imagine these surfaces, lines, etc., as physical counterparts. It seems that we distinguish the physical surfaces, lines, and points from their abstract mathematical idealisations. If Aristotle was right about his view, then this would undermine Plato's idea that forms are ontologically independent of physical objects.

We may observe from Aristotle that a type of *abstraction* process is being introduced. Let us consider a sphere-like physical object, an orange for instance. The concept of sphere, from an Aristotelian point of view, is obtained as a result of omitting the physical properties of the orange. An orange does not represent a perfect sphere as it may be uneven, the skin may be deeply textured, etc. Let us omit these faults. Furthermore, we omit its flavour, colour, weight, and similar physical properties. If we continue to carry out this abstraction process sufficiently long, we should eventually be able to reach to an abstract idealised sphere. So in Aristotelian realism, mathematics, to some extent, is ultimately built on sensory experience.

Perhaps Aristotle's one of the biggest contributions to philosophy is his distinction of *potentiality* and *actuality*. The concept of potentiality, originally translated from the ancient Greek word *dunamis*, generally refers to any "possibility" that a thing can be said to have. The notion of potentiality has been used throughout the history of science, particularly in classical mechanics by Newton. Apart from its physical meaning, the same concept has been used in mathematics to emphasise the type of treatment of the notion of infinity. Aristotle's philosophy of mathematics suggests that the concept of infinity

---

[11] Aristotle, Physics, Book II, 193b23–194a12.

should be rather taken as a potential instead of a completed totality. We will discuss this issue in Chapter 9 for its appropriateness. But let us now look at how objects of geometry come into being from an Aristotelian viewpoint.

> It is an activity also that geometrical constructions are discovered; for we find them by dividing. If the figures had been already divided, the constructions would have been obvious; but as it is they are present only potentially. Why are the angles of the triangle equal to two right angles? Because the angles about one point are equal to two right angles. If, then, the line parallel to the side had been already drawn upwards, the reason would have been evident to any one as soon as he saw the figure [...] Obviously, therefore, the potentially existing constructions are discovered by being brought to actuality; the reason is that the geometer's thinking is an actuality; so that the potency proceeds from an actuality; and therefore it is by making constructions that people come to know them.[12]

This is a clear indication that Aristotle's philosophy of mathematics—geometry, in particular—resembles that of Kant's as we will see. According to Lear, the virtues of Aristotle's philosophy of mathematics are also seen by comparing it with Hartry Field's nominalism.[13] For a detailed account of Aristotle's philosophy of mathematics, we refer the reader to Apostle [7] or the paper by Lear [176] for a general review.

3.4 Summary

Platonism, in summary, can be said to be a philosophical position which claims that the (mathematical) reality is independent of the mind, language, physics, etc. Mathematical objects are non-physical, non-spatiotemporal, and non-mental abstract entities. The truth should be seeked within nothing but the realm of being. For Plato, physical objects are the reflections and imperfect manifestations of the abstract forms. Of course, there is an ontological assumption behind this view that forms do exist and they are independent of us. On the one hand, there are *particulars*, i.e., objects of the physical world. On the other hand, there are *universals*, that is, forms which exist eternally and independent of the physical reality. The distinction between the two worlds should be clear that physical objects are subject to change and deteriorate, whereas forms are abstract, static, and eternal. For Plato, the only way to find the truth is through the knowledge of forms by applying the

---

[12] Aristotle, Metaphysics, Θ9, 1051a21–31.
[13] Lear, p. 187, 1982.

dialectical method. In *The Republic*, Plato says that it is the soul through which we come in contact with the forms since the soul is what resembles the forms most. Platonism argues for that true knowledge, i.e., *episteme*, is solely acquired by reasoning. Any other experience remains as an opinion or, in other words, *doxa*. Rationalism is the belief that only (or mostly) reasoning gives us true knowledge. Since mathematics is a gateway to the world of forms from the physical universe, Plato even emphasised on the importance of mathematics to the point that the entrance of the *Academy* was engraved with inscription "Let no one ignorant of geometry enter here". On ontological realism, mathematical objects exist independently in their own Platonic universe. Similarly, epistemological realism claims that every mathematical proposition has an independent absolute truth value regardless of the mind, language and the physical universe. Hence, a realist merely discovers what is true about the relationship between the mathematical objects in the Platonic universe.

Aristotle, on the other hand, agrees with Plato on the idea that forms exist, yet he sets himself apart from his mentor by disagreeing with the assumption that they are independent from the physical objects. This objection makes Aristotle one of the initiators of *empiricism*. According to Aristotle, form is the *essence* of a physical substance, contained in the object as a potential, that makes the object what it fundamentally is. Forms and physical objects are ontologically dependent on each other. Had we destroyed all tables in the physical universe, there would be no table form.

In the later history of philosophy, realism was adopted by many great thinkers such as Descartes, Leibniz, Frege, and Gödel. In addition, W.V.O Quine and Penelope Maddy were partly influenced by the realist philosophy. As a matter of fact, most philosophers had their own interpretations and so they introduced their own versions of realism. We will particularly study Quine, Gödel, and Maddy separately in Chapter 13.

Discussion Questions

1. What might be the main problems of ontological and epistemological realism?

2. Can we say that the idea of equivalence classes is implemented in the theory of forms? If so, to what extent universals and equivalence classes resemble each other?

3. Discuss how we can separate mathematical objects from their forms in Plato's divided line analogy. Does this mean that the "form of circle" has a higher sense of reality than an arbitrary circle in the general mathematical sense?

4. Do you think there is any motivation in believing the existence of inconsistent objects in the Platonic realm?

5. On Aristotelian realism, forms cannot exist without an object it can manifest itself within. If we had destroyed all tables in the universe, there would be no "form of table". Does this classify Aristotelian realism as a type of time-dependent realism?

6. Discuss that in what aspects Aristotle's philosophy of mathematics influenced the philosophers who came after him.

4

Intuitionism

"Mathematics is nothing more, nothing less, than the exact part of our think-
ing." — L. E. J. Brouwer, 19XX (From the MacTutor History of Mathematics
archive, University of St. Andrews)

In this chapter, we will study a philosophical position called *intuitionism*.
We shall first discuss Immanuel Kant, one of the most influential thinkers of
modern age philosophy. The reason why we introduce Kant first is because his
work is usually regarded as an early version of intuitionism in philosophy of
mathematics. We will later talk about L. E. J. Brouwer, who is known for his
constructivism, established in the early 20th century, which was influenced by
Kant's work.

4.1   Kant

Immanuel Kant (1724–1804), for his revolutionary works, is regarded as one
of the most greatest philosophers in Western philosophy ever since Plato and
Aristotle. He criticised and abandoned the classical conception of metaphysics
and epistemology, and with an analogy, made the Copernicus revolution of
philosophy. The Copernican turn was based on the idea that how human *mind*
(or the *subject*) was central to the theory of knowledge. Prior to Kant, it was
assumed that the object was central in attaining knowledge. Whereas for Kant,
knowledge is acquired through synthesis of the object and the subject. For
his emphasis on the mind, and a subject-central ontology and epistemology,
Kant is regarded as an idealist philosopher which separates him from the
philosophers who lived before him.

Kant's contribution in philosophy of mathematics starts by adapting Aristotle's seeds of empiricism to mathematics. His ideas on this subject can be
found, for the most part, in the *Critique of Pure Reason* [161]. Although it will
not be possible in our book to examine this great philosopher's work, which
was a turning point in the history of metaphysics, we will try to summarise
some of the parts in the *Critique* in order to understand Kant's philosophy of
mathematics. Readers may refer to the *Critique of Pure Reason* for a detailed

DOI: 10.1201/9781003223191-4

account. We will use the standard reference format A/B for the *Critique*, respectively denoting the first/second edition.

According to Kant, there are two types of knowledge, namely *a priori* and *a posteriori*.[1] In Kant's epistemological philosophy, *a priori* statements are universal in the sense that they are necessarily true or false. Moreover, they do not need sensory experience for justification.[2] They are not required to be justified by any empirical evidence to determine the truth value as it can be realised solely by reasoning or conceptual analysis. An example to an *a priori* proposition would be the statement "all bachelors are unmarried" or "a triangle is a three sided polygon". In the first statement, the concept of "bachelor" is in fact synonymous to "unmarried". One can be obtained from the other by conceptual analysis. Apart from this, *a priori* propositions also include statements of mathematics and geometry such as "$7 + 5 = 12$". On the other hand, *a posteriori* propositions are those whose truth value is justified by sensory experience. The truth value of an *a posteriori* proposition can only be attained as a result of an observation in the physical world. Examples of *a posteriori* propositions include statements such as "there are three apples in the bag" or "it is raining outside".

Kant further gives two types of judgments, namely *analytic* and *synthetic* propositions. According to his definition, *analytic* propositions are statements in which the subject is covertly contained in the predicate.[3] Otherwise, the proposition is called *synthetic*. In the examples we just gave above, "all bachelors are unmarried" and "a triangle is a three sided polygon" are both *a priori* propositions. Moreover, for Kant, they are analytic since the subject is, in fact, implicit in the predicate.

In his *Prolegomena* [162], Kant also studied the criteria for a proposition to be considered as analytic. He gives the example "gold is a yellow metal" to explain that the predicate does not *extend* the meaning of the subject.[4] That is, for Kant, one criterion for analyticity is that the proposition does not extend or add anything to our knowledge. As it can be seen, the concept of "yellow metal" does not extend the meaning of "gold". Kant calls analytic judgments in the *Critique*, *judgments of clarification*, and describes synthetic propositions as *judgments of amplification*. In the former, through the predicate, analytic judgments do not add anything to the concept of the subject, but only break it up by means of analysis into its component concepts. As for the latter, synthetic judgments add to the concept of the subject a predicate that was not thought in it at all and could not have been extracted from it through any

---

[1] In epistemology, *a priori* and *a posteriori* knowledge roughly mean, respectively, *non-empirical* and *empirical*. However, the term *a priori* literally means *prior (to experiment)*, and *a posteriori* means *post* (experiment).

[2] Kant, Critique of Pure Reason, A2.

[3] Kant, A7/B11.

[4] Kant, Prolegomena, §2 (b).

analysis.[5] As the third criterion for analyticity, Kant claims that the truth value of analytic propositions must always be able to be cognized sufficiently in accordance with the *principle of non-contradiction*, which can be written in the language of propositional logic as $\neg(p \wedge \neg p)$.[6] Consequently, the truth value of an analytic judgment, using the principle of non-contradiction, can be affirmed from its negation as the proposition itself and its negation cannot be simultaneously true. All tautologies, for that manner, are analytic.

Taking every combination of the types of knowledge and judgments, yields four cases one of which will be impossible.

1. Analytic *a priori*.

2. Synthetic *a priori*.

3. Synthetic *a posteriori*.

4. Analytic *a posteriori*.

Now, there is no such knowledge as analytic *a posteriori*. Any *a posteriori* knowledge must necessarily be synthetic since analytic propositions are necessarily *a priori* and hence, they do not need any sensory evidence for justification. Eliminating this case will leave us with three different types of judgments.

It is widely believed that mathematical knowledge is *a priori*. The fact that mathematical knowledge admits logical necessity is not the case for natural sciences or other *a posteriori* disciplines which are based on contingency. Only *a priori* knowledge gives us such a necessity. So there is a consensus that mathematics is *a priori*. Yet it is not very clear whether it is analytic or synthetic. In fact, we can rephase this problem as a question in terms of the relationship between axioms and their consequences: Do theorems add any knowledge to the axioms? But first and foremost, how is synthetic *a priori* knowledge possible? As we shall discuss shortly, Kant developed his most celebrated doctrine of *transcendental idealism* in order to find an answer to this question.

Synthetic *a priori* propositions, just as analytic judgments, are necessary and universal. In contrast to analyticity though, synthetic *a priori* judgments do extend our knowledge. Now *a posteriori* propositions are those which are justified by empirical evidence. On the other hand, analytic propositions are justified by conceptual analysis or logical reasoning. Then, as Kant puts it, we need a *third thing* to justify synthetic *a priori* propositions.[7] That "thing" is the result of a synthesis, attained from the pure space and time forms which exist *a priori* in the mind as pure representations of our senses, called *intuition* (German: *Anschauung*).[8] On this ground, Kant claims that the geometrical

---

[5] Kant, A7/B11.

[6] Kant, A151/B190.

[7] Kant, A154/B194.

[8] The word *anschauung* roughly means "view", "mental image", or "mental construct".

objects have representations in the pure space, whereas arithmetical entities have representations in the pure time form in the mind.[9]

An immediate difference between Kant and Aristotle, as well as Plato, is that Kant claims objects neither exist in an independent Platonic universe nor in the physical world with the absence of subject. For Kant, objects exist in the mind and that they are perceived through the concepts within the judgments. So in transcendental philosophy, objects cannot be independent from judgments, hence from the mind.

In Kant's transcendental philosophy, it is impossible to know a thing-in-itself (German: *Ding an sich*). These are objects, which Kant calls *noumena*, that just exist at face value. We attain the information about the object and perceive it only through a synthesis within the pure space and time forms that exist *a priori* in the mind. The primary requirement for an object to exist is that if it can be perceived within the pure space and time forms through their *a priori* representations, which are known by using intuition, i.e., mental constructs. In Kant's transcendental idealism, knowledge is merely obtained from judgments. Concepts and objects are not independent from judgments. Since all judgments reside in the mind, for if the mind ceased to exist, there would be no judgment, hence no object or concept. According to Kant, to extract any knowledge about an object, we must refer to its *a priori* representation which is separated from all sensory aspects of the object, in the transcendental pure space and time forms. Kant says:

> So if I separate from the representation of a body that which the understanding thinks about it, such as substance, force, divisibility, etc., as well as that which belongs to sensation, such as impenetrability, hardness, color, etc., something from this empirical intuition is still left for me, namely extension and form. These belong to the pure intuition, which occurs *a priori*, even without an actual object of the senses or sensation, as a mere form of sensibility in the mind.[10]

At his point, we may ask how the relationship between pure concepts of understanding, i.e., *categories*, and appearances is established. The interconnecting bridge between these two notions is formed by which Kant refers to as the *transcendental schema*. Transcendental schema is the mediating representation of the pure time and space form which makes an application of categories to appearances possible.[11] Through the transcendental schema, we form the *a priori* conditions to interconnect categories and appearances in order to unify pure understanding and sensibility which leads us to attain knowledge about an object. A detailed account of the transcendental schema

---

[9] Kant, Prolegomena, §10.
[10] Kant, A21/B35.
[11] Kant, A138/B177.

is given in the *Transcendental Aesthetic* part of the *Critique* to which we shall refer the reader for an extensive study.

For Kant, pure space and time are the *a priori* representations of our pure form of sensations, and they constitute a fundamental basis for (the existence of) knowledge. This includes the knowledge of mathematics and as well as geometry. The pure space and time forms are the intuited basis for both *a priori* and *a posteriori* knowledge. Any knowledge we learn about the objects must conform to their representations in the pure space and time forms and to the *a priori* conditions in the transcendental schema. In fact, the study of what these *a priori* conditions should be is the subject matter of what Kant calls *transcendental logic*. He says:

> I call a science of all principles of *a priori* sensibility the *transcendental aesthetic*. There must therefore be such a science, which constitutes the first part of the transcendental doctrine of elements, in contrast to that which contains the principles of pure thinking, and is named transcendental logic.[12]

Prior to Kant, there was no such investigation or an attempt to define such ontological and epistemological bases. Needless to say, Kant was opposed to the traditional Aristotelian metaphysics and argued for that objects should be in accord with our knowledge rather than vice versa.

It is worth noting that, for Kant, knowledge is not entirely subjective. The purpose of transcendental logic is, after all, to investigate the *a priori* grounds of objective elements of knowledge and the concept of pure understanding. Let us also discuss how transcendental logic differs from traditional logic.[13] Traditional logic abstracts the content and merely considers the form of the judgment. By doing this so, in traditional logic, not only do we cut off the relation between the judgment and *a posteriori* sensations, but we also eliminate empirical judgments from any pure forms of intuition. This is due to the fact that intuition can be empirical, but as well as purely be based on *a priori* evidence. Kant's revolution in epistemology is directly linked to this matter. According to Kant, transcendental logic investigates the *a priori* conditions for an object to be cognized and represented within the pure space and time forms, i.e., pure forms of intuition.[14] So transcendental logic, in a sense, unifies spatiotemporality with logic.

We said that intuition could be empirical, as well as purely *a priori*. In fact, according to Kant, without pure forms of intuition, empirical synthesis is impossible, that is to say, at the base of empirical observations lies pure form of intuition.[15] The separation made between pure and empirical intuition plays an important role in Kant's philosophy of mathematics.

---

[12]Kant, B36.
[13]This is extensively covered in A50-57 and B74-82 sections of the *Critique*.
[14]Kant, A11/B25.
[15]Kant, Prolegomena, §11.

For Kant, thoughts without content are empty, intuitions without concepts are blind. It is as necessary to make the mind's concepts sensible (i.e., to add an object to them in intuition) as it is to make its intuitions understandable (i.e., to bring them under concepts).[16] Thoughts that do not develop into knowledge is akin to doing mathematics without knowing what the ontological background of mathematical objects is. Without knowing it so, mathematical propositions would not extend our knowledge. As a matter of fact, by claiming that mathematics is synthetic *a priori*, Kant showed that the ontological ground of mathematical objects was defined in such a way to ensure that mathematical judgments would extend our knowledge. He manages to demonstrate this through transcendental logic, as a result of unifying pure space and time with traditional logic.

Now let us discuss how mathematical knowledge is synthetic *a priori* for Kant. The question whether mathematical knowledge is empirical or not has been debated for centuries. One of the greatest achievements of Kant, regarding philosophy of mathematics, so as to reconcile rationalism and empiricism, is his claim that mathematics is synthetic *a priori* as opposed to analytic *a priori*. According to Kant, propositions such as "7 + 5 = 12" or "From two points passes a line" are comprehended through the representations in pure forms of *intuition*. The same approach is thought to be valid for arithmetic just as for geometry. The formation of numbers, from this point of view, is only possible by giving units at distinct time periods in the transcendental representation of the pure time form of our intuition. For example, the number 7 exists via its representation, through the transcendental schema, in the form of seven distinct and consecutive units in the pure time form. In a famous paragraph in the *Critique of Pure Reason*, Kant explains how arithmetic operates in the following manner:

> To be sure, one might initially think that the proposition "7 + 5 = 12" is a merely analytic proposition that follows from the concept of a sum of seven and five in accordance with the principle of non-contradiction. Yet if one considers it more closely, one finds that the concept of the sum of 7 and 5 contains nothing more than the unification of both numbers in a single one, through which it is not at all thought what this single number is which comprehends the two of them. The concept of twelve is by no means already thought merely by my thinking of that unification of seven and five, and no matter how long I analyse my concept of such a possible sum I will still not find twelve in it. One must go beyond these concepts, seeking assistance in the intuition that corresponds to one of the two, one's five fingers, say, or five points, and one after another add the units of the five given in the intuition to the concept of seven. For I take first the number 7, and, as I take the fingers of my hand as an intuition

---

[16] Kant, A51/B75.

for assistance with the concept of 5, to that image of mine I now add the units that I have previously taken together in order to constitute the number 5 one after another to the number 7, and thus see the number 12 arise. That 7 should be added to 5 I have, to be sure, thought in the concept of a sum $= 7 + 5$, but not that this sum is equal to the number 12. The arithmetical proposition is therefore always synthetic.[17]

What Kant refers to as *five fingers* (or *five points*) corresponds to the representation in the mind of the concept of "five" in the pure space and time forms of our pure sensations.

The claim that mathematics is synthetic *a priori* implies that mathematical judgments extend our knowledge as in empirical knowledge, but they are universal and necessary as in analytic knowledge. Let us give a short note here. Despite the fact that Kant argued for that mathematical and geometrical judgments were synthetic *a priori*, he did not think the same for philosophy. What separates philosophy from mathematics is that philosophy is carried out by conceptual analysis, whereas mathematics is done by an interaction of empirical evidence and *a priori* conditions of cognition which involve things beyond conceptual analysis, e.g., intuition, construction. Regarding this, Kant says:

Philosophical cognition is rational cognition from concepts, mathematical cognition that from the construction of concepts. But to construct a concept means to exhibit *a priori* the intuition corresponding to it. For the construction of a concept, therefore, a non-empirical intuition is required [...].[18]

For Kant, mathematical knowledge is a type of knowledge attained by the *constructions* of concepts in the pure space and time forms of pure sensation. The construction of a concept corresponds to its spatial and temporal *a priori* representation in the pure forms of intuition. Let us consider an example from Euclidean geometry to elaborate on this matter. Consider the statement "Straight line passes from two points is the shortest line". The notion of "straight line passing from two points" and "shortest line" cannot be reduced to one another solely by conceptual analysis. This is due to the fact that, for Kant, the predicate contains an *additional* information that is not implicit in the subject; hence, the statement is synthetic. Now let us demonstrate geometrically that the sum of the interior angles of a triangle is 180 degrees.

As seen in Figure 4.1, we extend the edge AB from point B. From point B we inscribe a line parallel to the edge AC. Now, we know that the existence of these auxiliary lines follows from the axioms of Euclidean geometry. The

---

[17] Kant, B15.
[18] Kant, A713/B741.

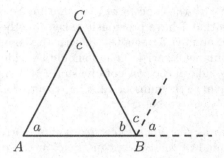

FIGURE 4.1
The sum of the interior angles of a triangle.

angle between BC and the line parallel to AC is same as the angle c. The angle between the line parallel to AC and the line that straightly extends AB is equal to the angle a. Then, we see that the sum $a + b + c$ is equal to 180 degrees which gives us the sum of the interior angles.

In the Kantian sense, the auxiliary lines are drawn with the assistance of our "intuition". Kant explains intuition as follows:

> In whatever way and through whatever means a cognition may relate to objects, that through which it relates immediately to them, and at which all thought as a means is directed as an end, is intuition.[19]

As we said earlier, mathematical knowledge and objects are acquired by way of their *representations* in the pure space and time forms of our intuition. Particularly, the objects of geometry are represented in the pure space form, whereas arithmetical objects have representations in the pure time form. Within the pure time form, if we consider counting, for example, the concept of "five" is represented in the mind by consecutive units consisting of *five points, five fingers*, etc. The Kantian view holds that mathematical and geometrical constructions are based on intuition in the sense of the example given above. The mathematician, by adding auxiliary lines or extending them, using intuition, actually displays certain properties that are "not implicit" in the concept of triangle. The challenging question is of course whether these properties really extend our knowledge or they are already implicit in the concepts and defintions.

Kant claims that mathematical judgments extend our knowledge and so they display new information. In the proposition "$7 + 5 = 12$", for instance, the concept of 12 cannot be derived solely from 7 and 5. Conceptual analysis

---

[19]Kant, A19/B31.

or logical derivation, does not introduce "new" knowledge that is not already implicit in the subject of the proposition. It rather gives the meanings which are covertly present in the subject. For this reason, from the Kantian point of view, since mathematical propositions do introduce new concepts and extend our knowledge, as given in the geometrical example, they cannot be analytic *priori*. Hence, they must be synthetic *a priori*.

The groundwork of the ideas introduced in the *Critique of Pure Reason* were, in fact, given in Kant's earlier works. Kant published an essay in 1768 on different space conceptions of Newton and Leibniz, called *Concerning the Ultimate Ground of the Differentiation of Directions in Space* [160]. In the essay, Kant presents a thought experiment, which he calls *incongruent counterparts*, that separates two space conceptions. The two space conceptions are called *absolute space* and *relative space* conceptions. Newton was an adherent of the absolute space, whilst Leibniz was known to support the relative space conception. Let us explain the difference between these two conceptions with an analogy. Suppose that all objects in our infinite universe were moved in the same direction and with same amount. In this case, has any change occurred in the system? According to the Newtonian interpretation, there is indeed a change. However, the Leibnizian interpretation tells us that, since the location of no object is changed relative to another, the state of the system remains the same as before. In an absolute space model, objects are defined on an existing space. On the contrary, relative space model argues for that the objects, in fact, define the space.

Kant's conception of space forms a substantial principle for his transcendental philosophy. In his 1768 essay, he holds a position in favour of Newton's interpretation of absolute space, where he gives the left/right hand argument to support his view. Consider a right hand and a left hand with the same properties, drawn on a plane, perhaps in an abstract form such as a pair of gloves. Then, two hands are expected to coincide with one another and so occupy the same space. However, due to the fact that they are in different "directions", despite their similarities, they do not coincide. This subtle difference can only be explained by means of how the objects are related to the space, rather than the implicit properties of the objects. But then this gives us to the absolute space interpretation. The moral of the story is the following: An object can only be cognised and perceived through its relation with the space, not by its own intrinsic properties. The emphasised point is that an object exists in the mind solely as a representation obtained via the *a priori* space and time forms of our pure intuition. Objects that are not processed through this synthesis of the pure intuition of the mind, e.g., *things-in-themselves* (noumena), cannot be entirely known. In Kant's transcendental philosophy, once the object is synthesised in the transcendental schema, it becomes an entity in the mind as a transcendental representation, which we may now call it a *phenomenon*. Considering this space conception, which constitutes one of the fundamental principles in the *Critique of Pure Reason*, mathematical knowledge was claimed to be synthetic as it is neither based on conceptual analysis nor logic, while it requires a kind of intuition; nevertheless, it should be noted that since knowledge of geometry is *a priori* and that it is universal and necessary, the intuition of this kind must rely on *a priori* factors.

There is still more to say, of course, about Kant's transcendental philosophy and the analytic/synthetic distinction, however it is beyond the scope of our book to cover them all together. For this reason, we shall move on to Brouwer's constructivism.

4.2   Brouwer and Constructivism

Kant was predominantly interested in how and on which grounds mathematical knowledge was attained, rather than the practical applications of the logical method. For his attempts to establish an idealistic ontology and epistemology of mathematics, he is considered as an early intuitionist. L. E. J. Brouwer (1881–1966), a leading figure in intuitionism, applied Kant's intuition-based approach on the mathematical method, particularly on logic. His works concerning the philosophy of mathematics can be mostly seen in his works [42] and [45].

Despite the fact that 20th-century intuitionists had some mutual disagreements, they were interested in the problem of how mathematical activity should be carried. More specifically, they focused on the rules of logical inference and reasoning. Mathematicians and philosophers, who adopted Kant's intuitionistic approach, were mainly concerned about the way classical logic was used in mathematics and they suggested that the laws of thought should be revised.

Classical laws of thought were introduced by Aristotle, long before Kant and Brouwer, to formulise the basis for rational discourse. The rules are imagined on purely logical, non-spatial, and non-temporal grounds. Classical laws of thought consist of three rules.

Law of Identity: Everything is identical to itself.

Law of Non-Contradiction: A proposition p cannot be true and false simultaneously.

Law of Excluded Middle: A proposition is either true or false, i.e., there is no third possibility or any partial truth. Symbolically $p \vee \neg p$.

In intuitionism, the *Law of Excluded Middle* (LEM) is not accepted as an axiom. This is not to say that it accepts the negation of LEM as an axiom.

Intuitionists just do not use the third law in their inferences. So modern day *intuitionism* is defined as the view which adheres for logical systems in which LEM is not included. The logical system in which all the classical laws of thought are accepted is called *classical logic*, and the corresponding mathematics that uses these rules is called *classical mathematics*. Whereas the the logical system in which LEM is not included as an axiom is called *intuitionistic logic*, and the mathematics based on this set of rules is called *intuitionistic mathematics* or sometimes *constructive mathematics*. The rejection of LEM precludes the mathematician from using certain rules and properties. One of these properties concerns the interpretation of the negation ($\neg$) operator. In classical logic, from $\neg\neg p$ we can deduce p.[20] This is due to the fact that there is no third possibility, as LEM puts it. If something is not false, then the only possibility remains is that it must be true. However, intuitionistic logic does not allow us to make this inference. That is, in intuitionistic logic, $\neg\neg p$ does not necessarily imply p. Of course if p is true, then $\neg\neg p$ must also be true. Also, just because the double negation elimination rule cannot be used in intuitionistic mathematics, does not mean that we are left with an infinite chain of negations. Brouwer proves in his [44] that 'absurdity of absurdity of absurdity is equivalent to absurdity', i.e., $\neg\neg\neg p \leftrightarrow \neg p$.[21] Let us quickly prove this. Assume that p and q are two propositions such that $p \to q$. Then $\neg q \to \neg p$. From this it follows that $p \to \neg\neg p$ implies $\neg\neg\neg p \to \neg p$. To prove the other direction, since $p \to \neg\neg p$, notice that, in particular, $\neg p \to \neg\neg\neg p$. Therefore, $\neg\neg\neg p \leftrightarrow \neg p$ as desired.

So given a proposition with $n > 2$ many negations, it is either reducible to the negation of the proposition or to the negation of its negation. Given the fact that $\neg\neg\neg A \leftrightarrow \neg A$, one may try to see if the double negation elimination rule could be derived from Brouwer's theorem. Let B be defined as $\neg A$. Now substitute B into the equality and get $\neg\neg B \leftrightarrow B$, and voila! But this only applies to B when it is defined in a specific way. So for an arbitrary proposition, the double negation elimination rule still does not hold in intuitionistic logic.[22]

The interpretation of the negation operator, as a matter of fact, depends on whether the ontology of the logical universe under consideration is binary or n-nary, for $n > 2$. Consider a binary logical universe, i.e., two-state logic, and let us call these states D_1 and D_2. Now in this case, D_1 and D_2 must complement each other since there is no third state. The negation of D_1 should be D_2, and the negation of D_2 should be D_1. Consider, however, a ternary universe of logical states and let us denote them by $D_1, D_2, and\ D_3$. If we ask what the negation of D_1 should be, we are somewhat faced with an ambiguity as to whether the negation of D_1 is D_2 or D_3. In fact, without further assumptions, the negation of D_1 is not uniquely defined, but it is rather the disjunction of D_2 and D_3.[23] Let us try to express this symbolically. Now in a binary

---

[20]This is called the *double negation elimination rule*.
[21]Brouwer, p. 253, 1925.
[22]Thanks to İbrahim Altun for pointing out this case.
[23]We will see later that this problem is actually related to the Axiom of Choice.

logical universe, it must be the case that $\neg D_1 \equiv D_2$ and $\neg D_2 \equiv D_1$. Then, $\neg(\neg D_1) \equiv \neg D_2 \equiv D_1$. However, in a ternary logical universe, we cannot map $\neg D_1$ to a unique logical state, so $\neg D1$ must be either D_2 or D_3. That is, in our ternary state logical universe we have

$$\neg D_1 \equiv (D_2 \vee D_3).$$

Similarly, it must be that

$$\neg D_2 \equiv (D_1 \vee D_3),$$
$$\neg D_3 \equiv (D_1 \vee D_2).$$

In such a logical universe, we can see why $\neg(\neg D_1)$ does not necessarily imply D_1. We have that

$$\begin{aligned} \neg(\neg D_1) \quad &\equiv \quad \neg(D_2 \vee D_3) \quad \equiv \quad \neg D_2 \wedge \neg D_3 \\ &\equiv \quad (D_1 \vee D_3) \wedge (D_1 \vee D_2). \end{aligned}$$

From this we can see that on an n-ary logical universe, if $\neg\neg D_1$ is true, it does not necessarily follow that D_1 is true. This is simply due to that the truth value of D_2 and D_3 should be known in advance. That is to say, since the negation operator has an effect on the rest of $n - 1$ states, $\neg\neg D_1$ is not necessarily logically equivalent to D_1.

Let us now turn to existence proofs. Existence and uniqueness proofs are quite common in many areas of mathematics. In particular, existence proofs may sometimes require complicated and long constructions. It is often more easier, for that matter, to prove the existence of an object rather indirectly. Unlike classical logic, proving the existence of a mathematical object or establishing the truth of a proposition indirectly is not an acceptable approach in this philosophical position called *constructivism*, which is based on intuitionistic logic. In classical logic, to prove the existence of an object, we may first assume that it does not exist. If this assumption leads to a contradiction, using LEM, we may conclude that the object must exist. Recall that there is no third possibility: an object either exists or not. So in classical logic, if it is not the case that something does not exist, then it must necessarily exist. This is in contrast with intuitionistic logic where one must prove the existence of an object by constructing it or defining it directly. From classical predicate logic we know that, for any relation R, if it is not the case that $R(x)$ holds for all x, then there must exist some x such that $R(x)$ does not hold. Symbolically, $\neg\forall x\, R(x) \rightarrow \exists x\, \neg R(x)$ is a valid statement in classical logic. This is not the case in intuitionistic logic as this implication is non-constructive. The distinction between a *constructive proof* and *non-constructive proof* is critical in intuitionism. Let us give an example regarding this to understand the distinction better. First we start with a straightforward example. Suppose that we want to prove there exists a prime number greater than 10. We can constructively prove this by explicitly stating that 11 is a prime number. A non-constructive proof would, for instance, first assume that there does not exist a prime number greater than 10 and then use Euclid's theorem that there are infinitely many prime numbers to derive a contradiction. Since it is impossible that there is no prime number greater than 10, we know by LEM that there must exist a prime number greater than 10. The claim is proved.

We now give a well-known theorem for which we give two proofs, respectively, constructive and non-constructive.

Theorem 4.1 There exist irrational numbers a and b such that a^b is rational.

Constructive proof. We give a and b explicitly. Let $a = \sqrt{2}$ and $b = \log_2 9$. Then, $a^b = 3$ and this is rational. Now we already know that $\sqrt{2}$ is irrational, and so is $\log_2 9$. If it were rational, then, for some natural numbers m and n, we would have that $\log_2 9 = \frac{m}{n}$. Followed from the property of logarithms, we would have $9^n = 2^m$. But 9^n is always an odd number, whereas 2^m is always even. $\qquad\square$

Non-constructive proof. We know that $\sqrt{2}$ is irrational. Let $q = \sqrt{2}^{\sqrt{2}}$. Now q is either rational or irrational. If q is rational, there is nothing to prove, we can simply let $a = \sqrt{2}$, $b = \sqrt{2}$, and the theorem will hold. If q is not rational, let $a = q$ and $b = \sqrt{2}$. We show that the theorem still holds in this case. Substituting the terms, we have that $a^b = \left(\sqrt{2}^{\sqrt{2}}\right)^{\sqrt{2}} = \sqrt{2}^{(\sqrt{2}\cdot\sqrt{2})} = \sqrt{2}^2 = 2$. Therefore, a^b is rational. $\qquad\square$

In the non-constructive proof, we showed in either case that there exist a and b which satisfy the theorem. What matters for the realists is that there are such numbers. But then what exactly is the solution? Is it that

$$a = \sqrt{2} \text{ and } b = \sqrt{2} \text{ ?}$$

Or is it

$$a = \sqrt{2}^{\sqrt{2}} \text{ and } b = \sqrt{2} \text{ ?}$$

As a matter of fact, what a and b are depends on whether or not q is rational. What makes the latter proof non-constructive is where we propose two possible solutions, without telling explicitly what they are, from the two possibilities of the status of q as to whether or not it is rational. We implicitly use LEM here, and so the proof is inevitably non-constructive.

The reason why Brouwer rejects LEM is because of his profound belief that mathematical activity should be grounded on constructive methods. Suppose that P is a property about the natural numbers. Then, the constructive proof of that $P(n)$ holds for some natural number n must be given by explicitly introducing a natural number n for which $P(n)$ holds. It is important to emphasise this point that, in constructivism, one must explicitly construct the object in order to prove its existence. In contrast with the common practice of classical logic, constructivists cannot deduce the existence of an object from the impossibility of its non-existence.

So why does intuitionism reject LEM? The problem is that we make metaphysical commitments. According to Shapiro, LEM is considered as a consequence of realism and that it indicates a belief in the independent existence of mathematical objects.[24] That is to say, LEM implicitly presupposes that mathematical reality is independent of our conceptions and capabilities. In actuality, intuitionism makes far less use of metaphysics than realism does.

Following Kant's footsteps, Brouwer as well believed that conceptual analysis would not suffice to establish the truth value of mathematical propositions and that we would need to refer to our intuitions. So according to Brouwer, mathematical knowledge is *synthetic*. Furthermore, he too agrees that mathematics is *a priori*. Just as Kant attempted to, Brouwer as well did try to reconcile rationalism and empiricism.[25]

We now turn to the problem of *impredicativity*, which is quite an interesting and deep problem in the ontology of mathematics. Impredicativity can be thought of as a generalised form of circularity. An *impredicative* definition of a mathematical object is a type of definition in which the whole or a part of the definiendum is referred in a totality that is *a priori* assumed to exist. Prima facie, impredicativity may be thought of merely as a some sort of circular schema. But it is in fact more general than circularity. Of course, if the definition is explicitly circular, then it is automatically impredicative. However, the other way around is not quite obvious. One example of impredicativity concerns the definition of *greatest lower bound* of an ordered set. Let A be a set. An element $y \in A$ is called the *greatest lower bound* of A if it satisfies the following properties:

(i) For every $x \in A$, $y \leq x$.

(ii) For every $x \in A$, $z \leq y$ whenever $z \leq x$.

If we look at part (ii) of the definition, the variable z quantifies over a totality containing *all* lower bounds of A, one of them being the greatest lower bound. Then, the definition in actuality refers to the definiendum in an existing completed totality. What is the justification that such a totality exists? For constructivists, an object cannot be claimed to exist without explicitly defining it. Constructivists rightfully believe that a totality of this kind has no ontological basis or justification.

The opposite of impredicativity is called *predicativity*. A predicative definition would only use terms or entities which have previously been defined. This enforcement, as it might be expected, does not usually yield a rich math-

---

[24]Shapiro, p. 174, 2000.

[25]As we explained earlier, empiricism claims that true knowledge is attained by sensory experience, whereas rationalism claims that it is attained by reasoning.

ematical system. For a detailed survey on predicativity, we refer the reader to Feferman's [86] paper.

For a mathematical realist, it may be reasonable to endorse impredicative definitions. There is no problem for a realist whatsoever in defining an independently existing object in terms of its smaller parts, or even via its equivalents for that matter, as long as the definition is internally consistent. Nevertheless, an intuitionist generally finds such definitions rather circular, hence problematic. For this reason, intuitionists avoid defining a object in terms of entities that have not yet been demonstrated to exist. There is a subtle point in their rejection of impredicativity. Allowing impredicative definitions on infinite domains, falsely permits the intuitionists to define completed infinite totalities, which, in turn, implies the existence of absolute infinities and enables them to use these objects in mathematical theories. However, the concept of absolute infinity is not regarded as a primary motivation in Brouwerian constructivism. In fact, absolute infinity is not a legitimate notion for Brouwer. We will come back to the debate between absolute and potential infinity, but first we give Brouwer's opinion on completed totalities. Brouwer raises an objection to the idea that some entities are, as realists and formalists claim, rather completed totalities. In regard to this, he says:

> [...] the formalist introduces various concepts, entirely meaningless to the intuitionist, such as for instance "the set whose elements are the points in space", "the set whose elements are the continuous functions of a variable", "the set whose elements are the discontinuous functions of a variable", and so forth.[26]

Brouwer's point is that since these entities contain infinitely many objects, possibly unordered, the intuitionist will never get to finish constructing these sets. Hence, the set of such objects cannot exist for the intuitionist.

Another controversial problem in constructivism is the status of the Axiom of Choice. Let X be a collection of non-empty sets. If f is a function such that $f(x) \in x$ for every $x \in X$, then f is called a *choice function* for X. The *Axiom of Choice* says that every family of non-empty sets has a choice function.

We are not concerned with the applications of this axiom in mathematical practice, as we shall leave that discussion for Chapter 12. Let us explain though that this axiom relies on non-constructive principles. As a matter of fact, there is no need to use the Axiom of Choice in every situation. If X is finite, we should have no problem in defining a choice function for X. In this case, we choose a member from each element of X. Since X is finite, the description of the choice function f will be finite as well. Unfortunately, it gets a little more complicated if X is infinite. If every element of X has some common property P, then we can indeed define a choice function by ourselves which would simply select the element with property P from every member of X. For instance, if we have infinitely many pairs of shoes, we may always choose,

---

[26]Brouwer, p. 82, 1912.

say, the left shoe from each pair. Having a "left pair" is a common property for every pair of shoes. In general, if we can somehow distinguish the elements of the members of X, then we can automatically choose some distinguished element without invoking the Axiom of Choice. But what should we do if there is no uniform way of distinguishing the elements of members of X, or there is no such common property that characterises them? Bertrand Russell gives a nice example that explains this case. Consider now that X consists of infinitely many pairs of socks rather than shoes. Now ideally socks in a pair cannot be distinguished from one another. In this case, we may use the Axiom of Choice to separate the socks in each pair and choose one. As the axiom states, such a choice function exists. But how exactly this function is defined is completely out of the question. The axiom asserts the existence of a choice function, yet it does not provide its description. So we really do not know in such cases what exactly the choice function might be. In fact, the mathematician often does not care how it is defined, if she is aware at all that she invokes the axiom. The constructivist is more peculiar that she believes the only way an object exists is if it is constructible. For if a choice function exists, its definition must be given explicitly. Asserting the existence of a mathematical entity is one thing, defining it explicitly is another thing. In constructivism, positing the existence of a choice function for the set X is meaningless unless it is explicitly constructed. For this very reason, the use of the Axiom of Choice has been a controversial issue in intuitionism. Nonetheless, according to Michael Dummett [81], the reason why intuitionists are biased against the Axiom of Choice is because of their expectation for the function, posited to exist, to be constructible while interpreting the quantification in the realist sense, as if the quantification is made over *all* functions, irrespective of them being constructible or not. For Dummett, as long as we assume that the quantification is made over *constructible* functions, i.e., when the domain of discourse is considered to be the class of *all* constructible functions, then the function whose existence is asserted in the axiom must also be constructible.[27] From this point of view, Dummett argued for that the Axiom of Choice could be, in fact, accepted as an axiom in intuitionistic mathematics.

We now look at the interpretation of real numbers in constructivism, in particular from Brouwer's point of view. Let us consider real numbers in the $[0, 1]$ interval. Of course by real numbers, generally we mean numbers that are irrational. From the point of view of classical mathematics, a real number is imagined to be a number with an infinite decimal expansion. Moreover, it is imagined as a completed number, as if the decimal expansion is a completed infinite totality. A mathematical realist would take and use this number as it is, literally as an infinite totality. For Brouwer, the existence of an arbitrary point in the real line is not something to be taken granted for. In fact, for a constructivist, there is no *a priori* reason that an arbitrary real number exists. Recall that in constructivism, an object must be constructed before we use it.

---

[27]Dummett, pp. 37–38, 1977.

According to Brouwer, the only way to posit the existence of a real number is to construct its decimal digits following a finite and well defined rule.[28] Normally, we may define real numbers as the set of limits of a special kind of sequences of rational numbers, what we call *Cauchy sequences.*[29] From the Brouwerian point of view, legitimate real numbers are those which can be represented by computable Cauchy sequences. Examples of such numbers include π, e, etc. The decimal digits of these numbers are still being computed to our day. They are, for this reason, *computable real numbers.* That is, for any given natural number n, we can algorithmically determine their nth digit of the decimal expansion.

Brouwer [45], in his later writings, introduces what he calls *free choice sequences* to support the notion of potential infinity in defining infinite sequences. One can have two types of treatments of the notion of infinity: *absolute infinity* and *potential infinity.* Actual infinity can be realised as a completed totality, for example, the use of the set of natural numbers $\mathbb{N} = \{0, 1, 2, 3, \ldots\}$, as an object on its own, enforces us to accept the notion of actual infinity. The *set* of natural numbers, for a Platonist, is an (actual) infinite object; it is a completed totality and it manifests the infinite in itself. Brouwer rejects this kind of treatment of the infinite. For him, the notion of infinity can never exist as a completed entity, but can merely remain as a potential that is yet to be completed. Potential infinity can be thought of as a never-ending *process* or *sequence* of objects where one is capable of introducing arbitrarily many new elements. The use of the set $\mathbb{N}$, as an object, makes reference to actual infinity, whereas the never-ending sequence $0, 1, 2, 3, \ldots$ is just a *process* of potentially becoming infinite but never a completed totality. We are not taking this sequence as a totality, but we rather treat it as an everlasting and on-going process.[30] In such a counting list, the next element after n is determined by the formula $n + 1$. Since n is arbitrary, we may always define a new number greater than any given number. Sequences of this kind, where the elements are determined uniformly, is called *ruled sequences.* In constructivism, infinite objects such as irrational numbers are defined via ruled sequences. For a mathematical realist, it is perfectly acceptable to claim that infinite objects exist as completed totalities and even that some sequences may be defined without any rule. The realist is not obliged to justify the existence of such *unruled sequences*, by any method whatsoever, for their reasoning that

---

[28] Brouwer, p. 85, 1912.

[29] Let $a_1, a_2, \ldots$ be a sequence of rational numbers. We call this sequence a *Cauchy sequence* if for every rational number $\epsilon > 0$ there exists a natural number N such that for every natural numbers n and m, $|a_n - a_m| < \epsilon$ whenever $n, m > N$. Furthermore, if $a_1, a_2, \ldots$ is a computable sequence (computable in the sense that there exists an algorithm that determines the nth element of the sequence, given any natural number n) and the values that N can take is bounded by a number, then we call it a *computable Cauchy sequence.* Now any convergent Cauchy sequence defines a real number. But we still have to be careful about that there may not necessarily exist a unique Cauchy sequence for every point in the real line.

[30] The concept of potential infinity is not something that was emerged from the works of Brouwer. The concept goes as far back as to Aristotle (see Chapters 3 and 9).

they are what they are. Since the realist adheres that all mathematical objects exist independent of us, she is not required to provide a construction. In some sense, free choice sequences reconcile ruled and unruled sequences. An irrational number with infinite decimal expansion could be defined up to some finite initial segment following some rule; but later, the proceeding digits may just be defined arbitrarily, one at a time. By "arbitrarily", we do not mean "all at once" in the Platonist sense, as if the digits existed before us. After defining a finite initial segment of the object, and in this case, say, of an irrational number, Brouwer leaves the rest of the digits to be determined by the "creative subject" through which we would be able to produce further digits as long as we desire. In free choice sequences, the digits of a number are not required to be determined by an algorithm. For instance, each next digit could be determined by the result of a toss of a fair coin. What we know about a free choice sequence is limited to what we have defined up to that stage since the information we have at any stage, when defining a free choice sequence, is just its initial segment of finite length. It should also be noted that we cannot be sure if the obtained potentially infinite sequence, in the end, could have been defined by a rule; and in this case, we may not know if we would have ended up with a rule-governed sequence. From this point of view, since any construction of a real number must be defined in stages, what we have so far, say by the end of some finite stage, must be a finite information about that number. Hence, for a constructivist, any mathematical truth about that real number must follow from a finite set of information.

Let us repeat the difference between the classical realist interpretation of real numbers and free choice sequences. The realist believes that any point on the continuum exists as it is. That is, all real numbers exist by virtue of themselves. The nth digit of the decimal expansion of a real number is fixed as it is, as if it has been already defined before us. For constructivists, this is too much of a metaphysical faith that only a realist would believe. In free choice sequences, after constructing a finite initial segment of a real number, by a pre-determined rule, we potentially leave the rest of the digits to be defined by a creative subject. Then, Brouwer's idea in defining a real number relies on the use of potential infinity for being able to continue indefinitely to write its digits, as much as one desires. The constructivist considers this as a never-ending construction process, not as a completed totality.[31]

A formalisation of intuitionistic logic was provided by Brouwer's student, Arend Heyting [144], who replaced *truth conditions* of classical mathematics with *proof conditions*. Heyting gives what is known today as the *Brouwer-Heyting-Kolmogorov semantics* as a set of proof conditions for intuitionistic mathematics. These rules are given as follows.

(i) A proof of $\varphi \wedge \psi$ is a proof of φ and a proof of ψ.

(ii) A proof of $\varphi \vee \psi$ is either a proof of φ or a proof of ψ.

---

[31] See Fletcher [94] (2020) for a survey on the relationship between choice sequences and the creative subject.

(iii) A proof of $\neg\varphi$ is to transform the proof of φ into absurdity.

(iv) A proof of $\varphi \to \psi$ is to transform the proof of φ into a proof of ψ.

(v) A proof of $\exists x\ \varphi(x)$ is to specify an object a and a corresponding proof of $\varphi(a)$.

(vi) A proof of $\forall x\ \varphi(x)$ is to find a procedure that, given any n, produces a proof of $\varphi(n)$.

Note that in this semantics, LEM is abandoned, as the semantics is incapable of establishing that every propositions must be either provable or refutable. Influenced by Brouwer, Heyting states that mathematics is mind-dependent. In [145], he writes:

> [...] [M]athematics is a production of the human mind [...] we do not attribute an existence independent of our thought, i.e., a transcendental existence. [...] Faith in transcendental [...] existence must be rejected as a means of mathematical proof [...] this is the reason for doubting the law of excluded middle.[32]

We said earlier that classical mathematics relies on metaphysics mostly due to the usage of LEM, as the law presupposes an ontological and epistemological reality independent of the mind. For Heyting, *to exist* is synonymous with *to be constructed*. He says:

> If 'to exist' does not mean 'to be constructued', it must have some metaphysical meaning. It cannot be the task of mathematics to investigate this meaning or to decide whether it is tenable or not. We have no objection against a mathematician privately admitting any metaphysical meaning he likes, but Brouwer's programme entails that we study mathematics as something simpler, more immediate than metaphysics. In the study of mental mathematical constructions 'to exist' must be synonymous with 'to be constructed'.[33]

The intuitionist interpretation of the negation operator leads us to revise the concept of counterexamples. Classically, to disprove a proposition, one finds an instance for which the proposition does not hold. We call this instance a *counterexample* for the proposition. For instance, a counterexample for the statement "All apples are red" would be to say that "There is an apple which is not red". As we know from classical logic, if p is not true, then p is simply false. We interpret the negation operator $\neg$ slightly different in intuitionism, for if the negation of p is true, this merely shows that p is refutable, i.e., a counterexample for p can be found. Of course, if there is a proof of p then

---

[32] Heyting, pp. 52–53, 1931.
[33] Heyting, §2, 1956.

proving that p does not have a proof is impossible. However, in intuitionism, showing that p does not have a counterexample, does not suffice to conclude that p is true. In fact, by this interpretation, we understand that p is stronger than $\neg\neg p$ in intuitionistic logic.

The earliest constructivist doubt of the validity of LEM as a principle of mathematics, when used in infinite domains, is found in Brouwer's doctoral dissertation [41]. In his [43], Brouwer manages to give one of the first counterexamples to the Platonist fact that the continuum is ordered (also known as the *Law of Trichotomy*), i.e., $\forall x, y(x < y \lor x > y \lor x = y)$. He claimed that this property about real numbers is false. As quoted in van Dalen [74], Brouwer says:

> Let d_v be the v-th digit in the decimal expansion of π and $m = k_n$ if in the growing decimal expansion of π at d_m it is the case that for the n-th time the part $d_m d_{m+1} \ldots d_{m+9}$ of this decimal expansion forms a sequence 0123456789. Let further $c_v = \left(-\frac{1}{2}\right)^{k_1}$ if $v \geq k_1$, otherwise $c_v = \left(-\frac{1}{2}\right)^v$. Then the infinite sequence $c_1, c_2, c_3, \ldots$ defines a real number r, for which neither $r = 0$, $r > 0$ or $r < 0$ holds.[34]

Another counterexample involves the construction of a π-like number, call it $\hat{\pi}$, using the decimal expansion of π. Let p denote the statement "In the decimal expansion of π, there exist 100 successive zeros". Suppose $\neg p$ denote the contrary. From the realist point of view, $p \lor \neg p$ is a true statement. But from the constructivist point of view, this is simply false due to a mistaken concept that the entire decimal expansion of π already exists as a completed totality. But as we said earlier, all the constructivist knows about π is the set of digits that have been constructed *up to* a certain stage. We define $\hat{\pi}$ as follows. Compute the decimal expansion of π until we find 100 successive zeros. The digits of $\hat{\pi}$ is same as π up to the point where the sequence of zeros starts. Suppose that n is the position of the first zero of the sequence. If n is odd, we let $\hat{\pi}$ terminate in its nth digit. If n is even, we let $\hat{\pi}$ have a 1 in the $n + 1$st digit and then terminate the process. Note that we do not know if such a natural number n exists. We know that $\pi = \hat{\pi}$ if n does not exist. But if n exists and if it is even, then $\hat{\pi} > \pi$. If n is odd, then $\hat{\pi} < \pi$. Let us now define $Q = \hat{\pi} - \pi$. Is q positive, negative or zero? Now the realist will say that q is necessarily either positive, negative, or zero. But the constructivist raises an objection to this transcendental information the realist claims about π. For if since we do not know whether π contains 100 consecutive zeros, we cannot know if q is positive, negative, or zero. Now the precise information about exactly which of the three cases q satisfies is not important for the realist. Just the fact that q must have one of the three properties is enough. However, the constructivist will further want to know actually which one of the three properties holds. From this argument, the constructivist infers that

---

[34]Brouwer, p. 3, 1923. Also in van Dalen (1999), p. 306.

mathematical truth is time-dependent and subjective. The word 'time' here does not refer to any physical quantity though. What we mean by 'time' is the *stage* of construction for that mathematical object.

The basic idea in all these example is to reduce the solvability of a problem to some known unsolved problem. Using counterexamples, we may show that a given proposition does not have a constructive proof. For this, we find a non-constructive law that is necessary for the proposition we desire to show its non-constructive nature. That is, if p is a proposition that we wish to show for which there is no constructive proof, we consider a non-constructive law r, such as LEM or any other statement for that matter, and constructively prove that $p \to r$. By contrapositive, we conclude that since r has no constructive proof, neither p. An interesting example which may be of great interest to mathematicians is *Diaconescu's theorem* [79].[35]

Theorem 4.2 (Diaconescu, 1975) The Axiom of Choice is a sufficient condition for LEM.

Proof. We use the Axiom Schema of Specification and the Axiom of Extensionality. We give the proof which was presented in Beeson [25].[36] Let p be a proposition. Using the full version of the Axiom of Choice, we (constructively) prove $p \lor \neg p$. We define the sets

$$A = \{n \in \mathbb{N} : n = 0 \lor (n = 1 \land p)\},$$
$$B = \{n \in \mathbb{N} : n = 1 \lor (n = 0 \land p)\}.$$

According to this definition, A and B are non-empty. Now let f be a choice function such that $f(A) \in A$ and $f(B) \in B$. Since members of A and B are natural numbers either $f(A) = f(B)$ or $f(A) \neq f(B)$ must hold.

If $f(A) = f(B)$, then p is true. Therefore, by $\lor$-introduction in propositional logic, $p \lor \neg p$ is true.

If $f(A) \neq f(B)$, then $\neg p$ must hold. To see this, suppose that p is true. Then, by the Axiom of Extensionality, $A = B$. Then, we have that $f(A) = f(B)$ which contradicts our first assumption. Therefore, $p \lor \neg p$ must hold. $\square$

LEM, by its own nature, is a non-constructive law. Since it is a necessary condition for the Axiom of Choice, we now know that the Axiom of Choice cannot have a constructive proof. For if it had a constructive proof, then, by Diaconescu's theorem, LEM would too have a constructive proof. Another interesting result related to LEM and set theory concerns well-founded sets.

Theorem 4.3 The Axiom of Regularity is a sufficient condition for LEM.

Proof. We give the proof provided by Schechter [254].[37] In fact, the proof

---

[35] Also known as *Goodman-Myhill theorem* [112].
[36] Beeson, p. 163, 1985.
[37] Schechter, p. 138, 1997.

looks quite similar to the previous one. Let p be a proposition. Here, p could be a proposition whose truth value is unknown or, at least, not known yet, for instance, Golbach's conjecture. Define S as follows:

$$S = \left\{ \begin{array}{ll} \{\emptyset, \{\emptyset\}\} & \text{if } p \text{ if true,} \\ \{\{\emptyset\}\} & \text{if } p \text{ if false.} \end{array} \right.$$

In either case, S cannot be empty. The reason is that, in any case, we have $\{\emptyset\} \in S$. By the Axiom of Regularity, it is possible to know which element of S has an empty intersection with S. This, in turn, gives us, for any proposition p, either p or $\neg p$ holds. Therefore, $p \vee \neg p$ holds.   □

A weak form of presenting counterexamples in Brouwer's intuitionism is called *weak counterexamples*. Such arguments are used to prove that a proposition has no constructive proof only *conditionally*. A weak counterexample is given in the following manner. First, we consider a proposition which has not yet been resolved, say, the Riemann hypothesis. We then define a function f in the following way so as to embed the information of the Riemann hypothesis into f:

$$f(x) = \left\{ \begin{array}{ll} 0 & \text{if the Riemann hypothesis is false} \\ 1 & \text{if the Riemann hypothesis is true.} \end{array} \right.$$

Now, let us define a proposition p as "$f(0) = 1$ or $f(0) \neq 1$". Based on this definition, if p has a constructive proof, then so does the Riemann hypothesis. However, it is not known whether the Riemann hypothesis has a proof, let alone the proof being constructive. Then, under this *condition*, p cannot have a constructive proof either. These types of arguments are mainly used for determining the "hardness" of a problem, in terms of its constructability, and for establishing a hierarchy regarding the complexity of various problems. As given in the example, proving p is *at least hard as* proving the Riemann hypothesis for the fact that p contains the information of the Riemann hypothesis. By the same method, one can define a new proposition which is constructively even harder to prove than p. One can further investigate the degrees of the hardness of their constructive provability and define, perhaps, an equivalence class induced by this reduction of provability. But we are straying too far from our subject. Constructive mathematics is an attractive subject and should be pursued in earnest elsewhere.

Intuitionism is being studied today, for the most part, in the form of constructivism. Many research papers have been written in mathematical logic, particularly *proof theory*, on various kinds of intuitionistic logics and *constructive set theory*, which are all beyond the scope of this book. Along with Brouwer's and Dummett's works, for further elaboration of constructivism, the reader may refer to works by Arend Heyting [144] and Errett Bishop [35], two notable figures in constructive mathematics. The reader may refer to Troelstra and Schwichtenberg [287], or Takeuti [279], for a detailed account on proof theory; and Bell [26] on constructive set theory.

Discussion Questions

1. From the Kantian point of view, do theorems extend the content of axioms?

2. What is the importance of the absolute/relative space distinction for Kant's transcendental philosophy?

3. According to Kant, arithmetic is represented in the pure time form. Do you think it might be possible to represent arithmetical objects in the pure space form?

4. List some of the advantages and disadvantages of intuitionistic mathematics?

5. How plausible is to make mathematical truth time-dependent?

6. Suppose that the triple negation theorem of Brouwer were false, that is, suppose $\neg\neg\neg p \not\equiv \neg p$. What would be the epistemological consequences of this?

7. Do you think that the notion of absolute infinite and impredicativity are intertwined? Do we still run into impredicativity if we rely on potential infinity instead?

5

Logicism

Mathematics and logic, historically speaking, have been entirely distinct stud-
ies. Mathematics has been connected with science, logic with Greek. But both
have developed in modern times: logic has become more mathematical and
mathematics has become more logical. The consequence is that it has now be-
come wholly impossible to draw a line between the two; in fact, the two are
one [...] The proof of their identity is, of course, a matter of detail. −Bertrand
Russell[1]

In Section 4.1 of Chapter 4, we discussed Kant's two main ideas. The first,
and foremost, is that mathematical knowledge is independent of sensory experience. The second is the idea that mathematical knowledge is independent
of the *a priori* forms of pure intuition, i.e., pure space and time forms, that
mathematics is synthetic *a priori*. Undoubtedly, Kant's work on philosophy of
mathematics is of critical importance. However, just like in any other philosophical ideology, even Kant's work received objections. These objections can
be classified into groups. The first criticism was to reject the idea that mathematics is *a priori* and to advocate that mathematics is, in fact, empirical, i.e.,
a posteriori. Second criticism was to adhere for that mathematical knowledge
is analytic. The idea that mathematical knowledge is analytic was profoundly
supported by Leibniz. As a matter of fact, Kant developed his own views as
a reaction to Leibniz and his followers such as Christian Wolff. On the other
hand, John Stuart Mill, who came after Kant, was one of the leading philosophers who supported the first objection. Mill, in claimed that mathematics
was solely based on sensory experience and physics. The 19th-century philosophical and mathematical community leaned towards, however, the second
objection. This, in turn, manifested itself in the logical arena.

In this chapter, we will study one of the central views in the philosophy
of mathematics, introduced in the late 19th century, called *logicism*. Logicism
argues for that mathematics can be in principle reduced to logic. The main
idea is to express primitive objects of mathematics, such as the concept of
number, successor, etc., using logical concepts. Solely by using the logical
terminology, mathematical theorems would simply be a consequence of logic.
We will first discuss Gottlob Frege, widely regarded as the father of logicism,

---

[1]Introduction to Mathematical Philosophy, §18 (194–195), 1919.

DOI: 10.1201/9781003223191-5

81

and then talk about Bertrand Russell. In the final section, we will look at Rudolf Carnap and logical positivism.

5.1   Frege

Gottlob Frege (1848–1925) is considered to be one of the greatest logicians in the history of Western thought. As bold as this claim may seem to be, we will try to explain and see why Frege deserves his rightful place.

Works of Leibniz in the 16th and 17th centuries, particularly his ideal *characteristica universalis*, in many of his writings, can be said to lay the foundations of the logicist ideology. Leibniz believed that reasoning could be carried by some kind of a formal system having a *universal language* so that any scientific problem, by translating the problem into the language of the system, would be solved merely by a calculation. Hence, solving a scientific problem would become no different than, say, solving an equation. Accomplishing this so, one could transform any scientific problem, so to speak, to a calculation problem. Frege, in his first book on symbolic logic, called *Begriffsschrift* [97], attempted to tangibly realise the idea of Leibniz and initialised the logicist programme. The *Begriffsschrift* is widely considered as the foundation of symbolic logic and it is the basis of Frege's later studies, *Die Grundlagen der Arithmetik* (The Foundations of Arithmetic) [98] and *Grundgesetze der Arithmetik* (Basic Laws of Arithmetic) [99] [100].

We shall begin with explaining what analyticity means for Frege. Recall that for Kant, given a proposition of the "subject-predicate" form, if the subject of the statement is implicit in the predicate, then the proposition is analytic. In Frege, propositions are rather considered in the form of functions and their arguments. For him, analyticity is directly correlated with logical verifiability/provability. Hence, Frege's analyticity is more informative than that of Kant's criterion. In his *Grundlagen*, Frege says:

> These distinctions between *a priori* and *a posteriori*, synthetic and analytic, concern, as I see it, not the content of the judgment but the justification for making the judgment. Where there is no justification, the possibility of drawing the distinctions vanishes [...] The problem now becomes that of finding the proof of the proposition, and of following it back to the primitive truths. If, in carrying out this process, we come only on general logical laws and on definitions, then the truth is an analytic one. If, however, it is impossible to give the proof without making use of truths which are not of a general logical nature, but belong to [...] general science, then the proposition is a synthetic one.[2]

---

[2]Frege, Foundations of Arithmetic, §3, 1884.

According to Frege, all logically verifiable propositions are analytic *a priori*. In fact, the only analytic propositions are those which are provable from general logical laws or from definitions. Moreover, for Frege, any statement of arithmetic must be ultimately provable or refutable from a set of logical rules or definitions. But Frege makes a stronger claim that, in fact, every mathematical proposition (or its negation), in particular, an arithmetical statement, is provable. To show that statements of arithmetic are analytic, one first needs to reduce arithmetic to general laws of logic, and this is exactly what Frege attempted to accomplish.

The act of *counting* is perhaps one of the most basic elements of arithmetic. We are particularly interested in equipollency of two collections, i.e., collections being equal to each other quantity-wise. We know that two collections have the same amount of objects if and only if there exists a bijection between them. This gives us the notion of *equinumerousity*, which can be defined as follows: two concepts M and N are *equinumerous* if and only if there exists a one-to-one correspondence between the objects falling under the concept M and the objects falling under N. For example, on a car racing setting, say, the cars are equinumerous with the drivers if there is a one-to-one correspondence between the cars and the drivers. Note that we have not yet defined what a "number" is, but instead we defined "number-equivalency". Although the definition speaks about equinumerousity, we made reference solely to logical concepts and avoided the use of natural numbers. Frege managed to define equinumerousity by using *second-order logic* and *Hume's principle*.

Hume's principle: For any given concepts F and G, the number of F is identical to the number of G if and only if F and G are equinumerous.[3]

In this case, the "number of a concept" is a reference, expressed in natural language, made to a singular term. Prima facie, it might be thought that Frege did not believe in the existence of universals. On the contrary, Frege is known to be an ontological realist (Platonist). However, Frege's realism should not be confused with Plato's conceptual (Idea/Form) realism. For Frege, concepts are regarded as functions. Nonetheless, these functions are not considered to be completed and perfect objects as in Platonism. In Frege, functions (or predicates) are incomplete and, in fact, *unsaturated* objects, which is in contrast with Plato's Forms for which such an incompleteless is out of the question. As an example, suppose that the predicate $P(x)$ is true if and only if $x > 0$ is true. Now, $P(x)$ does not have a truth value yet. But as soon as we provide it with a fixed argument n, $P(n)$ will become *saturated* and come out either true or false. For this reason, $P(x)$, as an open sentence, remains unsaturated.

---

[3]The principle originally appears in the Section I of Part III of Book I of Hume's *A Treatise of Human Nature* [152] precisely as follows: *When two numbers are so combin'd, as that the one has always an unite answering to every unite of the other, we pronounce them equal; and 'tis for want of such a standard of equality in extension, that geometry can scarce be esteem'd a perfect and infallible science.*

It is worth noting that, for Frege, logical justification is what matters in determining whether a proposition is analytic. Predicates, given an argument, that are not logically provable (or refutable) are not considered as analytic statements.

How can we then reduce numbers to logic, just by using the logical terminology? The goal of the logicist programme was ultimately to reduce all objects of mathematics to logical concepts. A logicist would naturally want to begin with the most fundamental concepts of mathematics, for instance, natural numbers. Among natural numbers, it is most reasonable to start with defining what 0 corresponds to in pure logic. Let $Z(x)$ be the concept "x is not identical to itself". Which objects fall under this concept? In fact, none. Since every object is identical to itself, there is no such a satisfying $Z(a)$. In other words, $Z(a)$ is true for no a. In fact, the number 0 is defined as this absence, that is, the number of Z. What about the rest of the numbers? Let us define the concept $T(x)$ as "x is identical to 0". It is easy to see that there is only one x which is identical to 0 and that is the number 0 itself, as defined earlier. Therefore, $T(a)$ is true if and only if $a = 0$. So only *one* object falls under T, namely 0. Frege defines, then, the number 1 as the number of T. The number is 2 defined in a similar fashion except with a trick involving logical disjunctions. Frege defines the number 2 as the number of the concept "x is identical to 0 or x is identical to 1" which contains exactly two objects, namely 0 and 1. The rest is defined similarly. Frege also gives the *successor* relation between two numbers as follows. Let us denote the concept "falling under F but not identical to x" by K. A number n is the *successor* of the number m if and only if there exists a concept F and an object x falling under F such that the number of F is n and the number of K is m. So n is the successor of m iff there is a concept under which n many objects fall and when we subtract one object from that concept, m many objects remain.[4]

The definition given above might look a little complicated at first sight. It is often better understood in terms of sets in the following manner. For any set A, the set

$$S(A) := \{x \cup \{y\} : x \in A \wedge y \notin x\}$$

is the successor of A.[5] Then, the successor of A is obtained by adding a new single element to A. As a matter of fact, $S(A)$ may not be unique in this case, as y could be arbitrary as long as $y \notin x$. Recall from Chapter 2, however, that the successor of a natural number n, in terms of sets, is defined as the unique set $n \cup \{n\}$. As we shall mention shortly, the natural number n is defined as the collection of all sets X such that $|X| = n$.

It remains to define what a natural number is. For this, Frege introduced *extensions* of concepts or, in the set-theoretical jargon, *equivalence classes*.[6] The *extension* of a concept is the collection of all instances of that concept.

---

[4] "iff" abbreviates "if and only if".

[5] We use the symbol := to mean "is defined as". So $x := y$ means "x is defined as y".

[6] Frege, §68, 1884.

For example, the extension of the concept of *cup* is the collection of all cups. An extension can be thought of as defining a second-order object, a class, so to speak. Then, the number of the concept F is the extension of a concept which is equinumerous to F. We think of the number 2 as the extension of all concepts whose number is 2. In the set-theoretic setting, as mentioned earlier, the number 2 is defined to be the class of all sets containing two elements. The proceeding numbers are defined in a similar fashion. So what is meant by the "extension" of a concept is the collection of all objects which fall under that concept. As a result, an extension is a higher-order type object.

Using second-order logic and Hume's principle, Frege proved some basic theorems of arithmetic which later came to known as *Frege's theorem*. Hence, with his theorem, Frege managed to reduce arithmetic to logic and showed that arithmetic is analytic *a priori*. However, mathematics is much bigger than just arithmetic, as it has been a controversial problem whether the logicist programme could be achieved for other branches of mathematics.

One ontological argument against logicism would be to claim that logical objects do not really exist. In fact, one might think that the logical method or conceptual analysis would not introduce new objects and so would not extend our knowledge. An objection of this kind, in fact, goes back to Kant. Kant emphasised that conceptual analysis would not reveal new knowledge or objects. Since Frege's work reduces arithmetic to logic and that logical analysis do not introduce new objects, as the Kantian view supports, arithmetical activity does not extend our knowledge, and nor does it introduce new objects of arithmetic. Based on this objection, either logical objects must not exist or, if they do exist, the logicist mathematics is merely a closed-system activity of re-stating the same propositions, repeatedly, in different forms since conceptual analysis do not allow us to expand our mathematical realm and the mathematical universe. In Poincaré's [220] words, the syllogism can teach us nothing essentially new, and if everything must spring from the principle of identity, then everything should be capable of being reduced to that principle'.[7] For Frege, the ontological universe of the logical domain is non-empty. He assumes that concepts are in the scope of the logical language and that extensions and concepts are simply intertwined. Hence, Frege was an adherent of ontological realism, particularly in the logical domain.

Frege, when reducing the arithmetic to logic, solely used the logical terminology such as *concepts* and *extensions*. Interestingly enough, in contrast with his idea of the logical nature of arithmetic, he did not attempt to reduce geometry to a logical basis. In other words, we see that Frege is in complete agreement with Kant on the aspect that geometry is synthetic *a priori*. According to Dummett [82], and independently by Shapiro [258], Frege held that geometrical truths rest on intuition, and it has been concluded that he

---

[7]Poincaré, p. 1, 1902.

thought, like Kant, that space and time are *a priori* intuitions and that physical objects are mere appearances.[8]

5.2   Russell

Frege's next volume, *Grundgesetze der Arithmetik* (Basic Laws of Arithmetic),
has a tragic ending. The reason is that Frege's one of the laws introduced
in the *Grundgesetze*, when used with the Axiom Schema of (unrestricted)
Comprehension, which was later discovered to be false, caused a contradiction.
The problem was the infamous *Basic Law V* and how it allowed us to define
paradoxical objects using naive set theory. Let us first express what the Axiom
Schema of (unrestricted) Comprehension is. Before the axiomatisation of set
theory by ZFC, it was (mistakenly) thought that for any property φ, any
collection whose elements satisfy φ would constitute a set. The Axiom Schema
of (unrestricted) Comprehension says that, for any property φ, the collection

$$\{x : \varphi(x)\}$$

forms a set. Frege's Basic Law V can be expressed as follows. Let F and G
be two concepts. The extension of F is identical to the extension of G if and
only if

for every a, $F(a)$ holds if and only if $G(a)$ holds.

In other words, the extension of F is identical to the extension of G iff the
same objects fall under F and G. In 1902, Bertrand Russell, in his letter to
Frege, said that his Basic Law V caused a fatal paradox in naive set theory.
To express this paradox in terms of sets, let $R = \{x : x \notin x\}$. We ask
whether or not $R \in R$. If $R \in R$, then $R \notin R$ by definition. However, if
$R \notin R$, then, following the definition of R again, it must be that $R \in R$. In
either case, we get a contradiction. This is known as *Russell's paradox*, named
after Bertrand Russell.[9] Russell explains the paradox in his own words in the
following manner:

> The comprehensive class we are considering, which is to embrace
> everything, must embrace itself as one of its members. In other
> words, if there is such a thing as "everything", then "everything"
> is something, and is a member of the class "everything". But
> normally a class is not a member of itself. Mankind, for example,

---

[8]Dummett, p. 233, 1982. See also Shapiro, §1–2, 1991.

[9]This is not the first paradox about naive set theory however. The *Burali-Forti paradox*,
"the set of all ordinal numbers", introduced by Cesare Burali-Forti [49] in 1897, appeared
a bit more earlier.

is not a man. Form now the assemblage of all classes which are not members of themselves. This is a class: is it a member of itself or not? If it is, it is one of those classes that are not members of themselves, i.e. it is not a member of itself. If it is not, it is not one of those classes that are not members of themselves, i.e. it is a member of itself. Thus of the two hypotheses—that it is, and that it is not, a member to itself—each implies its contradictory. This is a contradiction.[10]

Despite the fact Russell showed a fatal flaw in Frege's Basic Law V, when looked carefully, it is reasonable to take the aforementioned law as the definition of *class*. Apparently, what Russell finds problematic in Basic Law V is that it leads to a fallacy. Recall from Chapter 4 that an impredicative definition of an object is one in which the whole or a part of the definiendum is referred in a totality that is *a priori* assumed to exist. That is, if the definition, by any means, invokes the definiendum, then it is an impredicative definition. The definition of the *greatest lower bound* we gave in Chapter 4, as it quantifies over a totality containing *all* lower bounds of the given set, including the greatest lower bound, constitutes a good example of an impredicative definition. The reason behind Russell's paradox was thought to be caused by allowing impredicativite definitions in our theories. This is explained by Shapiro as follows:

> The development of Russell's paradox runs foul of the 'vicious circle principle'. To generate the paradox, we defined a concept R which 'applies to an object x just in case there is a concept F such that x is the extension of F and $F(x)$ is false'. The definition of R refers to all concepts F, and R is such a concept F. Thus, the definition of R is impredicative. We derive a contradiction from the assumption that the definition of R holds of its own extension. The ban on impredicative definitions precludes even making this assumption.[11]

So the contradiction follows from the fact that Basic Law V assumes that it holds of its own extension and also that every concept has an extension. Therefore, there is no set which contains *all* sets. Otherwise, it would be a set containing itself.

To avoid the use of impredicativity and paradoxes that are caused by such definitions, Russell proposed a systematic treatment of the ontological universe by partitioning it based on the "type" of objects. His proposal was came to known as *type theory*. The way type theory is supposed to work is that all objects must be defined in terms of previously defined concepts so that every object in the universe is classified according to its type. An *atom*

---

[10]Russell, p. 136, 1919.
[11]Shapiro, p. 116, 2000.

is defined to be an object which is not a class. All atoms are type 0 objects. Classes of atoms form type 1 objects. Similarly, classes of type 1 objects form type 2 objects, and so on. Defining numbers in terms of classes may sound more natural and comprehensible compared to Frege's approach. According to Russell, for any given class C, the *number* of C is the class of all classes which are equinumerous with C.[12] Then, for example, the number of a type 1 object is a type 2 object, the number of a type 2 object is a type 3 object, etc. Hence, the number of a class C defines a higher type than the type of C. Type theory was first described in Whitehead and Russell's *Principia Mathematica* [293] which aimed to reduce all of mathematics to a well defined set of axioms so all true arithmetical propositions, and even more, could be formally proved from the axioms. Russell, in his later work [252], maintained that the primary aim of *Principia Mathematica* was to show that all pure mathematics follows from purely logical premises and uses only concepts definable in logical terms.[13] Readers who are familiar with mathematical logic will recall that the goal of having a complete and consistent formal system, on a par with Principia Mathematica, was doomed to fail due to the theorems of Kurt Gödel, who proved extremely significant results in 1931 about the metamathematics of formal systems. We will study Gödel's theorem extensively in Chapter 7.

We shall also look at Russell's ontological philosophy as his view slightly differs from Frege's. Russell developed and revised many of his ideas en route to *Principia Mathematica*. His "no-class" theory constitutes a major part of his ontological philosophy of mathematics. With the realisation of the "no-class" theory, Russell diagnosed that classes could not refer to individual names. By diagonalisation he concluded that there must be more classes of individuals than individuals. Due to this asymmetry in equipollency, classes could not denote individuals. The passage below summarises his views on this matter:

> [T]he symbols for classes are mere conveniences, not representing objects called "classes", and that classes are in fact, like descriptions, logical fictions, or (as we say) "incomplete symbols".[14]

As opposed to Frege, it appears that Russell is not in favour of ontological realism. However, it may also not be correct to claim that Russell is a nominalist. In his *Problems of Philosophy* [250], he criticised the empiricists, particularly Berkeley and Hume, regarding their assumption that universals do not exist. The problem is to determine whether a drawn shape is, say, a triangle. Russell thinks it is impossible to avoid using a sort of resemblance relation to determine, given a particular instance of a triangle as a reference, if the drawn shape constitutes a triangle. He says:

> If we wish to avoid the universals whiteness and triangularity, we shall choose some particular patch of white or some particular

---

[12]Russell, §2(18), 1919.
[13]Russell, p. 74, 1959.
[14]Russell, §18(182), 1919.

triangle, and say that anything is white or a triangle if it has the right sort of resemblance to our chosen particular. But then the resemblance required will have to be a universal. Since there are many white things, the resemblance must hold between many pairs of particular white things; and this is the characteristic of a universal. It will be useless to say that there is a different resemblance for each pair, for then we shall have to say that these resemblances resemble each other, and thus at last we shall be forced to admit resemblance as a universal. The relation of resemblance, therefore, must be a true universal.[15]

The resemblance relation can be thought of as an equivalence relation without which every instance of a quality would be non-identical to the other. For now, we shall leave our discussion on Russell and move on to Carnap.

5.3 Carnap and Logical Positivism

Despite the pitfalls of Basic Law V and Russell's paradox, logicism continued to spread across the European continent. We will now look at another philosophical view which can be regarded as a branch of the logicist programme, called *logical positivism*, also known as *logical empiricism*. To understand what logical positivism is, it is best to remind the reader what is meant by *positivism* in the general sense. Positivism is an empiricism-based philosophical position, primarily against metaphysics, where the source of certain knowledge is believed to be based on empirical evidences and their natural relationships. According to positivism, science should not pursue generalised or universal judgments through abstractions, but should be in accord with the *a posteriori* nature of the physical world and sensory experience, as experimental knowledge is limited to particulars; hence, it is impossible to test *all* cases of a universal judgment through experimentation. In fact, the correct way of establishing a universal statement is to look at the relationship between the outcomes of experimental observations. Positivism, however, claims that we should be focusing on asking *"how?"* instead of *"why?"*.

Rudolf Carnap, in his, *Empiricism, Semantics and Ontology* [55], claims that there are two kinds of questions: *internal questions* and *external questions*. According to Carnap, internal questions are those which are inquired within the system, whereas external questions are those which are inquired outside the system. For example, if we want to know whether or not a particular mathematical object exists, say a set, then there are two ways to answer that question. One is to answer within the system, and in this case, we tend

---

[15]Russell, p. 55, 1912.

to answer it only from the perspective of a closed system in which no meta-physical term is absolute. The second way is to answer from the outside of system, in which case the metaphyical terms should become absolute as well as the answer. We will give examples shortly.

Logical positivism, emerged in the early 20th century from the *Vienna circle*, argues for that the only meaningful statements are those which are logically or empirically verifiable. For logical positivists, mathematical knowl-edge is analytic, hence *a priori*. Any proposition which cannot be tested by "scientific" means is meaningless. Such statements, on logical positivism, can-not be true or false. Then, logical positivism finds the source of knowledge in judgments that are logically or empirically verifiable. Even if a proposition is false, on this view, it is still acceptable as long as it is scientifically justifiable. Even if we "intuitively" know that a proposition is true, it is not a completely meaningful statement on logical positivism unless it is verifiable by scientific methods. Then, the first criterion is not whether the judgment is true or false, but whether its truth value can be determined by logical or empirical meth-ods. Let us give an example to make this clear. For instance, the statement "All crows are black" is a meaningful statement in logical positivism. Despite that it might be a challenging task to verify this statement, there are finitely many crows in the world, hence it is possible in principle to check every single crow and see if each one is black. If this is the case, then we conclude that "All crows are black" is true. Otherwise, we say that it is false. But it is not a matter of being true or false. It is rather about whether the statement is scientifically verifiable or not. On the other hand, the statement "There exists a number" is not a meaningful judgment on logical positivism for the reason that it cannot be shown *a priori* or by empirical means that any number exists in the most general sense at all. Now clearly, within arithmetic, "There exists a number" is followed by the proposition "There is an even number greater than 0". But then how do we know if there is *really* an even number greater than 0? According to Carnap, statements that use metaphysical terms are not considered to be legitimate from the point of view of logical positivism, unless these terms are defined in a language framework. Regarding this, Carnap says:

> Are there properties, classes, numbers, propositions? In order to understand more clearly the nature of these and related prob-lems, it is above all necessary to recognise a fundamental dis-tinction between two kinds of questions concerning the existence or reality of entities. If someone wishes to speak in his language about a new kind of entities, he has to introduce a system of new ways of speaking, subject to new rules; we shall call this procedure the construction of a linguistic *framework* for the new entities in question. And now we must distinguish two kinds of questions of existence: first, questions of the existence of certain entities of the new kind *within the framework*; we call them *in-ternal questions*; and second, questions concerning the existence of reality *of the system of entities as a whole*, called *external*

questions. Internal questions and possible answers to them are formulated with the help of the new forms of expressions. The answers may be found either by purely logical methods or by empirical methods, depending upon whether the framework is a logical or a factual one. An external question is of a problematic character which is in need of closer examination.[16]

What we need to understand by "linguistic framework" is the actually a criterion for "existence", or other metaphysical terms, based on logical or empirical methods. That is to say, if we can test "existence" by means of logic or sensory experience, then such ontological questions we ask under this framework become internal questions. On the other hand, asking whether a number exists, in the general sense, is classified as an external question and this is beyond consideration of the logical positivist. Let us take ZFC set theory for instance (see section 2.3.1 of Chapter 2). Within the system, questions like "given a set x, does there exist a set y such that $|x| < |y|$?" or "does there exist a set which is the intersection of two sets whose elements satisfy the property P?" are all internal questions. By the Axiom of Power Set, the collection of all subsets of a set is actually a set. Furthermore, by Cantor's theorem (see Theorem 9.1), the fact that the power set of any given set has a larger cardinality, it turns out that the first question has a positive answer and so the statement can be justified solely based on logical methods. The question whether there exists a set at all can be answered positively *within* ZFC. For the justification, we may refer to existential axioms of ZFC. The Axiom of Infinity and the Axiom of Empty Set are postulates of this kind, claiming the existence of sets with no other ontological assumptions. However, if we use the term *set* as a general concept, outside of the context of ZFC, then according to logical positivism, the question "does there exist a set?" becomes an external question independent of the language of the system. For Carnap, questions of this kind are not meaningful in logical positivism.

Another central figure in the Vienna Circle was A. J. Ayer (1910–1989). His ideas on the philosophy of logic and mathematics can be found in his book *Language, Truth and Logic* [11], which he wrote at a young age of 26. Just like Carnap, Ayer was a logical empiricist. For Ayer, propositions of logic are free of factual content that analytic propositions do not 'provide any information about any matter of fact'.[17] In his explanation of the universality and necessity of mathematical knowledge, he does give credit to rationalism, where he admits that 'there are some truths about the world which we can know independently of experience; that there are some properties which we can ascribe to all objects'.[18] He argues that, to destroy the foundations of rationalism, either mathematical knowledge is not *a priori* or that they do not concern world facts. He states his solution as follows:

---

[16]Carnap, §2, 1950. Also in Benacerraf and Putnam (1983), p. 242.

[17]Ayer, p. 73, 1936. Page references belong to the (2001) reprint version.

[18]ibid, p. 66.

Experience gave us a very good reason to suppose that a 'truth' of mathematics or logic was true universally; but we were not possessed of a guarantee. For these 'truths' were only empirical hypotheses which had worked particularly well in the past and, like all empirical hypotheses, they were theoretically falliable.[19]

For Ayer though, truths of logic and mathematics being independent of sensory experience does not mean they are innate. We do not discover an already existing knowledge, but it is "learned" by an inductive procedure. It is very likely that 'the principle of the syllogism was formulated not before but after the validity of syllogistic reasoning had been observed in a number of particular cases'.[20] He holds that mathematical knowledge is *a priori*, though he draws an essential distinction between empirical generalisations and statements of formal sciences like mathematics and logic. He says:

We may come to discover them through an inductive process; but once we have apprehended them we see that they are necessarily true, that they hold good for every conceivable instance. And this serves to distinguish them from empirical generalisations. For we know that a proposition whose validity depends upon experience cannot be seen to be necessarily and universally true.[21]

It may not be wrong to say that logical positivism lost its popularity. The criteria and conditions for scientific "verifiability" is not very clear. Determining what these conditions might be, in fact, would, in turn, define what is meant by scientific observation. Logical positivism, prima facie, may seem plausible apart from that it is known to be self-contradictory to some extent. On logical positivism, propositions that cannot be logically or empirically verified are non-sensical. But then the latter statement as well cannot be verified by any scientific means.

Discussion Questions

1. Argue whether or not logicism is compatible with anti-realist views?

2. Why Frege might have regarded geometry as synthetic *a priori*? Do you think geometry can be reduced to logic?

3. To what extent do you think Russell's paradox concerns daily mathematical practice? Can there be Russell-like paradoxes in more concrete parts of mathematics?

4. Discuss about what kind of problems one might have with logical positivism.

---

[19] ibid, p. 67.
[20] ibid, p. 68.
[21] ibid, p. 68.

5. Can there be "absolutely external" questions that pass beyond all language frameworks we could ever put forward?

6

Formalism

We must not believe those, who today, with philosophical bearing and deliberative tone, prophesy the fall of culture and accept the ignorabimus. For us there is no ignorabimus, and in my opinion none whatever in natural science. In opposition to the foolish ignorabimus our slogan shall be: We must know, we will know! —David Hilbert, Radio Address, Königsberg, 1930.

When we look at how proofs in modern mathematics are carried out, we see that they rely on the axiomatic method. The mathematician claims a statement and tries to prove/disprove it based on a predetermined set of axioms. This method, as we know today, goes as far back as to Euclid's geometry, i.e., *The Elements*. The philosophical view we study in this chapter, although influenced by the axiomatic method, emerged much later around the late 19th and early 20th-century.

In philosophy of mathematics, *formalism* was first advocated by Eduard Heine [140] and Carl Johannes Thomae [283], whose ideas can also be seen in Frege's criticisms in the *Grundgesetze*, to avoid certain Platonist commitments about abstract concepts. However, today formalism is more associated with David Hilbert and his programme to rid the foundations of mathematics of all the contradictions and the controversies appeared in the foundational crisis period of mathematics, which lasted from the late 19th century till nearly mid-20th century.

In a nutshell, formalism resembles the axiomatic method but with one major difference: The formalist usually rejects the meaning. That is, formalism argues for that mathematics is merely a game of meaningless symbols, purely on a syntactic basis. Despite that there are different versions of formalist positions, ranging from modest to extreme, formalism in the radical sense often rejects semantics. Modest versions of formalism, on the other hand, can acknowledge that symbols may have meanings. Someone who is an adherent of a radical formalist philosophy would believe that mathematical propositions do not have a meaning or have to be related to things about the so-called mathematical realm. Formalism adopts the view that mathematics is nothing more than a symbol manipulation and transformation activity determined by a given set of formal manipulation rules.

We often encounter in mathematics with notions like negative numbers, imaginary numbers, infinitesimal number, which we know from the ϵ-δ definition of the limit, whose meanings are substantially open to interpretation. To some extent, it can be said that early versions of formalism were invented to patch these notions and avoid all kinds of ontological and semantical commitments, although this was not the intention of the Hilbertian formalism which we shall discuss shortly. From this point of view, a formalist would claim that the symbol manipulation rules that are valid for imaginary numbers are also valid for real numbers, as the entire mathematical activity consists of symbol manipulation and rules that generate them.

6.1   Term vs. Game Formalism

Formalism can generally be classified under two main positions, as described by Resnik [243] and Shapiro [260], namely *term formalism* and *game formalism*.[1] Heine describes term formalism in the following manner:

> To the question what a number is, I answer, if I do not stop at the positive rational numbers, not by a conceptual definition of number, for example the irrationals as limits whose existence would be a presupposition. When it comes to definition, I take a purely formal position, in that I call certain tangible signs numbers, so that the existence of these numbers is not in question.[2]

According to *term formalism*, every mathematical entity must have a name and so it must be referred by a term. For example, in the realist sense, we would know that $4i + 2$ refers to the actual imaginary number. Whereas, in term formalism, it merely denotes the string "$4i + 2$". Similarly, the number 5 never refers to the usual meaning in the realist sense, but rather denotes the character symbol "5". The truth or falsity of mathematical propositions depends on whether or not the given expression is correctly followed from the symbol manipulation rules. For instance, on term formalism, we could say that $5 = 3 + 2$ is true because we might have defined a rule in advance, stating that the character "5" could be replaced with the character string "$2 + 3$". Then, mathematical knowledge is really just the relationship between these symbols. How are we supposed to interpret an equation like $a = b$? If the equality symbol is interpreted in the usual manner, then we interpret this equation as *the object a is same as the object b*. But what does the notion of "sameness" correspond to in formalism? What about the interpretation of the equation $a = a$? Purely on a symbolic basis, prima facie, we may interpret it as that the symbol on the left-hand side is same as the symbol on the right-hand side. But unless we have an upper level abstraction which classifies these symbols to be the "same kind" of object, we will fail to establish the latter equality.

---

[1] Resnik, p. 54, 1980.
[2] Heine, p. 173, 1872. Also in Frege, §87, 1903.

We may then introduce two kinds of symbols in term formalism: *Tokens* and *types*. Tokens are physical characters, marked uniquely by pen or pencil as we draw them. For example, in $a = a$, each symbol on either side of the equality is a token. Although the reader might not notice any difference between them, if we carefully observe them maybe with a high-scaled magnifying glass, we would be able to see that they are not perfectly similar. Hence, physically speaking, the token a on the left hand side is not the same as the token a on the right hand side. However, their *types* are identical. Inspired by Plato's concept of *forms*, the *type* of a token can be thought of as an abstract symbol which bears the essence that all tokens of a particular kind have in common.[3] Types are rather abstract symbols. In other words, an abstract type symbol describes all tokens that physically appear to show similar characteristics as that of the type symbol. For example, consider the binary sequence 10100. This sequence contains five tokens; each character admits a unique token. Yet, it only has 2 types, namely 0 and 1. Mathematics, from the point of view of a term formalist, then, is about the relationship between types.

There is a small problem here though. Consider the equality $5 + 7 = 6 + 6$. By this we mean the character string "$5 + 7$" is same as the string "$6 + 6$". However, let alone the tokens, even their types are different. How can a term formalist interpret this equality? Frege suggests a possible solution by introducing a rule for *substituting* the term on one side for the other. This means that we can replace the string "$5 + 7$" with "$6 + 6$" while at the same time preserve the truth value. Then, a statement of a kind $x = y$ tells us that the type of x and the type of y are formally interchangeable, by which we settle the problem of the interpretation of the equality symbol.

A major problem in term formalism is how to actually represent real numbers, more specifically, irrational numbers. A real number is generally assumed to have an infinite decimal expansion. Unfortunately, we cannot name every real number. The problem is that there are uncountably many real numbers, whereas given a finite alphabet (a set of symbols), we can only write countably many words over that alphabet.[4]

Another kind of formalism is known as *game formalism*, which Thomae explains it in the following manner.

> For the formalist, arithmetic is a game with signs which are called empty. That means that they have no other content (in the calculating game) than they are assigned by their behaviour with respect to certain rules of combination (rules of the game). The chess player makes similar use of his pieces; he assigns them certain properties determining their behaviour in the game, and the

---

[3] In this case, one needs to adopt a realist point of view as we assume that *universals* exist.

[4] The set of words generated from a finite alphabet is always countable. We may, for example, list every possible word that can be written in the Latin alphabet in the lexicographical order starting from strings of length 1, and then of length 2, and so on. In the end, the set of all possible words over the Latin alphabet is countably infinite.

pieces are only the external signs of this behaviour. To be sure, there is an important difference between arithmetic and chess. The rules of chess are arbitrary, the system of rules for arithmetic is such that by means of simple axioms of numbers can be referred to perceptual manifolds and can thus make important contribution to our knowledge of nature.[5]

Game formalism can be regarded as a loose version of term formalism. It treats all of mathematics as a *game* with no commitment to an ontology of objects or properties. In contrast with term formalism, a game formalist may attribute a meaning to symbols to some degree. A radical formalist, as we said earlier, is prone to reject the meaning. If a term formalist derives something like $x = 5$, she may also claim to have concluded $x = 2 + 3$ or $x = 4 + 1$. Contrary to as in term formalism, one does not need to explicitly state a separate substitution rule of "$2 + 3 = 4 + 1$".

Fundamentally, both type and game formalisms agree on that mathematics has no meaning, that numbers do not exist, and that the methodology of mathematics is purely based on syntactic manipulation of strings. Particularly, avoiding to articulate what the ontological background of mathematical objects might be, would classify mathematics as nothing more than a formal game of symbol manipulation. The problem concerning the ontological basis of mathematical objects in the formalist philosophy is indeed an interesting one.

In Frege's criticism of game formalism, he points out that, given a set of axioms under an interpretation in which the axioms are true, the theorems derived from the axioms are also expected to be true under the same interpretation. Frege aimed to construct a firm system in which every derivation would follow from the given set of inference rules in such a way that there would be no ambiguity and gaps whatsoever in-between the derivations.[6] Frege says:

> It is designed [...] to be operated like a calculus by means of a small number of standard moves, so that no step is permitted which does not conform to the rules which are laid down once and for all.[7]

It is worth emphasising one more point. Whilst, in Frege's logicist philosophy, definitions actually describe mathematical entities, definitions in formalism merely produce syntax rules. As a matter of fact, the letters between Frege and Hilbert reveals how two mathematicians significantly disagree with each other in many aspects. The passage given above is to demonstrate that—in

---

[5]Thomae, §1–11, 1898. Also in Frege, §95, 1903.

[6]We shall for now use the term *derivation* to mean *proof*, as the notion of *derivation* refers to a more language-independent and general concept.

[7]Frege, §91, 1884.

determining what the methodology of mathematics could be—Frege's logicism conforms to what we will call later *deductivism*, except that derivations in deductivism are symbolic, whereas in Frege they conform to our logic.

Formalists ought to answer the following question: If mathematical activity is merely symbolic manipulation and that mathematics has no content, then why do we employ it in our scientific theories?

6.2   Hilbert

A proponent of formalism does not necessarily need to believe that the consequences of the axioms of arithmetic would remain true in the same interpretation in which the axioms are true. A logicist needs to ensure that, to have a full conception of arithmetic, the rules of arithmetic should take us from true axioms to true theorems, i.e., consequences. That is to say, the derivations must be truth preserving in a given interpretation. Frege says:

> Whereas in an arithmetic with content equations and inequations are senses expressing throught, in formal arithmetic they are comparable with the positions of chess pieces, transformed in accordance with the rules without consideration for any sense. For if they were viewed as having sense, the rules could not be arbitrarily stipulated; they would have to be chosen so that from formulas expressing true propositions one could derive only formulas likewise expressing true propositions.[8]

Hence, for Frege, the rules are not to be chosen arbitrarily so as to solely satisfy the axioms, but they are to be picked in accordance with the notion of logical implication. So the manipulation rules are expected to preserve the rules of logic whilst being independent of the natural language.

Mathematicians and philosophers for the most part would agree with Frege's idea that rules are meant to produce *true* statements via deductions. Some formalists, however, may oppose to the view that axioms or theorems need to be sound. For such formalists, syntactic correctness of the deductions is of higher priority. This is called *deductivism*. To understand what this means, we shall speak about the logical/non-logical language distinction. On the one hand, we have logical terms which consist of connectives like *and* ($\wedge$), *or* ($\vee$), *if-then* ($\rightarrow$), exists ($\exists$), for all ($\forall$). On the other hand, non-logical terms are associated with the object language. Notions like *point, set, element, inclusion* are some examples of non-logical terms. Apart from this distinction, we should also look at the meanings of these terms. Logical terms are interpreted in the usual way. For instance, the meaning of the logical connective "and" is not

---

[8] Frege, §91, 1903. Also in Shapiro, p. 149, 2000.

really open to interpretation. Particularly on deductivism, the meaning of the non-logical terms are left open. From the deductivist point of view, if P is a theorem of arithmetic, then this merely means that P is in the deductive closure of the axioms of arithmetic. In other words, we are not entitled to make a judgment about what P might mean. Then, on deductivism, the syntactic correctness of the derivations has a higher importance than its meaning. This is, in general, the primary goal of formalism; renovating mathematics based on a syntactic manner so as to eliminate the semantics.

The deductivist approach was also realised in David Hilbert's classic text *Grundlagen der Geometrie* (The Foundations of Geometry) [146] in which he revisited the Euclidean geometry in order to give an axiomatisation free of any meaning and spatial intuition. In the early pages of his work, Hilbert writes:

> We think of these points, straight lines, and planes as having certain mutual relations, which we indicate by means of such words as "are situated", "between", "parallel", "congruent", "continuous", etc. The complete and exact description of these relations follows as a consequence of the *axioms of geometry*.[9]

The difference between the Hilbertian and Euclidean treatments of geometry is that Hilbert does not require the objects of plane geometry to correspond to physical counterparts or statements to be associated with the observations about the physical space. Hilbert's axiomatisation of geometry may be considered, in Kleene's terms (see Chapter 1), as a type of formal axiomatics.[10] In [149], Hilbert mentions that in the axiomatisation of geometry, one must always be able to say, instead of 'points, straight lines, and planes', 'tables, chairs, and beer mugs'.[11] Regarding Hilbert's axiomatisation of geometry, Paul Bernays [34] summarises it as follows:

> It consists in abstracting from the intuitive meaning of the terms [...] and in understanding assertions (theorems) of the axiomatised theory in a hypothetical sense, that is, as holding true for any interpretation [...] for which the axioms are satisfied. Thus, an axiom system is regarded not as a system of statements about a subject matter but as a system of conditions for what might be called a relational structure.[12]

Although the initial axioms that Hilbert specifies rely on the spatial intuition, as Frege [102] notes, his plan was seem to 'detach geometry entirely from spatial intuition and to turn it into a purely logical science like arithmetic'.[13]

In the previous chapter, we discussed the constructivist's criterion for existence, which is simply constructing the object following some well-defined

---

[9] Hilbert, §1, 1899.
[10] Kleene, p. 198, 1967.
[11] Hilbert, p. 403, 1935.
[12] Bernays, p. 497, 1967.
[13] Frege, p. 43, 1980.

procedure. Hilbert's criterion of existence is intertwined with consistency. In a letter, he wrote to Frege, he states:

> As long as I have been thinking, writing and lecturing on these things, I have been saying the exact reverse: If the arbitrarily given axioms do not contradict each other with all their consequences, then they are true and the things defined by them exist. This is for me the criterion of truth and existence.[14]

In fact, Poincaré [220] had the same view in regards to the mathematical ontology. He said 'mathematical entity exists provided there is no contradiction implied in its definition, either in itself, or with the propositions previously admitted'.[15] There is another point that can be said about deductivism and that is the refutation of Kant's thesis about the epistemological nature of geometry. Kant believed that the knowledge of geometry was attained through *a priori* representations in the pure forms of intuition, i.e., pure space and time forms. Advancements in the foundations of mathematics in the beginning of the 20th century, however, indicated that statements of geometry could be logically derived from the axioms of set theory. The validity of Kant's thesis then becomes disputable. In fact, Gödel objects Kant's idea of that in the derivation of geometrical theorems we always need new geometrical intuitions, and so a purely logical derivation from a finite number of axioms is impossible. Gödel in [122] says:

> That is demonstrably false. However, if in this proposition we replace the term "geometrical" by "mathematical" or "set-theoretical", then it becomes a demonstrably true proposition.[16]

Since Gödel considers the knowledge of geometry as a provable set of statements from axiomatic set theory, as opposed to Kant, he claims that geometry is analytic *a priori*. At that time, in the 18th century, of course, there was no such concern about the foundations of mathematics or formalising the mathematical method. Mathematics was not put on a firm foundation back then. In the early 20th century, however, significant part of mathematics, if not all, was successfully formalised by ZFC set theory.

In 1900, at the International Congress of Mathematicians, held in Paris, Hilbert proposed 23 problems that are wished to be resolved by the end of the millennium. The first problem is the Continuum Hypothesis. The sixth problem is rather strange for the mathematical community. It is to axiomatise physical sciences in which mathematics plays an important role, so as to establish a firm foundation, a set of rules for a possible "theory of everything", so to speak, but particularly for the theory of probabilities and mechanics.

---

[14]ibid, p. 42..

[15]Poincaré, p. 44, 1902.

[16]Gödel, p. 385, 1961. Also in Tieszen [285], p. 74, 2011.

Early 20th century was also a time when philosophically minded mathematicians intended to solve the foundational crisis of mathematics. By proposing his *formalisation programme* in order to overcome the crisis, Hilbert quickly became one of the leading savior figures. Hilbert, in his 1925 paper [147], states that the goal of his formalisation programme is to 'establish once and for all the certitude of mathematical methods'.[17]

The Hilbertian idea of formalisation was to apply the axiomatic method and symbolic logic—whether contentful or not—in all areas of mathematics in order to put it on a firm foundation so as to rid mathematics of all the contradictions. One could then also study this formal system by mathematical means. A crucial element in Hilbert's formalisation programme is the idea of *finitary arithmetic*. Consider, for example, a simple statement like "$2^2 + 1^2 = 2 + 3$", or "$3^{9^3} + 1$ is a prime number". Both statements are, in fact, about fixed values of natural numbers. Whether "$3^{9^3} + 1$ is a prime number" is really true can be decided *algorithmically* in modern terms. Of course, the definition of "decidability" was not very clear back then. To understand this better let us give an example. Consider the following statements below.

(i) There exist three consecutive 7's in the first million digits of π.

(ii) There exist three consecutive 7's in the decimal expansion of π.

Both statements include quantification over the domain of digits of π, however there is a significant difference. The quantification in (i) has an upper bound; the first million digits of π, which is of finite length. We may call such quantifications *bounded*. The type of quantification that appears in (ii) has no bound whatsoever as it ranges over all digits of the number π. Let us call this type *unbounded* quantification. According to Hilbert, the only type of sentences which conform to his finitistic arithmetic programme are the ones that can be expressed with bounded quantifiers. Statements that contain unbounded quantifiers are not of finitary nature. An important point that needs to be addressed here is that a statement can be expressed with bounded quantification even though it appears to range over infinitely many elements. For instance, if S is the Fibonacci sequence and p is some large number, then the statement "The number p is in S" merely involves bounded quantification due to the fact that S is an increasing sequence. So if we see a number $q > p$ in S without encountering p, then we know *for certain* that p will not be in the rest of S. Hence, having an infinite domain does not mean the statement involves unbounded quantification, as the algorithmic structure of the domain plays an important role. Decidable statements and statements that conform to Hilbert's finitism are those which can be reduced to statements involving only bounded quantification. We can algorithmically decide the truth value of such statements. In case of otherwise, we may fail to determine uniformly their truth value.

---

[17]Hilbert, p. 184, 1925.

What do we make out of a general statement like $a + 1 = 1 + a$? This is a completely legitimate statement in Hilbert's finitistic arithmetic as every instance of $a + 1 = 1 + a$, when a is fixed, can be effectively calculated. After all, this equation tells us that the symbols $a + 1$ and $1 + a$ are interchangable, as suggested by Frege on behalf of the term formalists. The problem arises when we consider negation of these statements. In regards to this issue, Hilbert notes the following:

> By negating a general statement, i.e., one which refers to arbitrary numerical symbols, we obtain a transfinite statement. For example, the statement that if a is a numerical symbol, then $a + 1 = 1 + a$ is universally true, is from our finitary perspective *incapable of negation*. We will see this better if we consider that this statement cannot be interpreted as a conjunction of infinitely many numerical equations by means of 'and' but only as a hypothetical judgment which asserts something for the case when a numerical symbol is given.[18]

To prove the negation of $a + 1 = 1 + a$, one then needs to find the existence of a number for which the equation does not hold. Now if it were possible to reduce $a + 1 \neq 1 + a$ to a statement containing bounded quantifiers, then one could effectively decide the truth value of the negated statement. The statement $a + 1 \neq 1 + a$, however, is expressed by infinitely many disjunctions and so it does not conform to finitism.

Finitary arithmetic has, in fact, a physical meaning. For Hilbert, the subject matter of finitary arithmetic is concrete symbols whose structure is expected to be immediately clear and recognisable. Each number is associated with concrete symbols. He says:

> [...] the only objects were numerical symbols $1, 11, \ldots, 11111$.
> These alone were the objects of material treatment.[19]

It is implied that natural numbers do, in fact, correspond to material symbols representing quantities. Each natural number n is conceived as consecutively written n many 1's so that a statement of arithmetic like $1 + 2 = 3$ signifies the fact that the length of "1" and "11", when concatenated together, is equal to the length of "111". The notation for the symbolic representation of numbers is rather irrelevant as the numbers could have been symbolically denoted otherwise, possibly by another character than just "1".

Arithmetic, despite its great significance, is just a small portion of mathematics. Beyond arithmetic, there is real analysis, geometry, set theory, etc. Hilbert refers to these fields as a part of his *ideal mathematics* which he hoped to treat under finitary arithmetic.[20] The intention was to establish a system in which mathematical propositions are treated in a finitistic manner.

---

[18]ibid, p. 194.
[19]ibid, p. 195.
[20]ibid, p. 196.

Hilbert, in his formalisation programme, aimed to construct a formal system containing a sufficently extensive set of axioms, but furthermore, it was expected that the system would have the following properties.

(i) Consistency: The system is required to be *consistent*. Hence, no contradiction should be derived from the axioms.

(ii) Completeness: The system should be *complete*. That is, for any proposition φ, written in the language of the system, either φ or $\neg\varphi$ must be derivable from the axioms.

(iii) Decidability: Given a proposition φ, the system should be able to algorithmically decide whether φ is true or not.[21]

The third condition is, in fact, correlated with Hilbert's 10th problem. The 10th problem was to find integer solutions to a given *Diophantine equation* using, in Hilbert's own terminology, finitistic methods.[22] The undecidability of the 10th problem was later proved in 1970 by Yuri Matiyasevich [195].

If mathematics could be captured in a formal system based on finitary arithmetic, then the investigation of the formal system itself would be of great mathematical interest. Real numbers, functions, sets, relations, and similar entities are mathematical objects. Whereas proofs, formal systems, formal languages are metamathematical objects. The mathematical study of these entities was came to known as *metamathematics*. According to Kleene's definition [164], metamathematics is just proof theory.[23] Since proofs themselves are objects of a formal mathematical system, they must conform to finitary methods. The consistency of a given formal system, according to Hilbert, is handled easily. He says:

> The problem of consistency is easily handled in the present circumstances. It reduces obviously to proving that from our axioms and according to the rules we laid down we cannot get '1 $\neq$ 1' as the last formula of a proof [...] that '1 $\neq$ 1' is not a provable formula.[24]

It is implicitly assumed that, given a set of axioms, there is a uniform enumeration of all possible formal derivations. Call this enumerated set S. If "1 $\neq$ 1" is not the final statement of any proof in S, then we understand that our formal system is consistent.

Undoubtfully, one of the most desired goals of Hilbert's formalisation programme was to prove, using finitary methods, the consistency of formal systems. John von Neumann [205] was regarded as a strong adherent of the

---

[21] Hilbert himself did not use the word "algorithm", but he rather used "finitary method". As we will discuss later, mathematical models of algorithmic computation defined independently by Gödel, Church, Kleene, and Turing.

[22] For integers a, b, and c, equations of the form $ax + by = c$ are called *Diophantine*.

[23] Kleene, p. 62, 1952.

[24] Hilbert, p. 199, 1925.

formalist programme. For if it were possible to establish a Hilbertian formal system for *all* of mathematics, the gain from this enormous project would be invaluable. According to Neumann , the outcomes and purposes of a possible Hilbertian system can be listed as follows.[25]

(i) To enumerate (uniformly) all symbols in mathematics and logic.

(ii) To characterise unambiguously all the combinations of these symbols which represent statements classified as "meaningful" in classical mathematics (syntactically correct well-formed formulas). In this case, "meaningful" does not necessarily mean "true". Neumann gives the example that both $1 + 1 = 2$ and $1 + 1 = 1$ are meaningful, i.e., syntactically correct, independent of the fact that the former is true and the latter is false. On the other hand, combinations like $1 + \to = 1$ and $+ + 1 \to =$ are meaningless.

(iii) Given a proof, i.e., a sequence of formulas, to supply an algorithmic method to check whether or not the proof is correct. For instance, a derivation given as

"$(5 = 1 + 2 + 2)$, hence $(5 = 1 + 4)$, and therefore $(5 = 5)$"

is correct. In the end, we want to have an algorithmic procedure that performs proof checking.

(iv) To devise an algorithmic procedure that, when given a true statement φ of mathematics, produces the proof of φ within the intended Hilbertian formal system.

Now, the conditions (i)–(iii) can be easily accomplished. However, there is a slight problem with (iv).

6.3 Gödel's Impact

In 1931, a young logician called Kurt Gödel (1906–1978) proved one of the most astonishing theorems of the 20th-century mathematics. His seminal result has come to known as *Gödel's incompleteness theorem*, which actually consists of two theorems. We will study Gödel's incompleteness theorem and prove it in Chapter 7 in more detail, but we will give a prelude now. Not only does the incompleteness theorem concern the foundations and philosophy of mathematics, it also relates to the theoretical foundations of computer science.

Now let T be an axiomatic system that captures basic arithmetic. In axiomatic systems like T, we can uniformly find a sentence G, in the formal

---

[25]Neumann, p. 118, 1931. Trans. in Benacerraf and Putnam [30], p. 63, 1965.

language of T, such that if T is consistent, then neither G nor $\neg G$ is provable from T. We will call this Gödel's *First Incompleteness Theorem*. In fact, as we will see in Chapter 7, the statement G codes the self-referential statement "I am unprovable". We will show that such a statement can be expressed in the language of arithmetic. So G is indeed a statement of formal arithmetic which codes the self-referential statement saying that it is unprovable. The fact that the aforementioned self-referential statement is expressible in arithmetic is of particular importance, as Gödel's theorem would not be applicable to mathematics otherwise. So there is, in fact, a formal statement in first-order arithmetic saying "I am unprovable". Since T cannot prove G or $\neg G$, T is incomplete. What if we add G to the axioms of T? Then we would obtain a stronger system, call it T'. But since Gödel's theorem also applies to T', we will be able to uniformly find a new statement G' such that neither G' nor $\neg G'$ is provable from T'. Furthermore, Gödel showed that how the incompleteness phenomenon was derived from T. We can prove from T, the statement "If T is consistent, then G is unprovable from T". But the statement which says "G is unprovable from T" is actually G itself. Therefore, from T, we can prove "If T is consistent, then G is provable". Suppose that T is consistent. Then this would imply that G is provable from T. But this contradicts the First Incompleteness Theorem. Then, if T is consistent, T cannot prove its own consistency. We call this Gödel's *Second Incompleteness Theorem*. Hence, the formal system that Hilbert aimed to establish fails to have Neumann's fourth feature, i.e., the system being able to effectively prove true statements.

By Gödel's theorems, Hilbert's programme to formalise *all* of mathematics became an utopia. Nevertheless, this failure has some advantages and disadvantages. To begin with the disadvantage, if our formal mathematical system is consistent, then there are true statements which cannot be proved within the system. That is, any sufficiently strong formal system, if consistent, is incomplete. Then, mathematical *truth* is, in fact, something more than just *provability*. That is, the notion of proof is not a sufficient tool for capturing the concept of mathematical truth. In fact, if our formal system is consistent, then the statement which corresponds to "This system is consistent" is among one of these true-but-unprovable statements. Even if our system is consistent, we cannot prove this fact within the very same system. We should not be very pessimistic about Gödel's theorems though. A promising thing about the incompleteness theorem is the indication that it is impossible to computerise or automatise *all* of mathematics. The mathematician will never be fully replaced by a computer. The factor of creativity and mathematical intuition in the mathematical activity is a distinctive feature which separates the "formal theorem prover" from the "human mathematician". As we shall discuss this theme in the next few chapters, we emphasise that mathematical reasoning cannot be completely formalised. In fact, the incapability of formalisation cannot be specific to mathematical activity; any arithmetic-like discipline which involves the act of making definitions, creating objects, and so on, will fail to be automatised entirely, if not partially.[26]

On the other hand, the claim that our mathematical "intuition" cannot be formalised might just be false. It may be the case that mathematical activity is more than our mathematical intuition and what we can ever imagine it to be with our finite minds. We shall end this discussion for now with Gödel's following statement, as quoted by Robert Goldblatt in [109]: *Either mathematics is too big for the human mind, or the human mind is more than a machine.*[27]

Curry's Formalism. Gödel's theorem had a great impact on the future of formalism. However, it is wrong to say that formalism was completely abandoned after Gödel. Formalism post-Gödel was still advocated by Haskell Curry.[28] Curry's formalism resembles term formalism rather than game formalism. He insists that formal systems must be free of metaphysical assumptions. In his [70], he states that 'mathematics may be conceived as an objective science which is independent of any except the most rudimentary philosophical assumptions'.[29] In mathematics, the central concept for Curry is formal systems. Thus, 'mathematics is the science of formal systems'.[30] For Curry there is an objective criterion for truth, that any given proof can be checked objectively. On the other hand, he holds that there might be undecidable propositions. But Curry claims that mathematics is not confined to a single formal system due to that 'metatheoretic propositions are included in mathematics'.[31] By this way, he avoids Gödel's first incompleteness theorem. He proposes a quasi-truth condition, called *acceptability* of a formal system. For acceptability he gives three criteria:

(i) the intuitive evidence of premises,

(ii) consistency,

(iii) the usefulness of the theory as a whole.

Curry maintains that 'a proof of consistency is neither a necessary nor a sufficient condition for acceptability.'[32] Since Curry and his followers do not seek to prove the consistency of a formal system within itself, they bypass Gödel's second incompleteness theorem.

Discussion Questions

---

[26] Despite that ZFC is a well-defined successful formal system for ordinary mathematics, it does not formalise "true" arithmetic.

[27] R. Goldblatt, p. 13, 1979.

[28] See Curry [69] (1951) for a detailed account on his formalist philosophy.

[29] Curry, p. 202, 1954. Page references to the reprint version in Benacerraf and Putnam (1983).

[30] ibid, p. 204.

[31] ibid, p. 204.

[32] ibid, p. 205.

1. Argue about the differences between the axiomatic method and formalisation.

2. How flexible can one be with game formalism? Do you think there are universal laws for formalisation?

3. We know that Hilbert's ontological position is plentitudinous, that is, an object exists if it is consistent with our axioms. Do you think a formalist could also endorse other forms, particularly non-plentitudinous versions of realism?

4. For Hilbert, numbers actually correspond to concrete symbols. Does this make arithmetic somewhat relative and physical since concrete objects may not be in the form of perfect abstractions?

5. Do you think the study of metamathematics requires extramathematical methods? Discuss briefly.

6. Can we draw a firm line between Hilbert's notion of *finitistic method* and what we call today the *algorithmic method*? Do you think these two notions are equivalent or not?

7. We said that Haskell Curry did not worry about proving the consistency of formal system under consideration. According to this view, Curry and his followers seem to bypass Gödel's Second Incompleteness Theorem. What are the merits and defects of this position?

7

Gödel's Incompleteness Theorem and Computability

In the early 20th century, David Hilbert proposed an ambitious long-term formalisation project of mathematics for resolving the foundational crisis. This project was called *Hilbert's programme*. Shortly after Hilbert's Königsberg speech, however, Kurt Gödel announced his incompleteness theorems which puzzled the mathematical and philosophical community of the time.[1] Hilbert's project was doomed to fail. In this chapter, we will prove Gödel's theorems and discuss its philosophical consequences and reasons behind the incompleteness phenomenon. For some of the parts of the proof, particularly the part where we explain Gödel's numbering, we will follow Smith's [268] presentation.

A *formal axiomatic system* consists of (i) a formal language, (ii) a set of axioms, (iii) a set of derivation rules.[2] For our purpose, we will only consider systems with certain properties. We will, later on, discuss what these properties are.[3] The intention behind Hilbert's programme is to axiomatise *all* of mathematics by which one would be able to determine effectively, using finitistic methods in Hilbert's terminology, the truth value of a given statement in the language of the system. It was also expected that we would be able to prove the consistency of the system from the axioms of the system itself. In the previous chapter, we gave the features of the formal system that Hilbert ideally wanted to establish. In Hilbert's ideal formal system, it was required that the truth value of every proposition would be *algorithmically* decided.[4]

---

[1] I say puzzled because the theorems were not fully understood at the time.

[2] Generally, laws of logical inference are used as derivation rules. However, derivations do not need to be logical. Specifically, in formalism, derivations may be solely based on syntactic manipulations.

[3] Gödel's theorems hold for "sufficiently strong" systems of first-order logic (see section 7.4 of Chapter 7). One can always check, using the truth table method, whether or not a given statement in propositional logic is a tautology. However, we cannot use the truth table method in first-order logic.

[4] In this case, "algorithmically" means anything that conforms to Hilbert's idea of finitistic method. One of the most fundamental elements of Hilbert's programme is the *decidability* feature. The definition of decidability was defined later in the 1930's by various mathematicians. For example, Gödel presented *general recursive functions* as a model for computability. Alonzo Church introduced what we call *λ-calculus*, which constitutes the basis of functional programming languages today. Alan Turing, on the other hand, introduced a universally accepted model of computation, called *Turing machine*, to capture the notion of effective computability. Although we may see more models in the future, the intuitive notion of computability cannot be exactly defined. The question whether *Turing-computability*

DOI: 10.1201/9781003223191-7

Kurt Gödel, in his 1931 paper [115], called in English *On Formally Undecidable Propositions of Principia Mathematica and Related Systems*, showed that sufficiently strong formal systems are either incomplete or inconsistent. We call this the *First Incompleteness Theorem*. Moreover, if the system is consistent, the system cannot prove its own consistency. This was came to known as the *Second Incompleteness Theorem*, a consequence of the first theorem. Proving the theorem with all its technical details is beyond our scope, hence we will omit the proof of some lemmas. A detailed study on Gödel's theorems can be found in Smith [268] or [203]. However, we will give all the essential elements of the proof.

7.1   Arithmetisation of Syntax

The first step in proving Gödel's theorem is to find a statement in formal arithmetic that corresponds to the paradoxical statement "This statement is unprovable". Without finding a correspondence, we cannot claim that such a paradoxical statement is a part of arithmetical truth. For this, Gödel *arithmetised* the language of arithmetic so as to be able to form a genuine arithmetical statement corresponding to the aforementioned paradoxical sentence. The uniform method of arithmetising the syntax is called *Gödel numbering*. Basically, Gödel numbering is a uniform way of coding each statement of formal arithmetic by a natural number. The first step is to assign a code number for every symbol used in the language of arithmetic by which we should be able to form a correspondence between natural numbers and statements of arithmetc.

The system we will be working in is *Peano arithmetic* (PA). The language of PA, along with the logical symbols, consists of the constant symbol 0, the *successor function* symbol S, binary function symbols $+$ and $\times$, and a binary relation symbol $<$. It also contains countably many variable symbols $x, y, z, \ldots$. We do not have symbols for numbers greater than 0. So we denote them in the language of PA through the application of the successor function on the constant symbol 0. For example, we denote the number 2 in the language of PA by $SS0$. We denote 3 by $SSS0$, etc.

The axioms of PA are given as follows.

(i) $\forall x \, (0 \neq Sx)$

(ii) $\forall x \, \forall y \, (Sx = Sy \rightarrow x = y)$

(iii) $\forall x \, (x + 0 = x)$

"exactly" defines the intuitive notion of computability is more of a philosophical problem rather than mathematical. It is usually accepted today in the mathematical community that the *Church-Turing Thesis* (see Chapter 8) defines the notion of effective computability.

(iv) $\forall x \, \forall y \, (x + Sy = S(x + y))$

(v) $\forall x \, (x \times 0 = 0)$

(vi) $\forall x \, \forall y \, (x \times Sy = (x \times y) + x)$

(vii) (Axiom Schema of Induction): $\{\varphi(0) \wedge \forall x \, (\varphi(x) \rightarrow \varphi(Sx))\} \rightarrow \forall x \, \varphi(x)$.
Here, the variable x in the formula $\varphi(x)$ is a *free variable*, that is, it is not in the scope of a quantifier.

As a matter of fact, Axiom Schema of Induction consists of infinitely many axioms; one instance for each formula φ. Let us now code every symbol in the language of PA by the following coding scheme.

$\neg$	$\wedge$	$\vee$	$\rightarrow$	$\leftrightarrow$	$\forall$	$\exists$	$=$	(	)	0	S	$+$	$\times$	$<$
1	3	5	7	9	11	13	15	17	19	21	23	25	27	29

x	y	z	$\cdots$
2	4	6	

The order of the symbols in the scheme is irrelevant. Given the language of PA, how many sentences can we write using the symbols in this language? In fact, countably infinite. Most of them will possibly be meaningless, that is, syntactically incorrect. Others will be syntactically correct. For instance, the formula

$$\forall x \, \exists y \, (Sx = y)$$

is a syntactically correct formula. So is the statement $S0 + 0 = 0$, although it is false in natural numbers. The statement $1 \times 0 = 0$, although it is true, is not a syntactically correct formula for the fact that the symbol 1 is not in the language of PA. We are just concerned right now with syntactically correct sentences. The sentence $+0 <= 0$ is not a formula of PA, it is syntactically wrong. Any sentence which uses symbols not belonging to the language of PA is syntactically incorrect. Let us simply call syntactically correct sentences *formulas*. Fortunately, we can effectively decide whether or not a given sentence is a formula. Hence, it is easy to separate formulas from syntactically incorrect sentences. Based on the coding scheme given above, we will assign a natural number for every sentence written in the language of PA. Now let e be an arbitrary expression $s_0 \, s_1 \, s_2 \ldots s_k$ of $k+1$ symbols, written in the language of PA. Let π_i denote the $i + 1$st prime number. Then, the *Gödel number* of e, is defined as the number

$$\pi_0^{c_0} \cdot \pi_1^{c_1} \cdot \pi_2^{c_2} \cdot \ldots \cdot \pi_k^{c_k}$$

such that c_i is the code number of the symbol s_i, based on the given coding scheme. For example, the Gödel number of the expression S is just 2^{23}. The Gödel number of $S0$ is $2^{23} \cdot 3^{21}$. Consider the statement $\forall x \exists y (x < y)$ which is a syntactically correct statement in the language of PA. In the standard interpretation, this means that for every natural number there exists a larger number. In other words, it says that there is no greatest natural number. The Gödel number of this statement is

$$2^{11} \cdot 3^2 \cdot 5^{13} \cdot 7^4 \cdot 11^{17} \cdot 13^2 \cdot 17^{29} \cdot 19^4 \cdot 23^{19}.$$

Notice that there is a one-to-one correspondence between numbers and sentences. A sentence is coded by a unique Gödel number, and similarly, every Gödel number codes a unique expression. Hence, we can effectively obtain the Gödel number from the given expression, and vice versa. The fact that the mapping is a one-to-one correspondence is followed by the *Fundamental Theorem of Arithmetic*: *Every natural number greater than 1 can be uniquely written as the product of prime numbers.*

Since we managed to code sentences, now we are interested in expressing their truth values via arithmetical relations. Given a formal language, consider the property of being a *formula* in that formal language. We shall define a relation for determining whether a given sentence is, in fact, a syntactically correct statement and denote it by Formula(n). We say that Formula(n) is true if and only if the number n codes a Gödel number for a syntactically correct sentence, i.e., a formula. Other than coding sentences, one can also code proofs. The reader may recall from Chapter 1 that, formally, a proof is just a finite sequence of statements. If we can code every statement of the proof, then we can also code the proof itself following the same method of Gödel numbering, providing the prime numbers once again as the bases of exponents. Consider a sequence of statements $e_0, e_1, e_2, \ldots, e_n$. Suppose that this sequence constitutes a valid proof, meaning that every e_i is either an axiom or follows from at least one e_j such that $e_j < e_i$. To define the Gödel number of the proof

$$((e_0 \wedge e_1 \wedge \cdots \wedge e_{n-1}) \rightarrow e_n)$$

we first find the Gödel number of each statement e_i. Let us denote the Gödel number of e_i by g_i. We then define the Gödel number of the proof as

$$\pi_0^{g_0} \cdot \pi_1^{g_1} \cdot \pi_2^{g_2} \cdot \ldots \cdot \pi_n^{g_n}.$$

Now we can define a binary relation $\mathrm{Prf}(m, n)$. We say that $\mathrm{Prf}(m, n)$ is true if and only if the expression with the Gödel number m is a *proof* of the statement with the Gödel number n.

7.2 Primitive Recursive Functions

We need to ensure that the relations we have defined so far are effectively capturable inside PA, by which we mean the expressibility of predicates like

Formula, *Prf*, and so on. This is where we need the concept of *computability*. For these predicates to be effectively captured inside our formal system, they should conform to Hilbert's finitary arithmetic. Functions we use in ordinary mathematics, such as addition, multiplication, exponentiation, the successor function, absolute value function, etc., are all fundamental to our mathematical studies. These simple functions are all calculable in the sense of Hilbert's finitary methods. In other words, they are all algorithmically computable. Moreover, they describe the simplest type of computability as they are all straightforwardly computable in a finite number of steps and that we are guaranteed to produce a result for any given argument. We will refer to such functions as *primitive recursive functions*. More formally, we define primitive recursive functions inductively as follows.

Definition 7.1 1. The functions given in (a), (b), and (c) are called *initial functions* and they are primitive recursive.

   (a) The *zero function*:

$$\text{For every } n \in \mathbb{N},\ \mathbf{0}(n) = 0.$$

   (b) The *successor function*:

$$\text{For every } n \in \mathbb{N},\ Sn = n + 1.$$

   (c) The U_i^k *projection function*:

$$U_i^k(\vec{m}) = m_i, \text{ for each } k \geq 1, \text{ and } i = 1, \ldots, k,$$

   where $\vec{m} = m_1, \ldots, m_k$.

2. If $g, h, h_0, \ldots, h_l$ are primitive recursive functions, then so is f obtained from $g, h, h_0, \ldots, h_l$ by one of the two rules given below.

   (a) *Substitution rule*:

$$f(\vec{m}) = g(h_0(\vec{m}), \ldots, h_l(\vec{m})).$$

   (b) *Primitive recursion rule*:

$$f(\vec{m}, 0) = g(\vec{m}),$$
$$f(\vec{m}, n + 1) = h(\vec{m}, n, f(\vec{m}, n)).$$

The smallest class of functions which satisfy the definition is called the class of *primitive recursive functions*. Many functions we use in ordinary mathematics is primitive recursive. For example, we can easily show that the addition function $+$ is primitive recursive. This follows from the primitive recursion rule.

$$m + 0 = m,$$
$$m + (n + 1) = (m + n) + 1 = S(m + n).$$

Besides the addition function, one can show that the *predecessor* function we use for subtractions, the *exponentiation* function, the *absolute value* function, etc., are, in fact, all primitive recursive. Gödel proved that the predicates such as *Formula*, *Prf*, and a couple of other relations are primitive recursive. The fact that these predicates are primitive recursive allows us to define the paradoxical Gödelian statement primitive recursively inside PA. For Gödel's theorems, we give two propositions for which we shall omit the proof. But first we give a definition.

Definition 7.2 Let $A \subseteq \mathbb{N}$ be a set of natural numbers. If there exists an algorithm that determines whether or not $n \in A$ for any given $n \in \mathbb{N}$, then A is called a *recursive* (or *computable*) set. If there exists an algorithm which enumerates the elements of A, then A is called *recursively enumerable* (or *computably enumerable*).[5]

Every recursive set is by definition recursively enumerable. However, we will show that not every recursively enumerable set is recursive. First we need to state two theorems.

Theorem 7.1 For any computable set A, there exists a formula $\varphi(x)$ in the language of PA such that

$$n \in A \text{ if and only if } \varphi(n)$$

The n in the formula $\varphi(n)$ is of course described in the language of PA by repeatedly applying the successor function symbol S on the constant symbol 0, e.g., $S \cdots S0$. The reader should understand that any natural number is represented in the formal language of PA in this form. The theorem given above tells us that every computable set is captured in PA. Another required theorem is as follows.

Theorem 7.2 Every primitive recursive function is captured in PA. That is, if $f : \mathbb{N} \to \mathbb{N}$ is a primitive recursive function, then there exists a formula $\varphi(x, y)$ such that

$$\varphi(n, y) \text{ if and only if } f(n) = y$$

for every $n \in \mathbb{N}$.

Similar theorem also holds for n-ary relations. So all primitive recursive functions and relations can be captured in PA. In particular, the relations that Gödel used such as *Formula*, *Prf*, and *Diag*, which we will mention shortly, are all captured in PA.

It may be critical here to note that although primitive recursive functions contain many useful functions used in ordinary mathematics, it does not include *all* computable functions. There are functions which are computable yet

---

[5]The terms "computable", "recursive", "decidable" have the same meaning as they are occasionally used interchangeably. For functions, however, we do not usually use the term "decidable".

not primitive recursive. Each derivation step in the computation of primitive recursive functions may be thought of as a finite string of symbols. More specifically, primitive recursive functions are objects with finite descriptions. Therefore, in principle, we can algorithmically enumerate all primitive recursive functions. Let f_n denote the nth primitive recursive function. Let us define $g(x) = f_x(x) + 1$. Now the definition of g is legitimate since every primitive recursive function is *total*, that is, for every x, $f_n(x)$ is defined. All we do when defining g is to add 1 to the outcome of the xth primitive recursive function on argument x. Then, g must be computable yet it cannot be primitive recursive since, by definition, $g \neq f_x$ for every x.

7.3 Diagonalisation

The next step of the theorem is to express the Gödealian sentence "This statement is unprovable" in the language of PA. For this, we use a well-known method called *diagonalisation*. Essentially, it is very similar to Cantor's method in proving+ that the set $\mathbb{R}$ of real numbers is uncountable (see Theorem 9.3). Suppose now that $\varphi(x)$ is a formula where x denotes the free variable of φ. Then, there is a Gödel number of this formula. Let us denote the Gödel number of φ by $\overline{\varphi}$. The diagonalisation method is the step where we substitute the Gödel number of φ into the free variable of φ. That is, the diagonalisation of $\varphi(x)$ will produce the formula $\varphi(\overline{\varphi})$. The reason that we can, in fact, produce this relies on the *Diagonal Lemma* we give below. Let us use the notation $PA \vdash \varphi$ to mean "φ is provable from PA".

Lemma 7.1 (Diagonal Lemma) Suppose that $\psi(x)$ is a formula in the language of PA. Then, there exists a formula φ such that

$$PA \vdash \varphi \leftrightarrow \psi(\overline{\varphi}).$$

Proof. Let $f : \mathbb{N} \to \mathbb{N}$ be a function which maps the number $\overline{\phi(x)}$ to the number $\overline{\phi(\overline{\phi(x)})}$ and maps the rest of the elements in the domain to 0. As given in Corollary 7.1, it is easily seen that f is a primitive recursive function. Then, the function f can be captured in PA, that is, for every $n \in \mathbb{N}$ there exists a formula $\theta(x, y)$ such that

$$PA \vdash \theta(n, y) \leftrightarrow f(n) = y.$$

Now let us define $\chi(x) := \exists y \, (\theta(x, y) \wedge \psi(y))$ and let φ be the formula $\chi(\overline{\chi(x)})$. This completes the proof of the lemma. $\qquad \square$

Corollary 7.1 For any formula $\varphi(x)$, the diagonalisation of φ, that is, $\text{Diag}(\varphi) := \varphi(\overline{\varphi})$ is primitive recursive.

Proof. It is a primitive recursive process to find the Gödel number of a given formula and to find the formula when given a natural number. Similarly, substituting the Gödel number of a formula into the free variable is a simple primitive recursive function by definition.   □

Now we will use the Diagonal Lemma to prove the First Incompleteness Theorem. It is inevitable that we omit some technical details in the proof, as Gödel's Incompleteness Theorem is normally taught as a one semester course when fully covered.

Originally, Gödel proved his theorem relying on a stronger assumption than what is actually sufficient. But we will see shortly that this does not change the presentation we give here.

Definition 7.3 Let T be a formal system. If the axioms of T separately proves each of the statements $\varphi(0), \varphi(1), \varphi(2), \ldots$, in the language of T, but if T also proves $\exists x \, \neg\varphi(x)$, then T is called ω-*inconsistent*. If a system is not ω-inconsistent, then it is called ω-*consistent*.

Every inconsistent system is, by definition, ω-inconsistent. Also, every ω-consistent system is consistent, but not every consistent system needs to be ω-consistent.

We begin with using the Diagonal Lemma. Let us first consider the formula $\forall y \, \neg\mathrm{Prf}(y, x)$. Call the Gödel number of this formula m. This formula tells us that the formula whose Gödel number is x has no proof. Now let us diagonalise this formula and obtain the following sentence G

$$G := \forall y \, \neg\mathrm{Prf}(y, m).$$

We have just obtained the desired paradoxical statement. G says that G has no proof. For reductio, suppose that G is provable. Let us denote the Gödel number of the proof by n. Then, $\mathrm{Prf}(n, m)$ must be true. Since we assumed that G was provable, we should be able to prove the statement $\forall y \, \neg\mathrm{Prf}(y, m)$. Since it holds for every y, it turns out that, in particular, $\neg\mathrm{Prf}(n, m)$ is provable, which is a contradiction.

Let us now suppose that $\neg G$ is provable. If PA is ω-consistent, then $\exists y \, \mathrm{Prf}(y, m)$ is provable from PA. In this case, for some n, $\mathrm{Prf}(n, m)$ is provable. But this gives us the proof of G, which is again a contradiction. Therefore, if PA is ω-consistent, we can neither prove G nor $\neg G$. This proves Gödel's First Incompleteness Theorem.

Barkley Rosser [246] managed to reduce the ω-consistency in the hypothesis of the theorem to plain consistency. Hence, Gödel's First Incompleteness Theorem can be stated as follows.

Theorem 7.3 (Gödel's First Incompleteness Theorem) If PA is consistent, then it is not complete.

7.4 Second Incompleteness Theorem

Next is to show that sufficiently strong formal systems cannot prove their own consistency. As a matter of fact, this is a consequence of the First Incompleteness Theorem. Define $\text{Prov}(n) := \exists m\, \text{Prf}(m, n)$. That is, $\text{Prov}(n)$ holds if and only if the statement with Gödel number n is provable. We consider PA as our basis system. PA is a sufficiently strong system.[6] We may restate the First Incompleteness Theorem as

$$\text{PA} \vdash G \leftrightarrow \neg\text{Prov}(\overline{G}).$$

We assume, of course, that PA is consistent. That is, we have $\text{PA} \nvdash \bot$, where $\bot$ denotes a contradiction. Now let us define $\text{Con}_{\text{PA}} := \neg\text{Prov}(\overline{\bot})$. Then, Con_{PA} indirectly states that PA is consistent. Recall that the First Incompleteness Theorem says that if PA is consistent, then G is unprovable in PA. That is, $\text{Con}_{\text{PA}} \to \neg\text{Prov}(\overline{G})$. In fact, we have

$$\text{PA} \vdash \text{Con}_{\text{PA}} \to \neg\text{Prov}(\overline{G}). \tag{*}$$

Moreover, in PA, we can prove

$$\text{PA} \vdash G \leftrightarrow \neg\text{Prov}(\overline{G}). \tag{†}$$

For reductio, suppose that $\text{PA} \vdash \text{Con}_{\text{PA}}$. Then, given that (*) holds, we get $\text{PA} \vdash \neg\text{Prov}(\overline{G})$. However, (†) says that $\neg\text{Prov}(\overline{G})$ and G are, in fact, provably equivalent in PA. That is to say, one can be derived from the other. Hence, we have $\text{PA} \vdash G$. But this contradicts the First Incompleteness Theorem and so our assumption must be false. We may then state the second theorem as follows.

---

[6] Now that the predicate Prov has been defined, it is worth discussing what is meant by a *sufficiently strong* system. The criterion is that the system must be able to prove every true predicate which is, computable complexity-wise, on a par with Prov. In computability theory, predicates that alike Prov have a special place in the hierarchy of degrees of unsolvability. Gödel's theorems rely on two basic assumptions: (i) The first-order theory in consideration is expected to be ω-*consistent* (or consistent), (ii) if R is a primitive recursive function/relation, any statement of the form $\exists x\, R(x)$ is expected to be provable within the system whenever it is true. We call statements of the form $\exists x\, R(x)$, in recursion theory, Σ_1^0 statements. Now if (i) is not satisfied, then our system will simply be inconsistent. So, by *ex falso* rule, anything can be proved, hence the system would be complete in the way that we do not want. If (ii) is not satisfied on the other hand, since the system will not be able to capture the notion of provability, there is no point in talking about the "provability" of G. It only makes sense when the system understands what provability is. Consider a system T which is not sufficiently strong, for example, any axiomatic system in propositional logic. Since no predicate on a par with Prov can be captured in T, the unprovability of G does not concern T. In fact, asking if G is provable within T would be non-sensical, similar to as we cannot speak about the truth of falsity of something when the system does not understand what the truth conditions are. Hence, sufficiently strong system means, in this context, that the system is able to prove every true Σ_1^0 statement. We call such systems Σ_1^0-*complete*. So we say that a system is *sufficiently strong* if and only if it is Σ_1^0-complete.

Theorem 7.4 (Gödel's Second Incompleteness Theorem) If PA is consistent, then PA cannot prove its own consistency.

Gödel's theorems are not limited to PA but they apply to any sufficiently strong system. As an immediate reaction, so as to try to deny the theorem, even though it is futile, one may try to add G as an axiom to our formal system. Unfortunately, this does not solve the problem. Even if we add G to our system, we automatically get a stronger system to which the theorem applies. We can again uniformly find another true-but-unprovable statement in this new system. No matter how large our set of axioms gets, we will never be able to avoid the incompleteness phenomenon. In some sense, not only did Gödel prove that sufficiently strong systems are incomplete, but he also proved that they are "incompletable".

One application of the Diagonal Lemma is Tarski's result on the *Undefinability of Truth* [280]. If M is a "structure" and φ is a statement which is "true" in M, then we denote this by $M \models \varphi$. Although we will discuss models and truth in Chapter 11, let us give a simple example to understand what we mean. Consider the statement

$$\forall x \, \exists y \, (y < x).$$

The statement above is false in the natural number "structure" due to the fact that there is no number in $\mathbb{N}$ strictly smaller than 0. On the other hand, the statement

$$\forall x \, \exists y \, (x < y)$$

is true in $\mathbb{N}$ since for every number in $\mathbb{N}$, there exists a larger number. We leave our discussion about models to Chapter 11 and now give Tarski's theorem.

Theorem 7.5 (Undefinability of Truth) Let $T = \{\overline{\varphi} : \mathbb{N} \models \varphi\}$ be a set of natural numbers. Then, there exists no formula $\psi(x)$, in the language of arithmetic, such that $n \in T \leftrightarrow \mathbb{N} \models \psi(n)$.

Proof. Suppose the contrary that there exists some formula $\psi(x)$ satisfying the condition of the theorem. From the Diagonal Lemma, it follows that there exists a formula φ such that PA proves the sentence

$$\varphi \leftrightarrow \neg\psi(\overline{\varphi}).$$

Then, $\mathbb{N} \models \varphi$ if and only if $\mathbb{N} \models \neg\psi(\overline{\varphi})$ if and only if $\mathbb{N} \models \neg\varphi$. A contradiction. $\square$

7.5   Speculations on Gödel's Theorems

Gödel's theorems should not be misinterpreted. The theorem does not say the followings.

(i) Mathematics cannot be formalised.

Which fragment and what kind of mathematics do we refer to? It depends on whether we mean ordinary mathematics or "true" mathematics. Ordinary mathematics can be formalised, as ZFC set theory is a clear example of this formalisation. If we mean formalising "true" mathematics, then it will be chimerical. This is due to the fact that the self-referential Gödelian statement, uniformly obtained within the system, is a true arithmetical statement and that we should accept it as a part of the mathematical truth. What "has been proved" can be formalised, yet what is "being proved" or even "yet to be proved" cannot.

(ii) No formal system can be complete and consistent at the same time.

Gödel's theorems do not apply to every formal system. For example, systems that are not Σ_1^0-complete (see footnote 6) can be simultaneously complete and consistent since such systems do not capture the Prov predicate, which is strong enough to compute all recursively enumerable predicates. An example of a complete and consistent system is the theory of dense linear orders without endpoints.[7]

(iii) Mathematics is inconsistent.

The incompleteness theorem does not show that mathematics is inconsistent. It merely says that sufficiently strong consistent formal systems cannot prove their own consistency. This is not to say that our system is inconsistent. If it is consistent, however, this very fact cannot be proved within the system. Consider, for example, ZFC set theory. If ZFC is consistent, then its consistency cannot be proved within ZFC. Consider, however, a stronger system, call it ZFC^+. Now if ZFC^+ is consistent, it can prove the consistency of ZFC. But then the consistency of ZFC^+ cannot be proved within the same system. No matter how much we enlarge our system, a sufficiently strong system cannot prove its own consistency. This implies that it is impossible to prove absolute consistency results, but we can merely prove relative consistency, assuming the consistency of stronger systems.

(iv) Gödel's incompleteness theorem is false and not accepted by the mathematical community.

The incompleteness theorem is, in fact, a legitimate theorem of mathematics. The only way to reject the incompleteness phenomenon is by denying the rules of classical logic. Although there may be mathematicians, particularly among non-logicians, who have not heard about the incompleteness phenomenon, those who know usually accept Gödel's theorems.

We will now discuss the philosophical consequences of this breakthrough

---

[7]See Hedman [138], §5.5, 2004.

result. Later we will introduce Turing machines and the halting problem to give a natural explanation as to why the incompleteness phenomenon exists. We will finally look at the notion of *algorithmic randomness* for further investigation concerning our search for reasons behind the incompleteness.

7.6   "Real" Mathematics vs. "Ideal" Mathematics

As a result of the incompleteness theorems, a distinction becomes more visible: the line between "real" and "ideal" mathematics; whilst the former refers to true statements of mathematics, the latter refers to the Hilbertian formalisation of mathematics. From Gödel's theorems, we understand that formal mathematics cannot *fully* capture mathematical truth. There are mathematical statements which are true in "real" mathematics, yet cannot proved in the "ideal" formal realm. When a formal system reaches to the point of being able to prove every true Σ_1^0 sentence, it allows us to uniformly generate, within the system, a statement like G which asserts an arithmetical truth in the language of the system. However, neither G nor $\neg G$ is provable from the axioms of the system. That is, the statement G becomes external to the formal system in consideration. Nonetheless, G is still a part of the mathematical truth, i.e., "real" mathematics. In other words, statements like G—although they do not belong to that particular ideal formal realm for the fact that it is an undecidable statement—contain information that escapes the power of ideal mathematics. As we said earlier, adding G as an axiom does not rid the incompleteness phenomenon. Regardless of how many axioms we add, we can still find a statement in the new system which is true but provable from the system. There are arithmetical truths beyond the provable facts of ideal mathematics. What is really required to capture true arithmetic? In fact, the answer to this question from the Kantian point of view can be said to hinge upon *intuition*. Kant claimed that mathematical knowledge is acquired through representations of mathematical objects in the pure forms of intuition, rather than solely by conceptual analysis. From this perspective, Gödel's theorems strengthen Kant's view on the perception of mathematical knowledge. The reason is that it does not seem Kant took the Hilbertian ideal mathematics as a basis when he explained the transcendental principles of perceiving mathematical knowledge. Instead, the type of mathematics Kant considers is more likely to be "real" mathematics whose knowledge is attained, according to his transcendental philosophy, through representations of objects in pure space and time forms. It is, therefore, hinted by Gödel's theorems that certain transcendental elements, such as intuition, representations, etc., are needed in order to capture true arithmetic. As a matter of fact, transcendental schematic elements like pure space and time forms cannot be expressed in the language of formal systems and *ipso facto* such components are not in the *a priori* sense contained in them. Frege [101], for example, disagreed with Kant regarding the nature of arithmetic. For Frege, pure thought, regardless of any content, is able to produce new judgments.[8] Logicism, however, could not be said to have completely succeeded due to Russell's paradox. Despite that Whitehead and Russell, in their *Principia Mathematica*, managed to rid mathematics of paradoxes, it is debated that to what extent it implemented the logicist programme. For example, in the Principia, how logical is the Axiom of Infinity? Although Principia cannot be considered as a complete formal system, it constitutes a foundation for a substantial part of mathematics. Nonetheless, ZFC set theory admits a stronger system. Needless to say, ZFC too contains contentious axioms as to whether they are "logical" or not. If neither Principia nor ZFC is purely logical, then which system should be adopted for mathematics? Which one of these systems is *true*? Both or neither? Or should we seek other systems which are more plausible? It is best if we presume that these theories do not posit absolutely true/false statements about the subject matter. We must divide mathematics into two in this case: the undisputable *real* mathematics, which contains "true" statements; and a vast area, called *ideal* mathematics, which contains the deductive closure of formally defined set of axioms that presumably bear relationships between certain structures inspired from the nature. On the one hand, there is real mathematics which is the universe of statements whose truth values are independent of any interpretation. On the other hand, we have ideal mathematics, i.e., the universe in which statements need not to have semantic content. However, the structures that the axioms of ideal mathematics exhibit would most certainly help the mathematician understand what lies behind real mathematics. The notion of truth, in ideal mathematics, is not an absolute concept but it is something interpreted merely in the formal system; to some extent, it resembles with the perception of truth from the deductivist point of view.

We earlier gave Gödel's objection to Kant's idea that the knowledge of geometry ultimately relies on intuition and representations in the pure forms of intuition. If the term "geometric" is interchanged with "set-theoretic", since every statement of geometry can be expressed in terms of sets, then every true geometric statement becomes provable in set theory. As we said earlier, however, what Kant must have considered, through the activity of perceiving objects through their representation in the pure forms of intuition, seems like "real" mathematics. Gödel's argument in reducing geometry to a set-theoretic basis stems from the fact that ZFC is the foundation of ordinary mathematics. But then again, ZFC cannot capture "real" mathematics. The idea that arithmetical and geometrical statements are perceived through their pure representations, via intuition, is closer to the Kantian position. From Gödel's theorems, though, we understand that formalisation has indeed failed to express "real" mathematics in a complete manner, yet it has succeeded to capture the ideal (formalisable) part of the mathematical realm.

---

[8]Frege, §23, 1972.

Consequently, mathematicians witnessed that the idea of formalisation was not a good candidate to fully capture the concept of mathematical reasoning. As a quirk of fate, the failure of Hilbert's formalisation programme led to an interesting turn of events, i.e., the invention of general purpose computers and programming. All programming languages are built upon formal systems. Hence, despite that the formalisation programme was not a success for mathematics, it gave birth to computer science and computer programming.

7.7   Reasons Behind Incompleteness

Now we shall try to explain the mathematical and philosophical reasons behind the incompleteness phenomenon. Why do we really have incompleteness in sufficiently strong formal systems? More precisely, why does there exist an undecidable statement like G? The answer to this question lies within the nature of the act of "proving". The critical reason behind the existence of undecidable statements is that the *activity of proving* a mathematical statement is generally an open-ended process, that there is no guarantee to find a stage after which we are certain that the statement turns out to be true or false. The proof activity can be thought of as an *open-ended search* process. Can we formalise this open-ended search procedure? So far, we have primarily talked about primitive recursive functions. These functions, however, cannot capture "all" *intuitively* computable functions or, in general, the notion of *algorithm*. Although primitive recursive functions contain all basic arithmetical functions used in ordinary mathematics, they do not cover *all* computable functions. Recall that we proved in section 7.2 that there exist computable functions which are not primitive recursive. In proving this result, we assumed that there was a uniform enumeration of all primitive recursive functions. Primitive recursive functions are *total functions*, i.e., they are defined on every argument. Functions that are not defined on every argument are called *partial*. But then, can we apply the same method to list all partial functions, diagonalise, and then claim that there are computable functions which are not partial? Unfortunately, we cannot apply diagonalisation on partial functions. Suppose we have an enumeration of all partial recursive functions. Given the nth partial recursive function f_n, is $f_n(n)$ defined? There is no guarantee for this, as $f_n(n)$ may well be undefined. Then, to obtain "all" computable functions, it is sufficient to consider partial functions, i.e., functions that do not need to be defined on every argument. The reason why we want to use partial functions to describe the intuitive notion of algorithmic computability is because of the fact that algorithms may not be defined or halt on every input. If we add the following feature to the definition of primitive recursive functions (see Definition 7.1), we obtain the class of *partial recursive functions*. If a function f is defined on argument n, let us denote it by $f(n) \downarrow$.

Definition 7.4 (μ-operator) If $g(\vec{n}, m)$ is a partial recursive function, then so is f given by

$$f(\vec{n}) = \mu m[g(\vec{n}, m) = 0],$$

where $\mu m[g(\vec{n}, m) = 0] = m_0 \iff g(\vec{n}, m - 0) = 0$ and for every $m < m_0$, $g(\vec{n}, m) \downarrow \neq 0$.

In other words, the μ-operator is a search process. This is exactly what we meant by "open-ended search", and we have just described it by the characteristics of the μ-operator.

In fact, we can also give a physical example of where we use the notion of open-ended searching. Imagine we are looking for a physical object in the entire infinite space. Provided that there is no time limit, if the object really exists, we should be able to find it after a finite amount of time, say in 10 years, or 1000 years, or maybe 10 billion light years. If we *ever* find it, it must have been found in a finite amount of time. If the object does not exist, in worst case, we keep the search process indefinitely and, futhermore, we cannot claim that the object does not exist with certainity since the universe is infinite and that there would still be an infinite space not yet been searched. In classic computability, the elapsed time before the algorithm (or the search process) terminates is always finite.

Before giving further reasons behind Gödel's incompleteness, it is worth noting one issue concerning not only just the incompleteness theorem but many antinomies that arise in the foundations of mathematics. It was mentioned in section 5.2 of Chapter 5 that Russell's paradox stemmed from the impredicative definition of the paradoxical collection R. The impredicative definition of R admits an *ontological circularity*, which is described as the ontological dependency of the existence of objects relative to each other. For example, if the existence of an object X is prior to the existence of an object Y, then X is ontologically prior to Y. If X is ontologically prior to Y and vice versa, then X and Y are ontologically dependent on each other. A natural manifestation of ontological circularity in the mathematical domain is impredicative definitions. It could be said that, to some extent, there is such an ontological circularity in formal systems to which Gödel's theorems apply. Due to the arithmetisation of syntax, which uses natural numbers, constructing the formal system and proving any statement of arithmetic within this system, thus, requires the existence of natural numbers. The representation of a formal system and the existence of natural numbers are, therefore, ontologically dependent on each other.

7.7.1 Entscheidungsproblem

We shall continue our discussion concerning why undecidable statements exist. The problem lies within, as we said earlier, the fact that algorithms are not guaranteed to halt on an arbitrarily given input. Although this is not the exact reason, we will delve into a deeper issue in the next subsection.

Hilbert and Ackermann [148], in their 1928 paper *Grundzüge der theoretischen Logik*, presented a decision problem.[9] *Entscheidungsproblem* is defined as a yes/no decision problem of finding whether or not a sentence in first-order predicate logic is valid. Alonzo Church [60] in 1936 showed that the *Entscheidungsproblem* is algorithmically unsolvable. In the same year Alan Turing [288] presented a concrete reason as to why this problem turned out to be unsolvable. Turing introduced a human-idealised computing model before actual computers were even invented. He calls this human-idealised machine literally a *computer*. Of course, Turing did not use this word in today's context of digital computers. Instead, by "computer" he depicted the formal mechanics of any "proving human mathematician". Hence, Turing's *computer* is really the ideal machine model of a mathematician. The modern usage of this term is, of course, very different. Turing first laid the foundations of algorithmic computability. He claimed that an algorithmic procedure is carried out by what we call today a Turing machine.

A *Turing machine* consists of an infinite tape divided into cells, a finite set Σ of symbols, called the *alphabet*, a tape head which reads the symbols on the tape and moves to left/right direction, a finite set Q of states, and a transition function δ such that

$$\delta : Q \times \Sigma \to Q \times \Sigma \times \{\text{left}, \text{right}\}.$$

That is, the transition function observes the current state of the machine, reads the symbol on which the tape head is placed and, if necessary, changes the state of the machine, writes an another symbol, and then moves the tape head to either left or right direction. The transition function does not need to be defined for every argument. As a matter of fact, the transition function is a set of rules that describes how the computation will be carried out by the machine. In other words, it is the set of instructions for the machine.

A Turing machine M starts its computation from the *initial state* $q_s \in Q$. Let $\Sigma^{<\omega}$ denote the set of all finite strings over the alphabet Σ.[10] At the beginning of the computation of the Turing machine, an input string $x \in \Sigma^{<\omega}$ is written on the tape, and the tape head is assumed to be placed on the leftmost symbol of x. The computation is carried out step by step, in accordance with the description of the transition function. If the machine ever enters any *halting state* $q_f \in Q$ at any stage of the computation, then we say the Turing machine M *halts* on input x and the output is defined as the tape content at that stage.

A Turing algorithm is, in fact, the transition function itself. The description of an algorithm is always a finite piece of information. Just as we cannot think of a food recipe of infinite length, we cannot conceive an algorithm with

---

[9]The decision problem was originally called in German, *Entscheidungsproblem*.

[10]By *string* we mean a sequence of symbols.

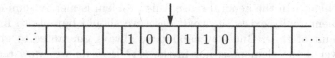

FIGURE 7.1
A Turing machine.

infinite description. The transition function of a Turing machine must be of finite length in the same sense and it must be written formally according to formal rules. Similar to that in primitive recursive functions, one can also uniformly enumerate all possible Turing algorithms. Consider now a uniform enumeration of all Turing machine descriptions (or Turing algorithms). Let us denote the ith Turing machine by M_i. We will now define a decision problem, known as the *halting problem*, that no Turing machine is able to solve.

Halting problem: Given a Turing machine M and an input x, decide whether or not M halts on x.

The halting problem was shown to be undecidable. To prove this, suppose for reductio that it is decidable. That is, assume that there exists a Turing machine that solves the halting problem. Call this hypothetical machine H. Let us now define a new Turing machine H_D in the following manner. The Turing machine H_D will take two inputs: the description of some Turing machine T and an input for n for T. We let H_D simulate T on argument n. Since we assume that the machine H solves the halting problem, we will use this machine in H_D. We continue to define H_D as follows. If the Turing machine T halts on input n (H can decide this), then we let H_D go into an infinite loop. Otherwise, i.e., if T does not halt on input n, we let H_D terminate. But then what should H_D do when given H_D as an input machine. As a matter of fact, we have a Russell-like paradox. H_D halts if and only if it does not halt. But then this is a contradiction. Hence, our assumption that there is a machine which can solve the halting problem must be false. Therefore, there exists no Turing machine which solves the halting problem.

7.7.2 Irreducible information

The reader should be able to understand the reason behind Gödel's incompleteness theorems better now. The concrete reason behind the incompleteness phenomenon is the halting problem. This stems from the possibility that, given an arbitrary set, the search problem of finding an element in that set may turn out to be an unbounded process. Any decision problem can, therefore, be reduced to a set membership problem, i.e., deciding whether or not a given element is a member of the given set (finite or infinite, depending on the problem). In the general sense, this problem is merely semi-decidable. If the element really exists, in worst case, we should be able to find it in a finite amount of time. But why cannot we always guarantee to give a negative answer if the element is not in the set? How do we know really know the element does not exist in the remaining part of an infinite set? Do we not spend enough time on solving the problem? Or is the problem genuinely unsolvable by Turing machines? Due to Gregory Chaitin [57], the reason behind the halting problem was claimed to be due to *algorithmic randomness* in mathematical structures.[11]

To see how the algorithmic content of a set of natural numbers is interrelated with the decidability of the set, consider a "random" subset S of natural numbers whose elements have no order between them whatsoever, at least as far as we know. Given a number n, we may not always be able to algorithmically decide whether or not n is in S. However, if S were the Fibonacci sequence, say, then we know for sure that the elements of S are strictly increasing. In that case, we search through S from the beginning of the sequence. If any point we encounter an element $m > n$ in the sequence before we see n, then we know *for certain* that n will not be in the remaining part of the sequence. It turns out that the algorithmic structure of an object is directly correlated to its solvability.

We know that we can code any mathematical structure by 0's and 1's. In general, a binary alphabet is sufficient for this purpose. The idea of mathematical reasoning is to reduce a mathematical information or a structure to a simpler form of explanation. For example, the infinite string

$$0101010101\ldots$$

seems less "random" than the string

$$10110101001\ldots$$

in terms of their algorithmic content. The first string can be easily defined as the infinite string which produces 0 for odd digits and 1 for even digits. We have just reduced a seemingly infinite structure to a finite description. Despite the fact that the first string can be in principle explained by simpler means than itself, for the second string it seems harder to make an analysis. We cannot say it *must* be "random". But if it really is random, it cannot be explained in simpler terms than itself. Of course, randomness is not an absolute concept and that there is no absolute mathematical definition of randomness, as well as of effective computability. There are definitions of randomness, however, in the way we *understand* the randomness concept.[12] A "random" structure cannot be reduced whatsoever to a simpler information than what it is. In other words, in order to determine the first n bits of a random 0–1 sequence, we need to know at least n bits of axioms.

---

[11] We may also call it *algorithmic unstructuredness*.

[12] See Downey and Hirschfeldt [80] (2010), pp. 225–323.

The traditional method of modern mathematics is based on the axiomatic method. The mathematician claims some hypothesis and attempts to prove it by reducing it to priorly shown theorems or axioms. Gödel's theorems indicate that the axiomatic method, that is, the idea of reducing mathematical statements to simpler propositions in order to establish their truth values, is insufficient to capture *all* mathematical truth. The halting problem is the genuine reason behind incompleteness. But why do we have the halting problem? According to Chaitin, it is due to the existence of irreducible information in pure mathematics, a real number that he calls Ω (also referred to as *Chaitin's constant* in the literature). This real number is defined in such a way that its digits are formed based on the halting probability of a given Turing machine. Before explaining how Ω is defined, it is best to define what makes a real number "computable".

A real number r is called *computable* if, given any natural number n, there exists an algorithm which determines the nth digit of r. For instance, the number π is a computable real since given any natural number n, we can algorithmically determine the n decimal place of π. In fact, the digits of π is still being computed to this day by computers. Since the description of a Turing machine—which includes the finite set of states, the alphabet, the tape content, the definition of the transition function—is finite, using Gödel numbering we can have an uniform enumeration of all Turing machines. Hence, there are (countable) infinitely many Turing programmes. But since the set of real numbers is uncountable, almost all real numbers must be uncomputable.

We first need to make some preperation to define Chaitin's constant. Let us denote the set of natural numbers by ω. Let σ be a string. The *length* of σ is the number of symbols in σ and it is denoted by $|\sigma|$. For example, if $\sigma = ab010c$, then $|\sigma| = 6$. The unique string of length 0 is called the *empty string* which is denoted by ϵ. The set $\{0,1\}^{<\omega}$ denotes the collection of all finite strings over the alphabet $\{0,1\}$. That is,

$$\{0,1\}^{<\omega} = \{\epsilon, 0, 1, 00, 01, 10, 11, 001, 010, 011, 100, 101, \ldots\}.$$

An *initial segment* of a string σ is the first n digits of σ for some $n < |\sigma|$. For instance, initial segments of 10111 are the strings $\epsilon, 1, 10, 101, 1011$, and 10111. Let σ and τ be two strings. If σ is an initial segment of τ, then we denote this relation by $\sigma \subseteq \tau$.[13] If neither $\sigma \subseteq \tau$ nor $\tau \subseteq \sigma$, then we call σ and τ *incompatible* and we denote this by $\sigma | \tau$. For two strings σ and τ, the *concatenation* of σ and τ is simply the string $\sigma\tau$. For instance, the concatenation of $\sigma = 00$ and $\tau = 1011$ is $\sigma\tau = 001011$. Note that concatenation is not commutative. That is, $\tau\sigma \neq \sigma\tau$.

The construction of Chaitin's constant is based on the concept of prefix-free sets. Let $A \subseteq \{0,1\}^{<\omega}$ be a set. If $\sigma | \tau$ for every σ and τ in A, then A is called a *prefix-free* set. Recall that the *domain* of a function f is the set of all

---

[13]Recall that the symbol $\subseteq$ is also used to denote the *subset* relation for two sets. In the domain of strings, the same symbol denotes the initial segment relation.

x's for which $f(x)$ is defined. If f has a prefix-free domain, then we call f a prefix-free function.

Let f be a computable function. Then, f is said to be a *universal function* if there exists a string w such that $f(wx) = g(x)$ for every computable function g and any argument x. Roughly speaking, the string w describes the definition of the function g. The function f, on the other hand, takes the definition of g as a prefix-free argument of f and then simulates g on the rest of the string as an argument.

Now that we know what prefix-free and universal functions are, next step is to define Chaitin's constant. Let P_f be the domain of some prefix-free universal and computable function f. Then, the number Ω with respect to f is defined as follows:

$$\Omega_f = \sum_{p \in P_f} 2^{-|p|}.$$

Clearly, Ω_f is a real number. We show that Ω_f is reducible to the halting problem. If we could compute the first n digits of this number, then we could also decide given any Turing machine M, whose description is of length at most n, whether M halts or not on any given argument. Let us give a short proof of this fact.

Consider a Turing machine programme p. Suppose that the description of p is of some length n. Before proceeding with the proof, let us first introduce a systematic method, used in computer science, for executing infinitely many programmes without falling into an infinite regress. Suppose that we have infinitely many computer programmes

$$p_1, p_2, p_3, \ldots$$

If we try to execute them in this order, given no time limit, we cannot wait indefinitely for the executed programme to halt since there is no guarantee that it would terminate. Recall that an algorithm admits a partial function and it might not terminate on every argument. If we start to execute a programme, since there is a possibility that it might not halt, the next programme might not take its turn for its execution. To avoid the infinite regress of executing a programme indefinitely long, we use a method called *dovetailing*. Dovetailing is essentially very similar to the *zigzag* method for proving that a finite union of countable sets is countable (see section 9.3 of Chapter 9). Using dovetailing, at the first stage, we execute p_1 just for one step. At the second stage, we execute p_1 for another step and execute p_2 for one step. At the third stage, p_1, p_2, and p_3 are executed for one step. In general, by the end of stage n, in total, p_n will be executed once, whilst p_1 will be executed n many steps, p_2 will be executed $n-1$ many steps, p_3 will be executed $n-2$ many times, etc. That is, by the end of stage n, p_m will be executed $n - m + 1$ many steps for

$m \leq n$. In overall, we have the following execution scheme:

Stage 1: p_1

Stage 2: p_1, p_2

Stage 3: p_1, p_2, p_3

Stage 4: p_1, p_2, p_3, p_4

$\vdots$

Stage n: $p_1, p_2, \ldots, p_n$

$\vdots$

In this fashion, even if a programme does not terminate, it is ensured that we will not be stuck with executing a particular programme. If we had executed p_1 indefinitely at stage 1 and put no limit on the execution time, p_2 would never be able to take its turn. We shall now continue the proof of that the halting problem is reducible to Ω. Roughly speaking, Ω is the halting probability of a universal prefix-free Turing machine. We first execute every Turing programme, in a dovetailing fashion, until a sufficient number of programmes terminate so that the halting probability they represent is within the first n digits of Ω. If a program p has not terminated yet, it will not halt. For if it did, it would contribute to the halting probability of p, this, in turn, would change the first n digits and so this would contradict the definition of p. This would then solve the halting problem of p.

Algorithmic randomness is a vast subject which led to a well-developed and rich theory. The subject itself constitutes a large research field in computability theory. Classifications and criteria of the concept of algorithmic randomness, order-theoretic and algebraic relationship between real numbers that satisfy these criteria are some of the topics that have been studied for many years in computability theory. For a detailed account of algorithmic randomness, we refer the reader to more comprehensive texts including Downey and Hirschfeldt [80], Nies [207] or Chaitin [57]. For computability theory, the reader may refer to Cooper [68], Soare [269], [271], or Davis [76].

Discussion Questions

1. Find the Gödel number of the following sentences using the given coding scheme in Section 7.1.

   (a) $\exists x \forall y (x < y)$.

   (b) $(x = 0) \to (\neg \exists y (y < x))$.

   (c) $SS0 = S(S0 + 0)$.

2. Show that the multiplication function is primitive recursive.

3. Show that the exponentiation function is primitive recursive.

4. Show that the predecessor function defined by

$$\delta(m) = \begin{cases} m - 1 & \text{if } m > 0 \\ 0 & \text{if } m = 0 \end{cases}$$

   is primitive recursive.

5. Let $A \subseteq \mathbb{N}$ be a set. Show that A is recursive iff A and $\overline{A}$ are recursively enumerable.

6. What might be the benefits of having an inconsistent but complete mathematics? How one can use this system in a productive way?

7. Do you think Gödel's Incompleteness Theorems are valid in intuitionistic logic? Discuss how.

8. What are some concrete practical implications of the halting problem?

9. Do you think absolute randomness exists? What are the implications of your answer?

8

The Church-Turing Thesis

This chapter is devoted to study of the definition of *computability*. For this, we will articulate what computability might or might not be. We will then argue that how the definition of computability leads to a new version of Pythagoreanism.

8.1 Minds and Machines

The connection between the mind and machines has been a topic of great interest in philosophy of computing. The question whether or not minds can be simulated by machines is one of the central problems in the field. It is well possible that this question can be settled by the theorems of mathematics. For example, can it be told that one of the consequences of Gödel's Second Incompleteness Theorem is that minds and machines are two different things? Smith [268] explains the relationship between minds and machines with an incompleteness argument in the following manner:

> Call the set of mathematical sentences which I accept, or at least could derive from what I accept, my *mathematical output O*. And consider the hypothesis that there is some kind of computing machine which can in principle list off my mathematical output - i.e., it can effectively enumerate the Gödel numbers for the sentences in O. Then O is effectively enumerable, and [...] it follows that there is a recursively axiomatized theory M whose theorems are exactly my mathematical output. Since I accept the axioms of Robinson arithmetic (a system weaker than Peano arithmetic) plus, [...] M is adequate (sufficiently strong). So I can now go on to prove that M can't prove its canonical Gödel sentence G_M. But in going through that proof, I will come to establish by mathematical reasoning that G_M is true. Hence M does not, after all, entail all my mathematical *output*. Contradiction. So no computer can effectively generate my mathematical output. Even if

DOI: 10.1201/9781003223191-8

we just concentrate on my mathematical abilities and potential mathematical output, I can't be emulated by a mere computer![1]

Although this seems to be a convincing argument, Smith immediately points out a possible problem with such an argument that it is rather not clear what is meant by my mathematical "output". The separation between the mind and machines was as well argued by Gödel in his work [121]. As a matter of fact, Gödel was also a supporter of this distinction. Referring to his Second Incompleteness Theorem, he says:

> It is *this* theorem which makes the incompletability of mathematics particularly evident. For, *it makes it impossible that someone should set up a certain well-defined system of axioms and rules and consistently make the following assertion about it: All of these axioms and rules I perceive (with mathematical certitude) to be correct, and moreover I believe that they contain all of mathematics.* If someone makes such a statement he contradicts himself. For if he perceives the axioms under consideration to be correct, he also perceives (with the same certainty) that they are consistent. Hence he has a mathematical insight not derivable from his axioms.[2]

As we said in section 7.4 of Chapter 7, although adding the self-referential Gödelian statement into our axioms might make that particular system look complete, it would yield a stronger system and so we are again faced with the same problem in the new system. One could perhaps approach this problem from a philosophical perspective by presuming that there are "absolutely" unprovable truths. Regarding this, Gödel puts his *disjunction* as follows:

> Either mathematics is incompletable in this sense, that its evident axioms can never be comprised in a finite rule, that is to say, the human mind (even within the realm of pure mathematics) infinitely surpasses the powers of any finite machine, or else there exist absolutely unsolvable diophantine problems of the type specified (where the case that both terms of the disjunction are true is not excluded, so that there are, strictly speaking, three alternatives) [...] where the epithet "absolutely" means that they would be undecidable, not just within some particular axiomatic system, but by any mathematical proof the human mind can conceive.[3]

We recommend the reader to refer to Feferman [87] for more elaboration on

---

[1] Smith, p. 259, 2007. Parentheses added.
[2] Gödel, p. 309, 1951.
[3] ibid, p. 310.

Gödel's disjunction. Koellner [167] investigates the same problem from a set-theoretic perspective. Horsten and Welch [151] published a book exclusively for this subject. For the relationship between Gödel's theorems and philosophy of mind, see Lucas [181] and Penrose [210].

8.2 Effective Computability

One thing that concerns the relationship between minds and machines is the definition of the concept of computability and what it really is. The notion of effective computability (or algorithm) is one of the fundamental concepts in the foundations of mathematics and logic. It is particularly a widely used concept in the epistemological and methodological philosophy of mathematics. Despite that the concept of algorithm has been widely used in mathematics and computer science today, its definition was not articulated in the mathematical community until mathematicians were engaged with Hilbert's formalisation programme. Classical properties of algorithms can be listed as follows.

1. Algorithms are finite procedures. In other words, the description of an algorithm must be of finite length.

2. Computations are performed in stages. That is, every step of the computation must be precise and clear so that it could be, in principle, carried out by a human being or a mechanical device.

3. In the classical conception, algorithms are deterministic. That is to say, each next step of the algorithm is uniquely determined by the previous step.

The characteristic conditions of algorithms were laid down more recently by Wilfried Sieg [263], who classified the principles of algorithmic computability in the following manner.[4]

Boundedness conditions:

(i) There is a fixed bound for the number of internal states that need be taken into account.

(ii) There is a fixed bound on the number of configurations a computer can immediately recognise.[5]

---

[4]Sieg, p. 390, 2002.

[5]By *configuration* we mean the content of the computation at a particular stage. If we take the Turing machine model, a *configuration* at stage s of a Turing computation consists of the following information: (i) The state of the machine at stage s, (ii) the tape content, (iii) the position of the tape head, i.e., which symbol to read next.

Locality conditions:

(i) Only elements of observed symbolic configurations can be changed.

(ii) The distribution of observed squares can be changed, but each of the new observed squares must be within a bounded distance of an immediately previously observed square.

Determinism condition:

The configuration at stage $n + 1$ is uniquely determined from the configuration at stage n.

An algorithm, therefore, is imposed to have the boundedness, discreteness, and determinism conditions. Nevertheless, there is no absolute definition of the concept of algorithm for the fact that the problem of specifying the characteristics of effective computability is rather a philosophical problem. In fact, there have been many proposals on what it might be. One particular mathematical schema which models the notion of effective computation is *μ-recursive functions*, proposed by Stephen Cole Kleene. Gödel, in his lectures at Princeton University, introduced *general recursive functions* for modelling effective computability. Alonzo Church, on the other hand, proposed *λ-calculus*, which is considered to be the basis of all functional programming languages that are being used today. One of the most "natural" models of computation, regarded in the modern day, is called a *Turing machine*, introduced by the British mathematician and computer scientist Alan Turing. Fortunately, all these models were shown to be equivalent in principle and that they all compute the same class of functions, namely computable functions. Hence, although the Turing machine model is usually regarded as the most *standard*, it is sufficient to consider just one model for our analyses. Before Turing machines were introduced, Church considered the λ-calculus model as the standard model of computation. Church [60], with regard to his own model, said:

> We now define the notion [...] of an *effectively calculable* function of positive integers by identifying it with the notion of a recursive function of positive integers (or of a λ-definable function of positive integers).[6]

Shortly after Turing machines were introduced, it was naturally wondered what the relationship between Turing machines and the notion of effective computability might be. This relationship is established by what is known today as the *Church-Turing Thesis*, which can be stated as follows.

The Church-Turing Thesis: Effective computability is Turing-computability.

---

[6]Church, p. 356, 1936.

The Church-Turing Thesis essentially says that any effectively or intuitively computable function is computable by a Turing machine. Many logicians today, including computability theorists such as Robert Soare [270], adopt the Church-Turing Thesis as *the* definition of computability.[7] The belief that Turing mechanical principles successfully model effective computability was also mentioned by Gödel [117], as he stated that "the correct definition of mechanical computability was established beyond any doubt by Turing".[8] The Church-Turing Thesis, of course, is a philosophical thesis rather than a mathematical one. On the one hand, we have an intuitive notion of effective computability. On the other hand, we have a definite mathematical object, i.e., Turing machine. Since this thesis is not a mathematical statement, it cannot be formally proved or disproved. However, we may of course present a convincing argument which constitutes an informal "proof". In fact, we already gave an example of this in section 7.7 of Chapter 7 when we showed that there is a computable function g which is not primitive recursive. Intuitively, g is observed to be a computable function, yet there is no formal definition of a function being "computable". The method used in the proof relies on Cantor's diagonalisation method and it is a highly convincing argument which implies that g is, observably, a computable function. As a matter of fact, we can simply consider that result as a theorem. Hence, similar approaches could be used to prove or disprove the Church-Turing Thesis. For example, Kripke [173] (2013) argues that the Church-Turing Thesis is a corollary of Gödel's completeness theorem for first-order predicate logic with identity. For Kripke, 'a computation is just another mathematical deduction, albeit one of a very specialized form'.[9] He also relies on the claim that the steps of any mathematical argument can be expressed in a language based on first-order logic.[10] Sieg also argues in his works [263] and [264] that the Church-Turing Thesis is susceptible to mathematical proof. So considering all these arguments, assuming that the thesis cannot be proved by *any* means would, therefore, be a mistake.

We shall also discuss what the Church-Turing Thesis could not be, as this thesis could be misinterpreted easily. First and foremost, the Church-Turing Thesis does not say that the human mind is a mechanical being. It merely says "*if* the human mind works solely based on algorithmic principles, then it could be simulated by Turing mechanical laws". However, we do not have any evidence whether the mind processes information in an algorithmic manner or not. Secondly, the Church-Turing Thesis does not put any constraint on what "machines" could compute. There may well be "machines" that model our intuitive conception of computability but which do not rely on Turing mechanics, yet that can solve classically unsolvable problems. If there is no such model of computation, its absence does not follow from the Church-Turing

---

[7]Soare, pp. 284–321, 1996.

[8]Gödel, p. 168, 193?

[9]Kripke, p. 80, 2013.

[10]ibid, p. 81.

Thesis. Finally, the Church-Turing Thesis does not advocate for the *computationalist* philosophy.[11] Piccinini [211], although stated that computationalism and the Church-Turing Thesis relate to each other, says that the thesis does not necessarily imply computationalism.[12]

Characteristics that are essential to the concept of computation can be found in various models. What makes a process to be regarded as computation? Is it merely physical? Or perhaps is it mental? What is the relationship between computable processes and scientific observation? Is every computable process scientifically observable? If we restrict the concept of computability to just physical processes, then the Church-Turing Thesis transforms into a claim about physics.[13] According to Piccinini, this splits into two cases.[14]

Physical Church-Turing Thesis (Modest): Anything physically computable is computable by a Turing machine.

Physical Church-Turing Thesis (Bold): Any physical process is computable by a Turing machine.

There is a significant difference in the assumption between the modest and the bold versions. The modest version assumes in the hypothesis that the physical process is computable. However, the bold version does not have this condition, as it is sufficient in the hypothesis to have any physical process, regardless of whether or not it is computable. Note, however, that the physical process is always assumed to have a mathematical model. For if it did not have any mathematical abstraction, it would not be possible to transform the problem of physics or the physical process to the mathematical domain of Turing machines.

An intriguing question that can be asked at this point is that if there exists a physically computable function/process which is not Turing-computable.

8.3   The New Pythagoreanism

The Church-Turing Thesis claims that the class of effectively computable functions is exactly the class of functions computable by Turing machines. If we adopt this thesis as the definition of computability, since any argument of a Turing machine is of finite length, intuitive notion of computability is automatically reduced to numerical computation, i.e., computations from the

---

[11]We call *computationalism* the view that cognition can be reduced to computation.
[12]Piccinini, p. 97–120, 2007.
[13]See Deutsch [78] (1985), p. 99, for the original statement of the physical version of the Church-Turing Thesis.
[14]Piccinini, p. 103, 2007.

domain ℕ to ℕ. In other words, assuming that the Church-Turing Thesis holds, Turing mechanical principles are sufficient to capture *all* intuitively or physically computable processes. According to the aforementioned thesis then, there exists no effectively computable process which is uncomputable by a Turing machine. Hence, any intuitive computation is bounded by the principles that Turing machines admit. It may be said that, for this reason, there appears a kind of "expressive completeness". Turing mechanical principles are sufficient to *completely* describe the intuitive notion of computability, solely via the relationship between natural numbers. A similar type of "expressive completeness", as it seems, appears in Pythagoreanism. In a nutshell, Pythagoreanism attributes describing the laws of the universe to natural numbers. Is the Church-Turing Thesis, then, a new form of Pythagoreanism in the arena of computability of the modern age? Recall again that Pythagoreanism in the broad sense says that anything in the universe, any relation between the objects and the harmony of the universe, so to speak, can be expressed with the natural numbers and their finite combination of ratios. Therefore, Pythagoreanism provides a complete description of the universe and the objects in it via natural numbers. There are two assumptions behind this view:

1. The universe is a closed system.

2. This closed system can be described in terms of relationships of the natural numbers.

Behind Pythagoreanism lies the idea of expressive completeness of natural numbers. Despite the fact that Pythagoreanism was later refuted by its own followers, the Church-Turing Thesis has survived and it still remains as a plausible philosophical claim that establishes a direct connection between Turing mechanical computations over the domain of natural numbers and intuitive computability. The discovery of *alogos*, that is $\sqrt{2}$, invalidated the Pythagoreanistic idea of the expressive completeness of natural numbers to describe the objects of the universe and, in particular, geometric measures. As a consequence, the Pythagoreanists had to find a resolution regarding the existence of measures or things in the universe which could not be expressed as the ratio of natural numbers. In this case, we are left with two choices. Either we must accept the existence of irrational numbers as a part of the reality, or we save the Pythagoreanistic doctrine by endorsing that irrational numbers are not related to the concepts and objects of the universe whatsoever. Are Turing mechanical principles really sufficient to describe what we call *computability*? According to the Church-Turing Thesis, this question has a positive answer, and hence Turing mechanical features are complete in the sense that they express the notion of intuitive computability. Should we then accept the completeness of Turing mechanics or try to find the "$\sqrt{2}$ of computability"?

Jensen, in his 1995 paper [157], speaks about the separation between "arithmetic" and "geometry", which has been present since the beginning of Pythagoreanism.[15] Whilst natural number arithmetic could be said to support some sort of completeness, knowledge of geometry indicated that we must go beyond what is comprehensible at the time in order to seek further objects. The adoption of $\sqrt{2}$ was a courageous attempt to go beyond the standard conception about the universe and indeed a critical point in the history of mathematics which led to the following question: Should we keep adding new axioms in mathematics? The quest of searching for things beyond what is already accepted resembles the discovery of $\sqrt{2}$ by the adherents of Pythagoreanism. So how does adding new axioms relate to the Church-Turing Thesis? We may say that the Church-Turing Thesis, until proven to be false, is the new Pythagoreanism in the domain of computing. Of course, this depends on whether the thesis is correct in reality or not. For many years, the majority of mathematicians, philosophers, and computer scientists regard the Church-Turing Thesis to be true. Hence, the thesis today is generally accepted as *the* definition of effective computability. However, if one wishes to break the Church-Turing barrier, it could be done in two ways:

(i) Find a Turing-computable function which does not seem to be intuitively "computable".

(ii) Observe a "computable" process which cannot be simulated by any Turing machine.

Now, (i) is very unlikely to happen. It is counter-intuitive to even think of a Turing-computable function which is not computable in the intuitive sense. Nonetheless, it may be possible to achieve (ii).

The attempt to find intuitively computable functions/processes which escape the expressive completeness of the Church-Turing Thesis and Turing mechanics has been a long time ambition of logicians and philosophers ever since the Church-Turing Thesis was first introduced. If we ever want to extend the Church-Turing Thesis so as to improve the notion of computability, it has to involve a phenomenon known as *supertasks*. We will study supertasks in Chapter 10; however, it is worth explaining this concept here shortly in order to understand how it is related to the Church-Turing Thesis. A supertask is a countably infinite sequence of tasks that occur sequentially within a finite interval of time. Let us now consider a hypothetical machine which works in the following manner. The machine performs its first operation in 1 second, and it performs each successive operation in half the time it performed the previous operation. Such a machine will be able to perform infinitely many operations in just 2 seconds. This would, in turn, solve many classically unsolvable problems by Turing machines, for instance, the halting problem or the Entscheidungsproblem. Therefore, an alternative to the Church-Turing Thesis would be to accept supertasks as our new conception of effective computability. Nevertheless, the logical and physical possibility, or otherwise, of

---

[15]R. Jensen, p. 393, 1995.

supertasks is a controversial problem in logic and philosophy. It is also debated whether or not supertasks should be regarded as genuine computations.

The question whether the Church-Turing Thesis is a neo-Pythagoeanistic claim, ultimately depends on the definition of computability. If our conception about effective computability is merely limited to Turing mechanical principles, then the Church-Turing Thesis can be seen as a new form of Pythagoreanism for the notion of effective computation, and so it exhibits the completeness of Turing machines in defining the concept of computability. The neo-Pythagoreanism we speak about here, though, should not be confused with the ancient Pythagoreanism, as the latter attempts to model the physical universe and all its objects, whereas the version we speak of is to model the notion of intuitive computability that occurs in the universe.

The concept of computability is fundamental to mathematics, and it is essentially concerned with the mathematical method. So does there exist an *absolutely* uncomputable function? Certainly, this is subject to our perception of computability and how we define it. If the Church-Turing Thesis is regarded as *the* definition of computability, then the answer to this question is positive indeed. Since the halting problem unsolvable by Turing machines, the thesis tells us that it must be absolutely unsolvable, as a Turing machine is all what we have for computability.

Discussion Questions

1. Do you think Gödel's Theorems put limit on machine intelligence?

2. Recall that Gödel's disjunction is the claim that either mathematics is incompletable and that it is too big or the human mind is more than a machine. Is there a third possibility?

3. Do you think the characteristic properties of algorithms should be revised so as to regard infinite procedures as legitimate algorithms?

4. How would you compare computationalism with formalisation?

5. Do you agree that the Church-Turing Thesis *is* the new Pythagoreanism? If not, why?

9

Infinity

Throughout the history of science and philosophy, the concept of infinity has been a mysterious, intriguing, and obscure notion. It has various sorts of applications and meanings in different disciplines. For the physicist, infinity might be the empty vast space, or the universe itself. For the mathematician, infinity could simply be the set of natural numbers. For the theologian, infinity may refer to God. For some, infinity can be defined, yet for others it cannot. Furthermore, for some, we cannot even speak about infinity. Despite that there is no consensus of opinion on what infinity might be, it could instead be said what it is not: The infinite is not finite. In any case, we shall try to give a philosophical and mathematical account on what infinity might be and how it manifests itself in the mathematical domain.

9.1 Infinity in Ancient Greece

Infinity, as a concept, was first used by the Milesian School, which was founded by Thales in the Ionian town of Miletus on the Aegean coast of modern Turkey. It was used for the sole purpose of understanding the ontological/epistemological basis of the universe, rather than for mathematical purposes.

In the history of philosophy, it is said that Anaximander of Miletus (B.C. 610–546), a student of Thales, was the first thinker in the pre-Socratic period who used the concept of infinity. Anaximander proposed that the principle of all things is that which is without any limit.[1] This limitless principle was referred to as *aperion* ($\alpha\pi\epsilon\iota\rho\text{o}\nu$).[2] For Anaximander, all opposites are coincided and transcended in *aperion*, and all conflict between created things is reconciled. Although Anaximander viewed *apeiron* as something *good, beautiful,* and *comprehensible*, the Pythagorean school, Pythagoras (B.C. 570–495), in particular, viewed *apeiron* quite the opposite way as Anaximander did. Hence, Pythagoras instead viewed *aperion* as *bad, ugly,* and *incomprehensible*. The reason why Pythagoras viewed *aperion* in that manner is because of how Pythagoreanists see the universe and how *aperion* lacks the "harmony", which

---

[1] In ancient Greek, the word limit is translated as peras ($\pi\acute{\epsilon}\rho\alpha\sigma$).
[2] The prefix $a-$ in Greek indicates a lack of feature in the suffix, e.g., moral/amoral.

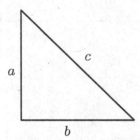

FIGURE 9.1
The Pythagorean Theorem states that $a^2 + b^2 = c^2$.

is something present in the cosmos. Recall that according to Pythagoreanism, anything in the universe and any relation between the objects can be expressed with the natural numbers and their finite combination of ratios. From their point of view, the universe (or the cosmos) is governed by the rules expressible through the ratios of natural numbers. The emanation of plurality from the singularity is akin to the construction of the rest of the natural numbers from the unity. For the Pythagoreanist, the principles of the universe is like a harmony of natural numbers. Ironically, they later came to conclude that their doctrine failed to express the reality of the universe, particularly the knowledge of geometry. The well-known *Pythagorean Theorem* tells us that if a and b are the lengths of the legs of a right triangle and c is the length of the hypotenuse, then the sum of the squares of the lengths of the legs is equal to the square of the length of the hypotenuse. As depicted in Figure 9.1, the Pythagorean Theorem says that $a^2 + b^2 = c^2$.

If the legs of a right triangle is of unit length each, then the length of the hypotenuse is equal to $\sqrt{2}$. It was proved that this number is not rational, that is, in modern terms, it is *irrational*.[3]

From the Pythagorenistic perspective, it can be said that natural numbers and their ratios have something *reasonable* about them, in the sense that they are easy to comprehend, more structured, and discrete as opposed to the nature of *aperion* for the Pythagoreanists. Entities that do not resemble the preciseness, simplicity of natural numbers are incomprehensible, chaotic, and in some sense "illogical". In ancient Greece, any geometric measure was thought to be a natural number. Natural numbers are quantities that admit properties of being odd/even, prime, etc. Geometric measures do not have

---

[3]The term *rational* is a Latin word originated from the word *ratio*. Although the word *ratio* means *rate* or *proportion*, *rational* means *logical* or *reasonable*. The meaning of the term *rational*, in fact, comes from the Greek word *logos* which translates to English as "word", "speech", but also as "thought". Ancient Greeks referred to the length of the hypotenuse as *alogos*, meaning "that cannot be reasoned about" or simply "illogical".

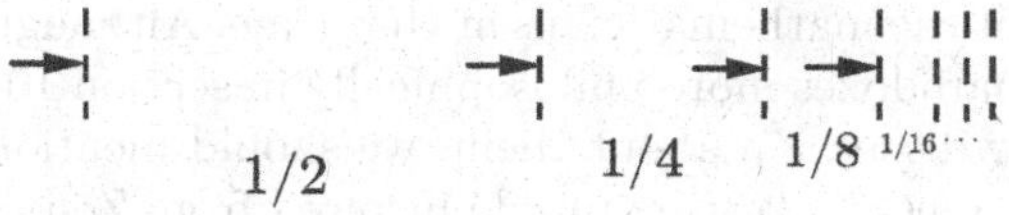

1/2   1/4   1/8 1/16...

FIGURE 9.2
Zeno's arrow paradox: Does the arrow reach its target?

these properties. For this reason, ancient Greeks separated the subject matter of arithmetic from that of geometry. The distinction between the "discreteness" of the natural numbers and the "continuity" of geometric measures had an effect on the relationship between the *finite* and the *infinite*. As a consequence of the Pythagorean Theorem, it was seen that the chaotic and incomprehensible *apeiron*, what Pythagoreanists perceived as, manifested itself in the continuous geometric measures.

Another influential philosopher who made major contributions in the understanding of infinity is Zeno of Elea (B.C. 490–430). To support the views of Parmenides that motion is just an illusion, Zeno introduced various paradoxes that demonstrate the conflicting interplay between discreteness and continuity. One of these paradoxes involves a flying arrow as to whether or not it can reach to a target (see Figure 9.2). Suppose there is a target at point B placed at a unit distance from point A, where the arrow is. For the arrow to reach the target at point B, it must first travel the half distance between A and B, i.e., $1/2$ unit length. Call this half point A_0. Now the arrow is at point A_0 and in order to reach point B, it must first travel the half way between A_0 and B, i.e., another $1/4$ unit length. Call this new point A_1. By the same argument, in order for the arrow to move from A_1 to point B, it must first travel the half way between A_1 and B, which is $1/8$ of the total distance, and so on. This will continue ad infinitum. Hence, by the end of n unit time, the total number of unit distance the arrow is supposed to travel will be

$$\frac{1}{2} + \frac{1}{4} + \frac{1}{8} + \cdots + \frac{1}{2^n} = 1 - \frac{1}{2^n}.$$

Note that the distance travelled is always less than 1. That is, no matter how large n gets, we always have $1 - \frac{1}{2^n} < 1$. Therefore, let alone the arrow hitting its target, it will never actually be able to move from one point to any point, since it requires infinitely many tasks. So Zeno comes to the conclusion that any kind of motion is an illusion.

Zeno's mind puzzling antinomy remained as a paradox for nearly 2000 years, until the discovery of calculus. Ancient Greeks did not really call this a contradiction outright, but they rather classified the problem as a puzzle with the hope that it would be settled in the future. It was not until calculus became a foundation that the mathematical community devised a convincing solution to Zeno's paradoxes. The tension between the finite and the infinite continued to haunt the mathematicians at that time. Although we shall also investigate Zeno's paradoxes more philosophically in section 10.1 of Chapter 10, before we end our discussion about them, we should mention another closely related antinomy. One interesting puzzle inspired from Zeno's paradoxes, was introduced by Lewis Carroll [56], the author of *Alice in Wonderland*, in the 19th century. Carroll published a short yet intriguing article titled *What the Tortoise Said to Achilles*. In this article, motivated from Zeno's paradoxes, Carroll claimed that one could never get to establish a conclusion in logic from a given collection of premises, that our reasoning would always end up in an infinite regress even if we attempted to make a single inference.

According to Carroll's argument, in order to infer any conclusion from a given hypothesis, we should first accept as a hypothesis the statement that our hypothesis implies the conclusion. But adding this implicational statement as a hypothesis yields a larger set of assumptions. We can then apply the same argument for the obtained set of assumptions, yielding an even larger set of assumptions, and so on. To derive any conclusion at all from a given set of assumptions, hence, requires one to make infinitely many inferences and accept infinitely many assumptions. Since the human mind can only conceive finitely many things, logical inference is impossible.

An immediate application of this argument on the inference rule *modus ponens* goes as follows: if we want to conclude q from a given hypothesis p, we should first accept the proposition that p implies q, i.e., $p \to q$. We then have the propositions p, $p \to q$ in the hypothesis. But in order to infer q from these premises, we should first accept as a hypothesis that our assumptions imply the conclusion. That is, we should first accept

$$(p \wedge (p \to q)) \to q \tag{i}$$

as a hypothesis. But to infer q from this, we should first accept the proposition

$$(p \wedge (p \to q) \wedge (\text{i})) \to q \tag{ii}$$

in the hypothesis. Again, to arrive q, we should first accept

$$(p \wedge (p \to q) \wedge (p \wedge (p \to q)) \wedge (\text{ii})) \to q \tag{iii}$$

in the hypothesis, and so on. The only possible way to arrive at q, as it seems, is to add infinitely many assumptions. Carroll then "arrived" at the conclusion that any logical inference was impossible. We shall end our discussion on Zeno's paradoxes for now.

The interplay between limited and unlimited also involved in the analysis of the relationship between *beings* and *becomings*, or *particulars* and *universals* in ontology. Parmenides argued for that the principle and the essence of all things was the One eternal, static, motionless being. Whereas for Heraclitus, all objects are subject to change and that "there is nothing permanent except change". On the one hand, we have the "Oneness" of the true eternal and static being. On the other hand, in the physical world, we have "many" physical instantiations of universals. The relationship between the One and the Many led to two different world views. The world which is static, unchanging, eternal, and infinite, and the world which is dynamic, mortal, and finite. Recall that Plato reconciles these two world views in his Theory of Forms, which we studied in Chapter 3.

Aristotle, as Plato's pupil, treated infinity rather much differently. According to Aristotle, infinity can only exist as a potential rather than as a completed actual totality. In fact, one of the most significant contributions of Aristotle in the history of philosophy is his idea of *potentiality* as another state of being, something between actual existence and complete absence. For Aristotle, the Form actualises itself in the matter. Without matter, the Form cannot transform into actuality from potentiality. Likewise, the matter also needs the Form to be conceived properly. Matter without any Form remains unconceivable. Aristotle's idea of potential infinity should not be conceived as a completed object, but as a never ending progress. For instance, the set $\mathbb{N} = \{0, 1, 2, 3, \ldots\}$ of natural numbers contains infinitely many objects. If it is used as a completed totality on its own, we are forced in this case to accept actual infinity. On the contrary, an endless sequence of natural numbers $0, 1, 2, 3, \ldots$, where we have the possibility to define a larger number for *any* number, may not be considered as a completed totality. The capability of *arbitrarily* defining a larger number to write the next element of the sequence each time, yet not seeing the sequence as a completed object is treating infinity in the potential sense. Besides the treatment of infinity, Aristotle also objects to the idea that continuity exists as an actual concept. For him, the reason behind that arrow does not hit its target in Zeno's paradox is due to the supposition that the space is "continuous". As a matter of fact, in the sixth book of his *Physics* [8], Aristotle argues for that the space is discrete and so the arrow, in reality, hits its target. In one passage, Aristotle states:

> Zeno's reasoning, however, is fallacious, when he says that if everything when it occupies an equal space is at rest, and if that which is in locomotion is always occupying such a space at any moment, the flying arrow is therefore motionless. This is false, for time is not composed of indivisible moments any more than any other, magnitude is composed of indivisibles.[4]

Aristotle's idea of potential infinity, had a major influence on philosophers in the later period. It can be rightfully said that the ancient Greek mathematician Eudoxus (B.C. 408–355) planted the seeds of calculus by introducing an approximation method which is called today the *method of exhaustion* to calculate the area of an inscribed circle. This method was later known to be used by Archimedes of Syracuse (B.C. 287–212). In the method of exhaustion, the area of a given circle is calculated as follows: inscribe a circle. Within the

---

[4] Aristotle, *Physics*, VI:9, 239b.

inscribed circle, inscribe a polygon with n many edges. Continue ad infinitum to inscribe polygons enlarging each time the number of edges.

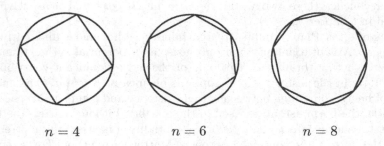

$$n = 4 \qquad\qquad n = 6 \qquad\qquad n = 8$$

FIGURE 9.3
Eudoxus' method of exhaustion.

As seen in Figure 9.3, inscribing a polygon inside the circle and arbitrarily increasing the number of edges at each step approximates to the area of the circle. The larger n gets, the better our approximation will be. Regardless of how large n gets, however, there will always be an error tolerance since we will never be able to calculate, by this method, the exact area of the circle. In principle, to calculate the exact area using this method, we should inscribe a polygon with "infinite" number of edges, which is impossible. Although the method of exhaustion is still being used today in numerical methods, it does not answer the most fundamental questions about geometry. If there is no absolute infinity, how come we accept the continuity in geometric measures on the Euclidean space?

Plotinus (A.D. 205–270), an influential philosopher of the Hellenistic period, was considered to be an adherent of what is known as *neoplatonism*. With Plotinus, we observe that the idea of actual infinity is introduced into consciousness in the Hellenic world as a positive force. For a number of similar reasons, he caused a significant impact on Christian thought. Influenced by Plato's works and in contrast with Aristotle, Plotinus defended that absolute infinity could exist in the metaphysical Platonic world. Nevertheless, for him it cannot exist by any means in any sensible realm.

9.2 Middle Ages, the Renaissance, and the Age of Enlightenment

The Middle Eastern world intellects in the medieval period with no doubt made significant contributions in mathematics, philosophy, medicine, and other sciences. Philosophers such as Al-Kindi, Al-Ghazali, Al-Farabi, Avicenna (Ibn Sina), Omar Khayyam, made enormous efforts by translating ancient Greek texts to Arabic, which was the language of science mostly back then. In particular, many arguments concerning infinite causality were presented by various theologists so as to strengthen the existence of Aristotle's "unmoved mover". One of these arguments was called *Burhan Al-Tatbiq* (proof of one-to-one correspondence), an argument to show that it is impossible to have an infinite descending chain of causality, and hence to prove that there must be a first cause and a beginning of the universe. The theological interpretation of Burhan Al-Tatbiq was also came to known as the "cosmological evidence". The idea behind this argument was to show the impossibility of having *al-tasalsul*, i.e., infinite descending causality chains. The word *tasalsul* in Arabic means *chain* or *sequence*, but in this particular case, it means *infinite descending chain*. The argument which we will present now later turned out to be mathematically invalid. The reason is that the mathematical theory of the infinite and infinite sets were not studied until the mid-19th century. Many of these arguments require a rigorous treatment of infinite sets and the definition of equipollency. Most of the paradoxes about infinite collections were due to the mistreatment of the theory of infinite sets.

Burhan Al-Tatbiq is a reductio argument that compares the length of two distinct infinite sequences for arguing that first cause must always exist. Consider two infinite descending chains A and B. We suppose that the elements of these sequences are associated with causality. So each sequence represents an infinite descending chain of events. Given two events x and y, if the cause of x is y, then let us denote this by $x \leftarrow y$. Let us now consider the following sequences of events:

$$A : x_1 \leftarrow x_2 \leftarrow x_3 \leftarrow x_4 \leftarrow \cdots$$
$$B : x_2 \leftarrow x_3 \leftarrow x_4 \leftarrow x_5 \leftarrow \cdots$$

Clearly, the sequence B starts 1 element behind A. The argument assumes for reductio that A is an infinite sequence. Then it continues to claim that B is "shorter" than A, and that the shorter one cannot be infinite. Hence, B is finite. But A is merely one element "longer" than B. Hence, A must be finite as well. This contradicts the initial assumption. Therefore, the argument concludes, there can be no infinite descending chain of events.[5]

Since the mathematical theory of infinity was not well developed back then, the fact that adding/subtracting a single element to/from an infinite set does not change the cardinality of the set was most likely overlooked. The arguments—predominantly proposed by theologians—concerning the impossibility of having an infinite descending chain of events were for the most part relied heavily on Aristotle's cosmological argument of the necessity of the *first cause*, i.e., the "unmoved mover", which can be found in Book Λ of his *Metaphysics*. In the sequences given above, for the event x_1 to happen, first x_2

---

[5]See Rahman [239] (1975), pp. 66–67, for details.

must happen; but for x_2 to happen, first x_3 must happen, and so on. Given an infinite decreasing chain of events, under the relation of causality, x_1 will never happen since we will never be able to get to the cause of x_1 in the first place. For if we did, the sequence which begins with the initial cause and ends with the final cause x_1 would have to be finite. Then, unless we transcend infinitely many causes, x_1 cannot exist. Something cannot come into existence if it requires an infinite number of other things existing before it. A typical objection to this might be to claim that "nothing comes from nothing" as Parmenides said, and that there is no such thing as "coming into existence". According to this view, every bit of atomic particle eternally exists, past, and present. But then it follows, in particular, that motion is impossible. Hence, if infinite descending chain of events (al-tasalsul) were possible, it would imply the impossibility of motion. To move from one point to another would actually be no different than immobility. Anyone who believes in the reality of motion and the concept of time would conclude that *al-tasalsul* is impossible.

In the Western world, various arguments concerning the infinite were presented in the Middle Ages. One of these arguments belongs to John Duns Scotus (1266–1308). Scotus gives a correspondence example to argue against the doctrine of indivisible parts and to support the contention that lines are not composed of an infinity of infinitesimal points (see Figure 9.4). The argument runs as follows. Consider a circle, around which there is another circle with a larger diameter with the same centre as the smaller circle.

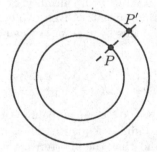

FIGURE 9.4
The diameter of the outer circle is larger, yet there exist a one-to-one correspondence between the points of two circles.

According to Scotus, the larger circle must contain "more" points than the smaller circle. However, we can define a one-to-one correspondence between the points of these two circles. But then they must contain the same number of points. This was considered to be a paradoxical example. This was yet another argument which misused the concept of cardinality.[6] A similar argument was

---

[6]The study of infinite sets became a mathematical study with Georg Cantor in the late 19th century. For the medieval period, it was fairly normal that scholars regarded this antinomy as a paradox.

proposed by Galileo [108]. On the one hand, we know that there is a one-to-one correspondence between natural numbers and perfect squares. On the other hand, the set of perfect squares is a strict subset of the set of natural numbers and so, in some sense, there are "less" perfect squares than there are natural numbers. This dilemma is known as *Galileo's paradox*. The argument was also given before infinite sets became a rigorous mathematical study. So Galileo's argument was considered as a paradox due to the limitations that the notion of cardinality was not properly introduced yet back then.

The Renaissance, a fervent period of the European history, was the beginning of the modern era that kindled the intellectual movements. Nicolas of Cusa (1401–1464) was a German cardinal, philosopher, and theologian, who played a crucial role in the development of the Renaissance movement. He made spiritual, political, and philosophical contributions. Most of his ideas were against the conservative views of his time. For example, based on various metaphysical principles, it was before Copernicus that Cusa proposed the Earth is not in the centre of the cosmos. Similarly, it was before Kepler that he observed planets do not have circular orbits. His philosophical writings differed from the classical views. Cusa's most completed proposals, including his philosophical ideas about mathematics, can be found in his *De docta ignorantia* [71]. The first book of this work is devoted to God, which he describes as the absolute "Maximum". For Cusa, God is the *coincidence of opposites*. In God, both the minimum and the maximum coincides in the infinite Oneness. The ultimate reality lies beyond the Law of Non-Contradiction. Rather in a mystical language, Cusa says:

> In God we must not conceive of distinction and indistinction, for example, as two contradictories, but we must conceive of them as antecedently existing in their own most simple beginning, where distinction is not other than indistinction.[7]

The incomprehensible infinite can only be "comprehended" through the Oneness in which all the opposites coincide. Cusa introduced a geometric, yet mathematically impossible analogy to elaborate on how a circle coincides with the line *in* the infinite. The illustration begins with imagining a circle. If the diameter of the circle continually increases, as in Figure 9.5, the circumference approaches the straight line and it will appear "less curved". In the infinite, the circle coincides with its tangent line.

Cusa makes bold cosmological claims contradicting the earlier conception of the universe. He says:

> [A]lthough the world (universe) is not Infinite, it cannot be conceived of as finite, since it lacks boundaries within which it is enclosed [...] Therefore, just as the earth is not the centre of the world, so the sphere of fixed stars is not its circumference.[8]

---

[7] Cusa, p. 29, 1997.
[8] ibid, pp. 158–159. Parenthesis added.

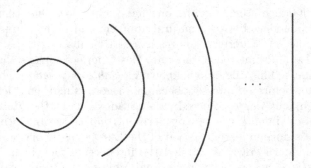

FIGURE 9.5

Cusa's analogy of how the circle coincides with the straight line *in* the infinite.

For Cusa, the universe has no centre. Since the universe is infinite, every point *relatively* defines a centre of the universe.

After the Pythagorean theorem and the discovery of that $\sqrt{2}$ is irrational, we see no evidence where irrational numbers were explicitly used in Europe until Fibonacci. Despite that irrational numbers were used after Fibonacci, it was debated whether or not they could really be considered as "legitimate" numbers. Michael Stiffel, a German reformist and mathematician, delivered in 1544 a criticism regarding this problem. In his *Arithmetica Integra*, he says:

> That cannot be called a true number which is of such a nature that it lacks precision [...] Therefore, just as an infinite number is not a number, so an irrational number is not a true number, but lies hidden in a kind of cloud of infinity.[9]

In order to reconcile geometry and algebra, René Descartes (1596 – 1650) and Pierre de Fermat (1601–1665) introduced a fascinating subject of mathematics called *analytic geometry*, by which all geometric objects on the Cartesian plane could be expressed in terms of algebraic equations. Any coordinate on the plane corresponds to a point expressed as a pair (x, y) of numbers. For instance, we know that a circle is a collection of points which is equidistance from a given central point. That is, if r is the radius, then the circle with radius r is defined as the set of coordinates (x, y) satisfying the algebraic equation $r^2 = x^2 + y^2$.

One fundamental assumption in analytic geometry is that there corresponds a number for every point in the continuous Cartesian plane. The Pythagorean theorem, though, implies that there are geometric magnitudes

---

[9]Kline [166], p. 251, 1972.

that correspond to no natural number unless it is a number with an infinite decimal expansion.

The discovery (or the invention) of calculus was a major breakthrough in the history of mathematics. It was emerged from the following question: *Given a curve and a point on the curve, what is the tangent line to the curve at that point?* In other words, given two quantities, we ask what the *rate of change* is with respect to each other. Calculus was systematically developed by Isaac Newton (1642–1727) and independently by Gottfried W. Leibniz (1646–1716). Generalising Eudoxus' method of exhaustion, calculus devised solutions to many problems including those which puzzled the mathematical community prior to the development of calculus, such as Zeno's paradoxes. However, it also led to philosophical controversies. Apart from the use of the irrationals and partial sums, Leibniz profoundly believed in the existence of an infinitesimal number. An *infinitesimal* is a positive value, denoted by ϵ, such that for every positive real number r, it holds that $\epsilon < r$. In other words, it is the "minimal element" of the real numbers above 0. The methods used in calculus were perceived as extremely obscure for some mathematicians and philosophers to the point that George Berkeley, a British theologian and one of the leading figures of British empiricism along with John Locke and David Hume, harshly criticised in *The Analyst* [32] the agnostic free-thinkers and mathematicians. He states:

> And what are these Fluxions? The velocities of evanescent increments? And what are these same evanescent increments? They are neither finite quantities nor quantities infinitely small, nor yet nothing. May we not call them the ghosts of departed quantities?[10]

Augustin Cauchy (1789–1857) and Karl Weierstrass (1815–1897) proposed to withdraw infinitesimals and the concept of absolute infinity from calculus and instead think in terms of quantities that can be *potentially* made arbitrarily small or arbitrarily large. The capability to assign quantities arbitrarily small (or large) values, relying on the use of potential infinity, was formalised in the conception of *limit*. The concept of limit may be thought of as an outcome of the process of imagining what the value would converge to if one could arbitrarily assign larger values indefinitely. For example, a sequence of polygons with increasing number of edges converges to a circle in the limit since we always have the facility to define a polygon arbitrarily close to a circle by selecting the number of edges to be sufficiently large. The concept of limit, thus, allows us to use infinity as a potential unlike the Leibnizian account of infinitesimals.

---

[10]Berkeley, *The Analyst*, §XXXV, 1734.

9.3　Cantor's Set Theory

Both analytic geometry and calculus need to answer the following question concerning their foundation: What is the legitimate basis for using objects, such as irrational numbers, which by any means cannot be finitely described? Richard Dedekind (1831–1916) is a well-recognised German mathematician for his definition of real numbers by a method called *Dedekind cuts*. He postulated that a *cut* separates the continuum into two subsets, say X and Y, such that whenever $x \in X$ and $y \in Y$, then $x < y$. If the cut is defined in such a way that X has a largest rational member or Y has a least member, then the cut corresponds to a rational number. However, if the cut is defined in such a way that X has no largest rational member and Y has no least rational member, then the cut corresponds to an irrational number. Although Dedekind defined irrational numbers by using sets of rationals, the cut itself corresponds to an infinite totality, thus, he makes reference to absolute infinity.

Georg Cantor (1845–1918) is known to be the first person who developed the mathematical theory of the infinite. Cantor brought an insightful perspective for the mathematical understanding of the infinite by introducing *transfinite (ordinal) numbers* as brand new mathematical objects. Cantor defines a transfinite "number" which is greater than all natural numbers

$$0, 1, 2, 3, \ldots$$

He denotes this transfinite number by ω. Thus, ω is the least transfinite number greater than every natural number. Then, we have

$$0, 1, 2, 3, \ldots, \omega$$

In fact, we can continue on defining new ordinals that follow ω. We get

$$0, 1, 2, 3, \ldots, \omega, \omega + 1, \omega + 2, \omega + 3, \ldots$$

After all these numbers, Cantor defines $\omega + \omega$, that is $\omega 2$. In a similar fashion, we get

$$\omega 2 + 1, \ \omega 2 + 2, \ \omega 2 + 3, \ldots$$

What comes after these numbers is $\omega 3$, and then if we continue we get $\omega 4, \omega 5, \ldots$, and finally ω^2. But why stop there? We may define $\omega^3, \omega^4, \ldots$. After all of these ordinals, we get ω^ω. As we continue to write, we have

$$\omega^\omega, \omega^{\omega^\omega}, \omega^{\omega^{\omega^\omega}}, \ldots$$

Later we get

$$\omega^{\omega^{\omega^{\omega^{\cdots}}}}$$

Let us call this number ε_0. That is,

$$\varepsilon_0 = \omega^{\omega^{\omega^{\omega^{\cdots}}}}$$

Now, ε_0 is a special ordinal number for the fact that it is the smallest ordinal α that satisfies the equation $\omega^\alpha = \alpha$. Of course, we may continue to define $\varepsilon_0 + 1, \varepsilon_0 + 2, \ldots$, ad infinitum.

According to Cantor, in order to know whether the cardinality of two sets are equal, there needs to be a one-to-one correspondence (bijection), between those sets. That is, given two sets A and B, we say that the *cardinality* of A and B are equal if and only if there exists a bijection between A and B. If the cardinality of A and B are equal, this is denoted by $|A| = |B|$. The cardinality of finite sets is just the number of elements in the set. The cardinality of the set $\mathbb{N} = \{0, 1, 2, 3, \ldots\}$ of natural numbers is denoted by $\aleph_0$ (Aleph zero), which is the least infinite cardinal. Cantor showed that for every set A, there is another set, namely the set of all subsets of A, whose cardinality is larger than that of A. We call the set of all subsets of A the *power set* of A and we denote it by $\mathcal{P}(A)$. We now give the following theorem.

Theorem 9.1 (Cantor's theorem) For every set X, $|X| < |\mathcal{P}(X)|$.

Proof. We show that $|X| \leq |\mathcal{P}(X)|$ and $|X| \neq |\mathcal{P}(X)|$. Suppose that $f : X \to \mathcal{P}(X)$ is an onto function (surjection). Consider the set

$$Y = \{x \in X : x \notin f(x)\}.$$

Now Y is a subset of X. But we claim that Y is not an element of the image of f.[11] Assume for a contradiction that $f(z) = Y$ for some $z \in X$. Then, we would have $z \in f(z)$ if and only if $z \in Y$ if and only if $z \notin f(z)$. A contradiction. Thus, f is not an onto function from X to $\mathcal{P}(X)$. Hence, $|X| \neq |\mathcal{P}(X)|$.

Now we show that there exists a one-to-one function (injection) $f : X \to \mathcal{P}(X)$. The function $f(x) = \{x\}$ is clearly a one-to-one function from X to $\mathcal{P}(X)$. Therefore, $|X| < |\mathcal{P}(X)|$. $\qquad\square$

We give some basic facts about cardinal arithmetic.

(i) $\kappa + \lambda = |A \cup B|$, where $|A| = \kappa$, $|B| = \lambda$, and A and B are disjoint,

(ii) $\kappa \cdot \lambda = |A \times B|$, where $|A| = \kappa$, $|B| = \lambda$,

(iii) $\kappa^\lambda = |A^B|$, where $|A| = \kappa$, $|B| = \lambda$.

We know that if $|A| = \kappa$, then $|\mathcal{P}(A)| = 2^\kappa$. Thus, Cantor's theorem can be formulated as

$$\kappa < 2^\kappa \text{ for every cardinal } \kappa.$$

Here are some more basic facts.

---

[11]The *image* of a function f is the set $\{y : f(x) = y\}$.

Theorem 9.2 (i) Addition and multiplication operators both are associative, commutative, and distributive.

(ii) $(\kappa \cdot \lambda)^{\mu} = \kappa^{\mu} \cdot \lambda^{\mu}$.

(iii) $\kappa^{\lambda+\mu} = \kappa^{\lambda} \cdot \kappa^{\mu}$.

(iv) $(\kappa^{\lambda})^{\mu} = \kappa^{\lambda \cdot \mu}$.

(v) If $\kappa \leq \lambda$, then $\kappa^{\mu} \leq \lambda^{\mu}$.

(vi) If $0 < \lambda \leq \mu$, then $\kappa^{\lambda} \leq \kappa^{\mu}$.

(vii) $\kappa^0 = 1$; $1^{\kappa} = 1$; $0^{\kappa} = 0$ if $\kappa > 0$.

We shall also note the fact that

$$\kappa + \lambda = \kappa \cdot \lambda = \max\{\kappa, \lambda\}.$$

When Cantor's theorem is applied on infinite sets, this leads to a significant philosophical and mathematical consequence: *There are larger infinities.* In fact, for any infinite cardinal, there exists a larger infinite cardinal. The cardinality of the set of natural numbers is $\aleph_0$. However, there are many sets whose cardinality is $\aleph_0$. As a matter of fact, there are too many of them that we cannot even count them in principle. If the cardnality of a set is $\aleph_0$, then we call that set *countably infinite*. All finite sets or countably infinite sets are countable. Sets that are not countable are called *uncountable*.

We may give many examples of countably infinite sets. The set $\mathbb{N}_E = \{0, 2, 4, 6, \ldots\}$ of even numbers has the same cardinality as that of $\mathbb{N}$. For this, we map every element n in $\mathbb{N}$ to $2n$ in $\mathbb{N}_E$. The set $\mathbb{Z} = \{\ldots, -2, -1, 0, 1, 2, \ldots\}$ of integers is countably infinite too, thus it is of cardinality $\aleph_0$. We can have a one-to-one correspondence between the set of integers and the set of natural numbers. We can map even numbers to non-negative numbers and odd numbers to negative integers. The same holds for the set $\mathbb{Q}$ of rational numbers. To prove that $\mathbb{Q}$ is countable, consider a matrix of fractions as in Figure 9.6. Thus, we know that the set of integers and the set of rational numbers are all countable. However, Cantor's another celebrated result shows that there are sets which are not countable.

Theorem 9.3 The set of real numbers is uncountable.

Proof. The proof of the theorem uses the *diagonalisation method*. For reductio, assume that the set of real numbers is countable. This would imply that the set of numbers in the $[0, 1)$ interval is also countable. If we could show that this interval is uncountable, then so would be the entire set of reals, and so this would suffice to prove the theorem. If the interval $[0, 1)$ is countable, then there must exist a one-to-one correspondence between the reals in $[0, 1)$ and natural numbers. Suppose we list the elements of $[0, 1)$ as in Figure 9.7 and that $\{n_i\}$ is an enumeration of this interval.

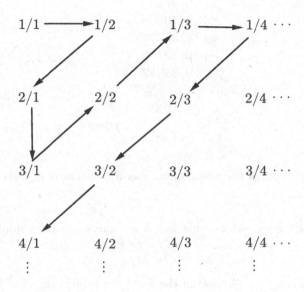

FIGURE 9.6
Using the zigzag method we count all fractions.

Let us now define a real number d which falls in the interval $[0,1)$ as follows: Suppose that $d(j)$ denotes the jth digit of d after the decimal point. Define $d(j)$ to be a number different than the jth digit of n_j after the decimal point. That is, the jth digit of d should be different than the diagonal entry of the real number that corresponds to n_j in the list. We look at the diagonal entries, shown in bold in Figure 9.7. The new number d is indeed a real number in the interval $[0,1)$. Then, d should appear somewhere in the list since we assume there is a one-to-one correspondence between $[0,1)$ and $\mathbb{N}$. Suppose that d is mapped to some natural number n_i. But then what is the ith digit of d after the decimal point? By construction, it must be different than whatever it is. This is a contradiction. Thus, our assumption that there is a one-to-one correspondence between the set of reals and the set of natural numbers must be false. Therefore, the set of real numbers must be uncountable. $\qquad\square$

The cardinality of the power set of natural numbers is, in fact, the same as the cardinality of real numbers, as they both have the cardinality $2^{|\mathbb{N}|}$, that is $2^{\aleph_0}$. We know that $\aleph_0$ is the smallest infinite cardinal. But what is the *next* smallest infinite cardinal after $\aleph_0$? Is it really $2^{\aleph_0}$? This is not an easy question

$$n_1 : 0, 46489 \cdots$$
$$n_2 : 0, 70936 \cdots$$
$$n_3 : 0, 90374 \cdots$$
$$n_4 : 0, 19495 \cdots$$
$$n_5 : 0, 85017 \ldots$$
$$\vdots \qquad \ddots$$
$$n_i : 0, 51408 \cdots (?) \cdots$$

*i*th digit

FIGURE 9.7

We define the digits of the new real number d from the diagonal entries of the table.

to answer at all. Suppose now that we list all infinite cardinal numbers in order

$$\aleph_0, \aleph_1, \aleph_2, \ldots$$

The question can be rephrased in the following form: is $\aleph_1 = 2^{\aleph_0}$? The claim that $\aleph_1 = 2^{\aleph_0}$ is known as the *Continuum Hypothesis* (CH).

Cardinal exponentiation has somewhat a strange behaviour. In fact, the CH stems from the properties of cardinal exponentiation, that, for a given cardinal κ, cardinal exponentiation does not exactly determine what 2^κ really is. The following proposition gives one of these features concerning exponentiation.

Proposition 9.1 For any infinite cardinals κ and λ such that $2 \leq \lambda < \kappa$, $\lambda^\kappa = \kappa^\kappa$.

Proof. Suppose that A is a set such that $|A| = \kappa$. Then, κ^κ is the cardinality of the set of functions $f : A \to A$. The graph of any function of this kind must be a subset of $A \times A$. Hence, $\kappa^\kappa \leq |\mathcal{P}(A \times A)|$. But since $|A| = |A \times A|$, we have that $|\mathcal{P}(A)| = |\mathcal{P}(A \times A)|$. Then,

$$2^\kappa \leq \lambda^\kappa \leq \kappa^\kappa \leq |\mathcal{P}(A \times A)| \leq |\mathcal{P}(A)| = 2^\kappa.$$

Therefore, $2^\kappa = \lambda^\kappa = \kappa^\kappa$. $\qquad\qquad\square$

If we try to calculate λ^κ satisfying the hypothesis of the proposition, we have that $\lambda^\kappa = 2^\kappa$. In fact, if $|A| = \kappa$, then 2^κ is the cardinality of $\mathcal{P}(A)$. We know then $\kappa < 2^\kappa$, but we still do not know what exactly 2^κ is. Since the cardinality of $\mathbb{R}$ is 2^κ for $\kappa = \aleph_0$, we cannot tell which aleph is equal to $2^{\aleph_0}$. In fact, is it consistent with ZFC to posit that $2^{\aleph_0} = \aleph_n$ for any natural number $n > 0$.

The CH says that there is no other infinite cardinal number strictly between $\aleph_0$ and $2^{\aleph_0}$. In other words, any infinite subset of the set of reals is either countably infinite or has cardinality $2^{\aleph_0}$. According to the CH, the cardinality of the set of real numbers is the least infinite cardinal larger than the cardinality of the set of naturals. One can also generalise this to other alephs. Generalised version of the CH can be stated as follows.

Generalised Continuum Hypothesis (GCH): For every α ordinal, $\aleph_{\alpha+1} = 2^{\aleph_\alpha}$.

We know at least that $2^{\aleph_0}$ is not equal to $\aleph_\omega$ due to König's theorem [172], thus claiming that they are equal to each other leads to a contradiction in ZFC. Kurt Gödel [116] proved that if ZFC is consistent, then so is ZFC+GCH. That is, adding GCH into ZFC does not contradict the axioms of set theory. Thus, Gödel showed that ¬CH cannot be proved from ZFC. Paul Cohen [62] , on the other hand, proved that if ZFC is consistent, then so is ZFC + ¬GCH. That is to say, adding ¬GCH does not lead to a contradiction. Then, GCH too cannot be proved from ZFC. From the results of Gödel and Cohen, we conclude that GCH is *independent* from ZFC. Gödel, in his proof, defined a model of set theory in which all axioms of ZFC hold. Furthermore, in this model, he showed that GCH is true. Cohen later introduced a novel method, which is widely used today for independence proofs, called *forcing*. Studying this revolutionary method is unfortunately beyond the scope of this book. For a detailed account on advanced topics in set theory and forcing, we encourage the reader to refer to Jech [155] or Kunen [175].

The first question arises from this independence result concerns the "reality" of the continuum and how cardinalities are really related to each other. How do we settle the Continuum Hypothesis? Considering Gödel's and Cohen's theorems together, the axioms of set theory cannot seem to decide the truth value of this simple looking yet deep problem. ZFC is not strong enough to settle the relationship between different sizes of infinities. We leave our discussion to later chapters regarding whether or not to accept the Continuum Hypothesis as an axiom.

The notion of actual infinity has been more apparent in mathematics since Cantor. It should be noted that the fact that whether one endorses the notion of potential infinity or actual infinity changes the mathematical practice dramatically. For example, one question is whether potential infinity is compatible with classical logic or requires a weaker logic like intuitionistic logic. It turns out that for a potentialist, all statements of first-order arithmetic are meaningful, yet she is not entitled to full second-order arithmetic.[12] This is just one example to show that how philosophical perspective effects the mathematical practice.

---

[12]See Linnebo and Shapiro [180] (2019) for further discussion.

Discussion Questions

1. The idea of convergence of the limit of an infinite series in calculus offers a solution to Zeno's paradoxes. But what about Lewis' paradox of infinite hypotheses? Can we apply the same idea to logical inference to settle this paradox? Why/why not?

2. Do you think that the usage of absolute and potential infinite makes any difference in mathematical practice?

3. Is Cusa's analogy of the coincidence of opposites merely geometrical? Can you think of other non-geometric analogies that explain this idea?

4. It is known that $0,\overline{9}$ is equal to 1. If we were to accept the existence of an infinitesimal number ϵ though, could not we define $0,\overline{9}$ otherwise? That is, could we somehow define the difference between 1 and $0,\overline{9}$ as the infinitesimal number ϵ?

5. Can you think of other applications of the diagonalisation method in more traditional areas of mathematics that similarly introduces new objects once applying it?

6. If the continuum hypothesis was proved false, how many different cardinalities do you think there would be strictly between $|\mathbb{N}|$ and $2^{|\mathbb{N}|}$?

10

Supertasks

A *supertask* is defined as a countably infinite sequence of operations performed in a finite interval of time. Computations involving supertasks are often called *transfinite computation* or *hypercomputation*. Theory of transfinite computability is a fascinating subject that intersects mathematics, computer science, and philosophy. The aim of this chapter is to concisely introduce supertasks, both mathematically and philosophically.

10.1 Transfinite Computability and Continuity

Although the term "supertask" was coined by the British philosopher J. F. Thomson in his work [284], the concept stems from the paradoxes of Zeno of Elea (see section 9.1 of Chapter 9). Zeno presented a series of paradoxes by which he argued for the impossibility of motion. To move from one point to another, one must first travel the $1/2$ way. But for that, one must first travel $1/4$ way. For this, one must first travel $1/8$ way, ad infinitum. Thus, to move from one place to another, one should perform infinitely many tasks. Zeno then claims, by his paradoxes, that motion is merely an illusion.

A modern variation of a similar paradox was introduced by Thomson. Imagine a hypothetical lamp which remains switched on for 1 minute. Then it is switched off for $1/2$ of a minute. After that, we switch it on for $1/4$ of a minute. After, we switch it off again for $1/8$ of a minute, and so on. We know from calculus that the summation of these time intervals converges to 2 minutes. But is the lamp on or off by the end of exactly 2 minutes? The literature of full of answers. However, Thomson's lamp can be seen in the form of *Grandi's series*, named after the Italian mathematician Guido Grandi, which is defined as

$$\sum_{n=0}^{\infty}(-1)^n.$$

Depending on how we take the parentheses, the outcome will either be 1 or 0. It will be 0 if we take the parentheses as

$$(1-1)+(1-1)+(1-1)+\cdots.$$

DOI: 10.1201/9781003223191-10

However, the outcome will be 1 if we take them as

$$1 + (1 - 1) + (1 - 1) + \cdots .$$

As these two outcomes conflict with each other, the series do not converge to a fixed value. For this reason, we say that Grandi's series is divergent.

The problem as to why Zeno's paradox is in contrast with the reality of motion hinges upon the physical conception of space. As Aristotle states as well (see section 9.1 of Chapter 9), Zeno must have made some fundamental assumptions about the physical space. Concerning this problem, Max Black [36] states the following.

> If a line in space actually consists of infinitely many points, no motion at all is possible, for the smallest shift of position would involve the crossing of infinitely many points, i.e., the actual performance of an infinite number of acts.[1]

Max's point makes perfect sense. If the space were continuous, it would be dense. Hence, for any two points on a line in space, there would be a point strictly in-between them. In order to move from one point to another, either the space should consist of indivisible parts or, in case of otherwise, one must perform infinitely many tasks. It is clear that Zeno must have assumed that the space is continuous. Nevertheless, even if we presume this so, any physical metric applied on space transforms the space into a spatial object of indivisible parts. The metric we apply in this case is the concept of physical motion.

Let us now turn to transfinite computations. One may naturally ask how powerful transfinite computional models are. In the analysis of determining the limits of computation, one is usually required to fix a "standard" model of computation, in this case, a model of transfinite computation. Before answering the latter question, we shall remind the reader the definition of Turing machine, which was introduced in section 7.7.1 of Chapter 7. A *Turing machine* consists of an infinite tape divided into cells, a finite set Σ of symbols, called the *alphabet*, a tape head which reads/writes symbols and moves to left/right direction, a finite set Q of states, and a transition function δ such that

$$\delta : Q \times \Sigma \to Q \times \Sigma \times \{\text{left}, \text{right}\}.$$

The machine receives the input string which is written on its tape, then reads a symbol from the tape, observes the current state, and then it changes the state of the machine, writes an another symbol, and finally moves the tape head to either left or right direction. The output of the machine is whatever written on the tape at the stage where the machine enters any of the halting states. A *configuration* is the information that contains the tape content, the current state, and the tape head position of the machine. In some

---

[1]Black, p. 90, 1959.

sense, a configuration can be thought of as the snapshot of the computational stage. Despite that the concept of algorithm or effective computability has no absolute mathematical definition, it is generally believed that the Church-Turing Thesis defines effective computability in the way we perceive it. Recall that the Church-Turing Thesis is a philosophical claim which says that effective computability is simply Turing-computability. Any algorithmically computable process is computable by Turing machines. The computational power of Turing machines is limited. For instance, given an arbitrary statement of arithmetic, there is no algorithm that decides whether it is true or false. We also showed in section 7.7.1 of Chapter 7 that the halting problem is unsolvable.

Every Turing-computable function admits a partial function. We call such functions *partial recursive*. Partial recursive functions are not necessarily defined on every argument. We assume that there is a uniform enumeration of all partial recursive functions. We shall denote the eth partial recursive function by φ_e. We will use this notation in the following sections.

10.2 Infinite Time Turing Machines

Despite that there are many transfinite computational models, similar to its classical counterpart, a "natural" model which led to a rich theory is called *infinite time Turing machines*, introduced by J. D. Hamkins and A. Lewis [133] in the early 2000's. In this section, we give some interesting results proved in their paper. An infinite time Turing machine has the same mechanics as that of the standard Turing machine. Just like the standard Turing machine, the configuration at stage $n + 1$ is uniquely determined from the configuration at stage n. An infinite time Turing machine has extra tapes though, namely an *input* tape, *scratch* tape, and *output* tape. All these tapes are assumed to be of infinite length. The alphabet is defined to be the set $\{0, 1\}$ for conventional purposes. The real difference, however, is when the computation of the machine transcends all finite stages. Infinite Turing machines are allowed to compute at transfinite ordinal number stages.

Let α be an ordinal such that $\alpha \neq 0$. If $\alpha = \beta + 1$ for no ordinal β, then α is called a *limit ordinal*. Otherwise, α is called a *successor ordinal*. The smallest infinite limit ordinal is denoted by ω and it is the well-ordering of the set of natural numbers. Note that

$$\omega, \ldots, \omega 2, \ldots, \omega 3, \ldots, \omega^2, \ldots, \omega^3, \ldots, \omega^\omega, \ldots, \omega^{\omega^\omega}, \ldots$$

are all limit ordinals. As a matter of fact, these ordinals are countable. The smallest uncountable ordinal is denoted by ω_1 and it is defined as the least upper bound, i.e., the supremum, of all countable ordinals.

Since infinite time Turing machines are supposed to extend the operation of stanard Turing machines to transfinite ordinal stages, it is a natural requirement to define a distinct *limit action* of the machine performed at limit ordinal stages. At every limit ordinal stage, the machine rewinds the tape head to the leftmost cell and it enters at a distinguished *limit* state. When a machine is in the limit stage, the limit of the cell values are taken as follows: if the values appearing in a cell have converged, that is, if it is either eventually 0 or 1 before the limit stage, then the cell retains the limiting value at the limit stage. Otherwise, in the case that the cell values have alternated from 0 to 1 and back again unboundedly often, the limit cell value is fixed as 1. Essentially, we take the limit supremum of the cell values appearing before that limit ordinal stage. This describes the configuration at any limit ordinal stage β. After that, the machine will continue its computation for stages $\beta+1$, $\beta+2$, etc., as usual. At stage $\beta+\omega$, the machine enters the limit state again and follows the same limit action procedure, and so on. If at any stage of the computation the machine enters any of the *halting* states, then the output of the machine is whatever written on the output tape at that stage.

Every infinite time Turing machine programme admits a partial function. Let us denote an infinite time Turing machine programme p with argument x by $\varphi_p(x)$. The domain of φ_p is the set

$$\{x : \varphi_p(x) \text{ halts}\}.$$

Generally speaking, the input of an infinite time Turing machine is an element of the Cantor space $2^{\mathbb{N}}$. In fact, we may see these inputs as real numbers in the form of binary strings. So the computations of infinite time Turing machines are defined over reals rather than naturals.

Functions that are computable by standard Turing machines, i.e., partial recursive functions, are defined over the natural numbers. For this reason, classic recursion theory concerns "computable" subsets of the set of natural numbers.[2] Let us give a few definitions before giving some results on infinite time Turing machines.

Definition 10.1 (Hamkins and Lewis, 2000) Let p be an infinite time Turing machine programme. If $f = \varphi_p$, then f is called *infinite time computable*.

This is the infinite time counterpart of the standard definition of partial recursive functions. The theorem given below shows that it is sufficient for us to take countable ordinals into account in transfinite computability.

Theorem 10.1 (Hamkins and Lewis, 2000) The length of every halting infinite time computation is countable.

Proof. This is proved by a simple cardinality argument. Suppose that the length of the computation has reached to an uncountable ordinal and it has

---

[2]For model theorists, though, recursion theory is often regarded as the study of "definable" subsets of the set of naturals.

not halted yet. Since the smallest uncountable ordinal is ω_1, the tape head of the machine at stage ω_1, by definition, is placed at the beginning of the tape. If we could show that the configuration at this stage has appeared in an earlier countable ordinal stage, then this would imply that the machine is in an infinite loop. Now if the cell value is 0, then it must have converged to this value at some countable ordinal stage before ω_1, it must have stabilised and not changed after that. If the cell value is 1, then there are two possibilities. Either the cell value has stabilised as 1 at some countable ordinal stage before ω_1, or the cell value has gone back and forth between 0 and 1 unboundedly often and so, by definition, the cell value is fixed as 1. Applying a simple cofinality argument, since there are countably many tape cells, by taking a countable supremum, we can find a stage α_0 at which the cells which eventually stabilise have, already all stabilised. After this stage, the only cells which change are the ones which will change cofinally often. Thus, there exists a countable sequence of countable ordinal stages

$$\alpha_0 < \alpha_1 < \alpha_2 < \ldots.$$

such that between α_n and α_{n+1}, all the cells which change at all after α_n have changed value at least once by stage α_{n+1}. We let $\delta = \sup_n \alpha_n$. Now δ is a limit ordinal, so the tape head of the machine at this stage is placed at the beginning of the tape. Moreover, the cells which have stabilised before stage ω_1 have, in fact, stabilised before δ. The cells which alternates between 0 and 1 unboundedly often has similarly changed their values unboundedly often before δ. Then, the configuration at stage ω_1 is same as the configuration at stage δ. Since we observe the same configuration at two distinct limit ordinal stages, we conclude that the machine must be in an infinite loop. $\qquad \square$

Definition 10.2 The *characteristic function* of a set A is defined as

$$\chi_A(x) = \begin{cases} 1 & \text{if } x \in A \\ 0 & \text{otherwise.} \end{cases}$$

Definition 10.3 Let A be a set of real numbers. If the characteristic function of A is infinite time computable, then A is called *infinite time decidable*.

The computational power of infinite time Turing machines is strictly greater than that of standard Turing machines. For instance, given a statement ψ in the language of arithmetic, an infinite time Turing machine can decide the truth value of ψ.

Theorem 10.2 (Hamkins and Lewis, 2000) The truth value of any arithmetical statement can be decided by an infinite Turing machine.

Proof. By induction on the complexity of formulas, we can show that any arithmetical statement is infinite time decidable. We give the proof of the existential form. Suppose that we are given a proposition in the form $\exists n\, \varphi(n, x)$.

Using an infinite time Turing machine, we decide the truth value of this statement by testing $\varphi(n, x)$ for every $n \in \mathbb{N}$. Checking all values of n is indeed possible for infinite time Turing machines. If $\varphi(n, x)$ holds for at least one n, then the machine halts and returns true, otherwise it returns false. We leave the proof of the statements of other forms to the reader. $\square$

An ordinal α is called *computable* if there is a computable well-ordering of a subset of the natural numbers with the order type α. The least uncomputable ordinal is called the *Church-Kleene ordinal* which is denoted by ω_1^{CK}. For an infinite Turing machine programme p, if $\varphi_p(0)$ halts in exactly α many steps, then α is called a *clockable ordinal*.

Theorem 10.3 (Hamkins and Lewis, 2000) Every computable ordinal is clockable.

A consequence of this theorem is that clockable ordinals extend at least up to ω_1^{CK}. In fact, they may be extended further.

Theorem 10.4 (Hamkins and Lewis, 2000) $\omega_1^{CK} + \omega$ is a clockable ordinal.

Another interesting result is the fact that there are *gaps* in-between clockable ordinals. That is to say, if α is a clockable ordinal, not every $\beta < \alpha$ needs to be clockable. The authors say:

> Any child who can count to 89 can also count to 63; counting to a smaller number is generally considered to be easier. Could this fail for infinite time Turing machines? Could there be gaps in the clockable ordinals—ordinals to which infinite time machines cannot count, though they can count higher? The answer, surprisingly, is yes.[3]

Note that a binary relation Δ on $\mathbb{N}$ can be coded by a real number x such that $n\Delta k$ exactly when $x(\langle n, k \rangle) = 1$, where $\langle ., . \rangle$ denotes some uniform pairing function $\mathbb{N} \times \mathbb{N} \to \mathbb{N}$. Let us write one more definition before giving another theorem.

Definition 10.4 For an infinite time Turing machine programme p, the real number that is outputted by the function $\varphi_p(0)$ is called a *writable real*. If the output of $\varphi_p(0)$ codes a well-order, the ordinal coded by this well-order is called a *writable ordinal*.

Theorem 10.5 (Hamkins and Lewis, 2000) There are no gaps in the writable ordinals.

Proof. Let α be a writable ordinal. Then, there exists a programme p such that $\varphi_p(0)$ outputs a real number x which codes a well-ordering R whose order type is α. For any $\beta < \alpha$, there exists a number n which gives the β-th

---

[3]Hamkins and Lewis, p. 18, 2000.

element of R. We can define an infinite time Turing machine programme p which writes the real number x on its tape and outputs the first n digits of x. This output then codes the well-ordering relation R up to n and this is of order type β. Therefore, β is a writable ordinal. $\qquad\Box$

Concerning the length of infinite time Turing machine computations, Philip Welch [291] showed that the halting times of infinite time Turing machines, coded as ordinals by sets of integers, are capable of being halting outputs of such machines. Welch also developed in his work [292] a degree theory of infinite time Turing machines in terms of some set-theoretic structures as a transfinite counterpart of the classical Turing degrees.

Another interesting question that can be asked about transfinite computability is whether the famous **P** versus **NP** problem could be formulated for infinite time Turing machines.[4] Indeed, a version of the **P** versus **NP** problem was introduced by Ralf Schindler [255], in which he showed that **P** $\neq$ **NP** for this particular version.[5]

Transfinite computing models are not limited to infinite time Turing machines. Ordinal Turing machines [171] and Blum-Shub-Smale machines [37] are other examples of transfinite computing models.

So far, we have discussed about the logico-mathematical groundwork of supertasks, or transfinite computability. Surprisingly, exciting studies have been made regarding even the physical realisability of supertasks. One hypothetical supertask model is a *real computer*, that is, a computer which physically computes with real numbers. Digital computers, as we know, are in principle standard Turing machines whose inputs and outputs are finite. In particular, standard computation is defined from naturals to naturals, but natural numbers can also be coded as finite binary sequences. Real computers, on the contrary, use real numbers in the input and output. However, whether real computers are feasible or not depends on the limits of physics and our measurements. The description of a real computer must be of infinite length. The reason behind this is that the complexity of a model which *uses* irrational numbers as auxiliary objects is expected to be at least as complex as irrational numbers. The existence of real computers, thus, depends on how real numbers are projected on the physical world.

Many transfinite computational models that have been proposed are based on calculations relying on the concept of *continuity*. We will now see how infinite amount of tasks, using models in theoretical physics can be in principle performed in a finite interval of time.

---

[4]The class **P** denotes the class of problems solvable by deterministic Turing machines in polynomial time, and **NP** denotes the class of problems solvable by non-deterministic Turing machines in polynomial time. It is one of the millenium prize problems in the field of theoretical computer science whether **P** equals **NP**.

[5]See Hamkins and Welch [134] (2003) for a solution of the general case of the same question for infinite time Turing machines. See also Deolalikar et al [77] (2005) which extends results of Schindler, Hamkins, and Welch.

10.3   Physical Realisations

Infinite time Turing machine is a mathematical model that is entirely independent of physics. Perhaps the most effective tool which allows supertasks on physical settings is *relativity*. Results on general relativity in the early 20th century indicates that spacetime is not rigid, but it is actually bendable. Researchers, including J. Earman [83], M. Hogarth [150], G. Etesi and I. Németi [85], made great efforts to show that general relativity can be used so as to perform supertasks in physical settings.

Definition 10.5 A *light cone* is a double cone centred at each event E in spacetime (see Figure 10.1). The *future light cone* contains all paths of light that begin from E and travel into the future, whereas the *past light cone* consists of all paths of light that end at E and come from the past. The *worldline* of an object is a sequence of spacetime events.[6]

Earman proposed certain spacetime models in which it is possible to perform supertasks. In such spacetime models, two parties are considered to complete a supertask co-operatively. The first party is assumed to exist indefinitely. On the other hand, the second party is assumed to perform the desired task within a finite time. However, the second party is situated in a spacetime such that its past light cone contains its entire worldline.

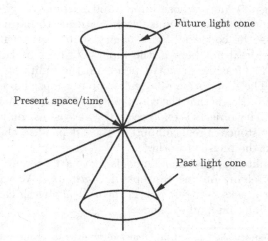

FIGURE 10.1
A light cone representing the three dimensional space and the time.

---

[6]Worldlines are timelike curves in spacetime.

Since the first party is present eternally and is able to perform its task indefinitely, i.e., in an open-ended manner, it can, in principle, compute Turing non-computable functions while staying in communication with the second party.

Relativistic computers use the concept of *time dilation* to perform hyper-computations. Consider the following example. Assume that a team of mad research scientists wants to find out if the decimal expansion of π contains three consecutive 7's. To find an answer, on a relativistic setting, our team travels to a satellite orbiting the Earth. The satellite's instantaneous *tangential velocity* is

$$v(t) = c\sqrt{1 - e^{-2t}},$$

where t denotes time in Earth's frame and c is the speed of light. If we denote the satellite's local time by τ, then we have

$$d\tau = e^{-t}dt.$$

Moreover, since

$$\int_0^\infty e^{-t}dt = 1,$$

when one second passes on the satellite, an infinite amount of time passes on Earth. While our research team orbits the planet on the satellite, their colleagues are examining the decimal expansion of π to see whether it has three consecutive 7's. Suppose that their colleagues ask their younger colleagues to continue their work, and so forth. If they ever encounter three consecutive 7's at any point, then this information is sent to the satellite. If the research team does not receive any message within a second, then they understand that the decimal expansion of π does not contain three consecutive 7's. The problem is that we must need a spacetime model in which we can carry out such supertasks.

A *Malament-Hogarth [M-H] spacetime* [153], [85] is a spacetime structure, consistent with general relativity, which allows us to perform supertasks in similar settings. The definition is known as follows.

Definition 10.6 *M-H spacetime* is a relativistic spacetime which has the following property: There exists a path λ on M-H spacetime and there is an event p such that all events along λ are a finite interval in the past of p. However, the proper time along λ is infinite. In this case, p is called the *M-H event*.

The basic idea is to set a Turing machine at some point before p, then run along λ. Then, we set the machine to signal to p at any stage of its infinite computation. The observer then travels through time to p in a finite proper time to pick the solution. Prima facie, such a setting might sound

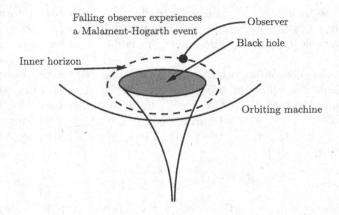

FIGURE 10.2

Setting of a relativistic computer on the event horizon of a black hole.

quite eccentric and fictional, but we should emphasise that the M-H spacetime structure is consistent with general relativity.

The next supertask model perhaps provides us the physical realisation of infinite time Turing machines. The model uses rotating black hole as a computer. Jack NG and Van Dam [206], and indepedently Brooks [40], argued for that rotating black holes are actually hypercomputer candidates. It is claimed that the exterior of a rotating black hole formed an M-H spacetime. Figure 10.2 is to show the setting of relativistic computing model on an rotating black hole.

Stephen Hawking [137] observed that rotating black holes emit a kind of thermal radiation what he called *Hawking radiation*. This radiation consists of particle and anti-particle pairs. The Hawking radiation is caused by the emission of particle and anti-particle pairs of the event horizon of rotating black holes.[7] It was also observed that one of these particles could fall into the black hole, whilst other staying outside. This arises the idea that black holes might be possible quantum objects. Hawking discovered that quantum information was preserved through the black hole. How large is the memory of a black hole and how fast can it compute? The answer to this question depends on the lifetime of the black hole.

Hypercomputers should not be confused with any form of digital computer. They are not quantum computers either. In fact, quantum computers [78], DNA computers [2], and similar models do not perform supertasks.[8] Quantum computers do not compute more functions in principle than what classical

---

[7]The event horizon is bounded by the spacetime that surrounds the edge of a black hole. It is important to note that no event occurs beyond the event horizon.

[8]See Hagar and Korolev [125] (2007) for a discussion about the relationship between quantum computers and hypercomputation.

computers do. They just solve some problems more efficiently. The speed-up is significant though. In fact, some problems are solved asymptotically faster by quantum computers when compared to classical computers. By classical computers, solving the integer factorisation problem, for example, which is commonly used in RSA cryprography, takes an exponential time. Whereas by quantum computers, the same problem, as proved by Peter Shor [262], can be solved in polynomial time. Another example concerns unstructured searching. Classically, given an unordered list of length n, the computational complexity of finding an arbitrary element in the list is in the order of n. Lov Grover [124], however, showed that a quantum algorithm could solve this problem in the order of $\sqrt{n}$ time complexity.

Hypercomputing models indeed break the Church-Turing barrier that is set by Turing mechanical principles. These models can compute Turing non-computable functions. They can solve, for instance, Hilbert and Ackermann's *Entscheidungsproblem.* They can solve the daunting halting problem. Nevertheless, it is not all sunshine and rainbows with hypercomputers, as they cannot solve every problem. The most straightforward example is that they cannot solve their *own* halting problem. That is, a hypercomputer cannot decide whether a given hypercomputer will halt on a given argument.

The concept of supertask still remains as a philosophical and logical interest. Some philosophers, including J. F. Thomson, argued against the possibility of supertasks, whereas others such as Paul Benacerraf [27] supported the idea. It was also argued that supertasks could not be considered as legitimate algorithms since it was thought that there is no "final" step in a supertask process. The problem is to decide then what kind of procedures would constitute an algorithm in the most general sense.

As a final note, one interesting type of procedure was proposed by Clark and Read [61], called *hypertasks*. A hypertask (not to be confused with hypercomputation) is peforming an *uncountable* number of operations in a finite interval of time. In this sense, hypertasks are quite similar to supertaks, but instead of performing countably many operations, we are supposed to perform uncountably many operations. Despite the fiery debates on the possibility of supertasks, it was claimed by Clark and Read that hypertasks are logically impossible. Nonetheless, it was shown by Al-Dhalimy and Geyer [4] that it is actually possible to perform hypertasks if one models time, i.e., the temporal continuum, with surreal numbers.

For a more detailed account on transfinite computability, the reader may refer to what we have also found very helpful, the work by Syropoulos [278], which includes the summary of many issues surrounding the concept of supertasks.

Discussion Questions

1. Let $\{\varphi_i\}$ be a uniform enumeration of standard Turing programmes, and

let

$$\text{Tot} = \{i : \varphi_i \text{ is total}\}.$$

Is Tot infinite time decidable?

2. If an infinite time Turing machine enters in the same configuration at two distinct limit stages, then we understand that the machine must have entered into a loop. Why is it not sufficient to look at two distinct arbitrary configurations instead?

3. Suppose that we had a hypothetical computer which is capable of processing with real numbers. What would be the physical requirements and consequences of having such a computer?

4. Discuss whether or not it would be possible to devise a logic in which we might be allowed to perform hypertasks.

11

Models, Completeness, and Skolem's Paradox

It is generally believed that mathematics is the most rigorous and precise of all sciences and that every notion in mathematics is absolute. Surprisingly though, even the most pure branches of mathematics involve some relativity to a certain extent. In this chapter, we will discuss one of these relativistic concepts that appears in set theory and its consequence, what is usually referred to as Skolem's paradox. Skolem's paradox is not, in fact, seen as an outright paradox in the traditional sense, but it is certainly considered as an antinomy. To understand the nature of this antinomy, we shall give some preliminary mathematical work aside from what we have given in Chapter 2. The primary focus of this chapter will be the notion of "model". Model theory is a beautiful subject with many philosophical bearings inside. Our concern is rather more restricted: to introduce Skolem's paradox and discuss its consequences. Readers who do not wish to go through the mathematical details about Skolem's paradox may skip to the discussion part of this chapter given in the end. For an extensive study on the philosophical aspects of model theory, we refer the reader to Button and Walsh's *Philosophy and Model Theory* [53].

Before giving any rigorous mathematical definition of "model", let us say roughly that a *structure* is a set together with function, relation, and constant symbols along with an "interpretation" of each symbol. An example of a structure would be the structure of natural numbers $(\mathbb{N}; +, 0, 1)$. The set $\mathbb{N}$ here denotes the *universe* (or the *underlying set*) of the structure, $+$ denotes the symbol for a binary function defined over the elements of the universe, 0 and 1 denote the constant symbols, where each of these symbols are interpreted in the usual manner.

We will consider first-order predicate logic for our purpose. As the reader might be already familiar, what makes a logic nth-order depends on the type of quantified objects. In first-order logic, the quantification is made merely over the *elements* of the universe. Second-order logic quantifies over the elements of the *power set* of the universe. In predicate logic, there are two types of quantifications. We use the symbol $\forall$ to denote "for all" and use $\exists$ to mean "there exists". A simple example of a first-order statement, in the language of arithmetic, would be something as

$$\forall x \, \exists y \, (y = x + 1)$$

which means in the standard interpretation "every number has a successor".

DOI: 10.1201/9781003223191-11

Or we may write something like

$$\forall x \, \exists y \, ((x = y + y) \vee (x = y + y + 1))$$

which means "every number is either odd or even". Let us give a few definitions for our notation.

Definition 11.1 A *formal language* (or simply *language*) $\mathcal{L}$ consists of the following elements.

(i) A set $\mathcal{F}$ of function symbols and a positive integer n_f for every $f \in \mathcal{F}$;

(ii) a set $\mathcal{R}$ of relation symbols and a positive integer n_R for every $R \in \mathcal{R}$;

(iii) a set $\mathcal{C}$ of constant symbols.

In the definition above, n_f and n_R denote, respectively, that f is a function of n_f variables and R is a relation of n_R variables. Note that $\mathcal{F}, \mathcal{R}, \mathcal{C}$ can be empty. All logical symbols are automatically assumed to be included in our languages. Along with the symbols $\neg$ (negation), $\wedge$ (and), $\vee$ (or), $\rightarrow$ (implies), $\leftrightarrow$ (if and only if), we also include the parentheses $(,)$, the equality symbol $=$, and the symbols $\forall$ (for all) and $\exists$ (exists) which are used exclusively in predicate logic.[1] The equality symbol may also be thought as a part of relation symbols. We should not forget that, for $i \in \mathbb{N}$, the variable symbols x_i are also included in the language. The cardinality of a language is the number of all symbols in it. So a countable language is a language which contains countably many symbols.

In symbolic logic, a formal language and formal grammar rules are essential to describe what constitutes syntactically correct expressions. Given a language, we construct expressions starting from the primitive elements and, by induction, form more complex ones. We shall first give the definition of "term", "atomic formulas" and proceed with "formulas". All of these will be finite strings over the symbols of the language. We begin with the following definition.

Definition 11.2 Let $\mathcal{L}$ be a language. We define *terms* of $\mathcal{L}$ as follows.

(i) Every $c \in \mathcal{C}$ is a term.

(ii) For $i \in \mathbb{N}$, each variable symbol x_i is a term.

(iii) If $f \in \mathcal{F}$ and $t_1, \ldots, t_{n_f}$ are terms, then so is $f(t_1, \ldots, t_{n_f})$.

Next we define formulas of $\mathcal{L}$.

---

[1] It is worth reminding that in predicate logic, the statement $\forall x \, R(x)$ is equivalent to $\neg \exists x \, \neg R(x)$. We shall use this equivalence in the proofs as well.

Definition 11.3 Let $\mathcal{L}$ be a language. If t_1 and t_2 are terms and if ϕ is any expression of the form $t_1 = t_2$ or of the form $R(t_1, \ldots, t_{n_R})$ such that $t_1, \ldots, t_{n_R}$ are terms and $R \in \mathcal{R}$, then ϕ is called an *atomic formula*. Every atomic formula is a *formula*.
We define $\mathcal{L}$-*formulas* as follows.

(i) If ϕ is a formula, then so is $\neg\phi$.

(ii) If ϕ and ψ are formulas, then so are $(\phi \wedge \psi)$, $(\phi \vee \psi)$, $(\phi \rightarrow \psi)$, and $(\phi \leftrightarrow \psi)$.

(iii) If ϕ is a formula, then so are $(\exists x_i \, \phi)$ and $(\forall x_i \, \phi)$.

Example 11.1 Let $\mathcal{L}_{or} = \{+, \cdot, <, 0, 1\}$ be a language. As noted earlier, we assume that all logical symbols, parentheses, and variable symbols are included in the language. We do not need to write the outer parentheses. For instance, $(\phi \rightarrow \psi)$ essentially means $\phi \rightarrow \psi$. The parentheses are just for readability and to specify the priority of the operations. The examples listed below are formulas in the language $\mathcal{L}_{or}$. That is, they are $\mathcal{L}_{or}$-formulas.

(i) $(x_1 = 0) \vee (0 < x_1)$;

(ii) $\exists x_2 \, (x_2 \cdot x_2 = x_1)$;

(iii) $\forall x_1 \, (x_1 = 0 \vee \exists x_2 \, (x_2 \cdot x_1 = 1))$.

If a variable of a formula is not in the scope of some quantifier, then we call it a *free variable*. Otherwise, we call it a *bound variable*. If we look at the examples given above, the variable x_1 appering in (i) and (ii) is free, but bound in (iii). A formula which does not contain any free variable is called a *sentence*. Every sentence is either true or false in a given structure. If M is the universe of a structure and if ϕ is a formula with free variables $x_1, \ldots, x_n$, then ϕ admits some "property" about the elements of M^n. A formula ϕ with its free variables is written as $\phi(x_1, \ldots, x_n)$.

Although we already gave some of the rules of logic and some logical tautologies (valid statements) in Chapter 2, since we will be using them shortly, let us remind them once again, particularly those which we shall refer to in the proofs.[2]

Rule A. Let P be a propositional tautology independent of the values of its variables $r_1, \ldots, r_n$. Then, the result of replacing each r_i in P by any proposition is a tautology.

Rule B (Modus ponens). If P and $(P \rightarrow Q)$ are tautologies, then so is Q.

Rule C.

---

[2]In predicate logic, usually the word "valid" is used to denote universally true statement, which is similar to the meaning of tautology in propositional logic. We will use both interchangeably.

(i) If c_0, c_1, and c_2 are constant symbols, then the following statements are tautologies:

$$c_0 = c_0,$$
$$(c_0 = c_1) \to (c_1 = c_0),$$
$$((c_0 = c_1) \land (c_1 = c_2)) \to (c_0 = c_2).$$

(ii) Let A be a proposition and let c_0 and c_1 be constant symbols. If A' represents A with every occurrence of c_0 replaced by c_1, then

$$(c_0 = c_1) \to (A \to A')$$

is a tautology.

Rule D. Let A and A' be two sentences as given in the previous rule. Then, $A \leftrightarrow A'$ is a tautology.

Rule E. For any constant symbol c, $\forall x\, A(x) \to A(c)$ is a tautology.

Rule F. Let B be a statement which does not involve c or x. If $A(c) \to B$ is a tautology, then $\exists x\, A(x) \to B$ is a tautology.

Rule G. Let $A(x)$ be a formula containing x as its only free variable and assume every occurrence of x is free. Let B be a statement which does not contain x. Then each of the following statements is a tautology:

$$\neg \forall x\, A(x) \leftrightarrow \exists x\, \neg A(x),$$
$$(\forall x\, A(x) \land B) \leftrightarrow \forall x\, (A(x) \land B),$$
$$(\exists x\, A(x) \land B) \leftrightarrow \exists x\, (A(x) \land B).$$

Definition 11.4 Let $\mathcal{L}$ be a language. An $\mathcal{L}$-*structure* $\mathcal{M}$ consists of the following elements.

(i) A function $f^{\mathcal{M}} : M^{n_f} \to M$ for every $f \in \mathcal{F}$;

(ii) a set $R^{\mathcal{M}} \subseteq M^{n_R}$ for every $R \in \mathcal{R}$;

(iii) an element $c^{\mathcal{M}} \in M$ for every $c \in \mathcal{C}$.

Here, M denotes the universe of $\mathcal{M}$. We say that $f^{\mathcal{M}}$, $R^{\mathcal{M}}$, and $c^{\mathcal{M}}$ are the *interpretations* in $\mathcal{M}$ of f, R, and c, respectively.[3] It should be understood that when we speak about a structure, we usually mean to refer to its underlying set. We may occasionally use, depending on the context, $\mathcal{M}$ and M interchangeably. However, unless stated otherwise, $\mathcal{M}$ denotes the structure, whereas M denotes its underlying set.

---

[3] For any $n \in \mathbb{N}$, $M^n = \underbrace{M \times M \times \cdots \times M}_{n \text{ times}}$.

Some structures can be contained in others as a "substructure". Let us now give a definition regarding this notion.

Definition 11.5 Let $\mathcal{L}$ be a language. Let $\mathcal{M}$ and $\mathcal{N}$ be two $\mathcal{L}$-structures with universes M and N, respectively, and let $(a_1, \ldots, a_n) \in M^n$. A one-to-one mapping $h : M \to N$ is called an $\mathcal{L}$-*embedding* if it satisfies the following conditions.

(i) For every constant symbol c, $h(c^{\mathcal{M}}) = c^{\mathcal{N}}$.

(ii) For every n-ary relation symbol R, $(a_1, \ldots, a_n) \in R^{\mathcal{M}}$ if and only if $(h(a_1), \ldots, h(a_n)) \in R^{\mathcal{N}}$.

(iii) For every n-ary f function symbol,
$h(f^{\mathcal{M}}(a_1, \ldots, a_n)) = f^{\mathcal{N}}(h(a_1), \ldots, h(a_n))$.

The embedding relation is denoted by $h : \mathcal{M} \to \mathcal{N}$.

Suppose that $M \subseteq N$. If $h : \mathcal{M} \to \mathcal{N}$ is an $\mathcal{L}$-embedding, then $\mathcal{M}$ is called a *substructure* of $\mathcal{N}$. For example, $\mathcal{N} = (\mathbb{N}; +, <, 0, 1)$ is a substructure of $\mathcal{R} = (\mathbb{R}; +, <, 0, 1)$ since $\mathbb{N} \subseteq \mathbb{R}$ and that for every n-ary relation symbol A in the language of $\mathcal{N}$ and $\mathcal{R}$, we have that $A^{\mathcal{N}} = A^{\mathcal{R}} \cap \mathbb{N}^n$. Similarly, for every n-ary function symbol f, we have that $f^{\mathcal{N}} = f^{\mathcal{R}} \upharpoonright \mathbb{N}^n$, where $f^{\mathcal{R}} \upharpoonright \mathbb{N}^n$ denotes the *restriction* of the domain of the function $f^{\mathcal{R}}$ by the elements of $\mathbb{N}^n$. The same holds for constant symbols, that is, for every constant symbol c, $c^{\mathcal{N}} = c^{\mathcal{R}}$ holds.

The cardinality of a structure $\mathcal{M}$ is defined as the cardinality of its underlying set M, and it is denoted by $|M|$. If $h : \mathcal{M} \to \mathcal{N}$ is an embedding, then $|M| \leq |N|$.

We shall now look at the conditions for a sentence to be "true" in a structure. This leads us to define the notion of model. Every sentence in a given language is either true or false in a structure having the same language. To understand what we mean intuitively, consider the $\mathcal{L}_{or}$-formulas we gave in Example 11.1. We see that some of those formulas hold in the structure $(\mathbb{N}; +, \cdot, 0, 1)$, whilst some do not. For example, (ii) is true in $(\mathbb{N}; +, \cdot, 0, 1)$ for $x_1 = 4$, yet false for $x_1 = 5$. Whereas, the sentence (iii) is true in the same structure. Let us now explicitly give the conditions for a formula $\phi(x_1, \ldots, x_n)$ to be true in $\mathcal{M}$. From now on, we will abbreviate an n-tuple $(a_1, \ldots, a_n)$ as $\vec{a}$.

Definition 11.6 Given a structure $\mathcal{M}$, let $\phi(\vec{x})$ be a formula in the language of $\mathcal{M}$ and let $\vec{a} \in M^n$. The relation $\mathcal{M} \models \phi(\vec{a})$ is defined by induction as follows.

(i) If ϕ is of the form $t_1 = t_2$ and if $t_1^{\mathcal{M}}(\vec{a}) = t_2^{\mathcal{M}}(\vec{a})$, then $\mathcal{M} \models \phi(\vec{a})$.

(ii) If ϕ is of the form $R(t_1, \ldots, t_{n_R})$ and if $(t_1^{\mathcal{M}}(\vec{a}), \ldots, t_{n_R}^{\mathcal{M}}(\vec{a})) \in R^{\mathcal{M}}$, then $\mathcal{M} \models \phi(\vec{a})$.

(iii) If ϕ is of the form $\neg\psi$ and if $\mathcal{M} \not\models \psi(\vec{a})$, then $\mathcal{M} \models \phi(\vec{a})$.

(iv) If ϕ is of the form $(\psi \vee \theta)$ and if $\mathcal{M} \models \psi(\vec{a})$ or $\mathcal{M} \models \theta(\vec{a})$, then $\mathcal{M} \models \phi(\vec{a})$.

(v) If ϕ is of the form $(\psi \wedge \theta)$ and if $\mathcal{M} \models \psi(\vec{a})$ and $\mathcal{M} \models \theta(\vec{a})$, then $\mathcal{M} \models \phi(\vec{a})$.

(vi) If ϕ is of the form $\exists x_i \psi(\vec{x}, x_i)$ and if there exists some $b \in M$ such that $\mathcal{M} \models \psi(\vec{a}, b)$, then $\mathcal{M} \models \phi(\vec{a})$.

(vii) If ϕ is of the form $\forall x_i \psi(\vec{x}, x_i)$ and if $\mathcal{M} \models \psi(\vec{a}, b)$ holds for all $b \in M$, then $\mathcal{M} \models \phi(\vec{a})$.

If $\mathcal{M} \models \phi(\vec{a})$ then $\phi(\vec{a})$ is said to be *true* in $\mathcal{M}$.

A set of sentences in a formal language is called a *theory*.[4] Let T be a theory and let $\mathcal{M}$ be a structure. If $\mathcal{M} \models \phi$ for all $\phi \in T$, then $\mathcal{M}$ is said to be a *model* of T. This relation is denoted by $\mathcal{M} \models T$.

Definition 11.7 Let S be a collection of statements. We say that a sentence A is *derivable* from S, if for some $B_1, \ldots, B_n$ in S, the sentence $(B_1 \wedge \cdots \wedge B_n) \to A$ is valid. If $A \wedge \neg A$ is not derivable from S for any statement A, then S is called a *consistent* theory. Otherwise, S is *inconsistent*.

Theorem 11.1 (i) If A is a valid statement, then A is true in every model.

(ii) Let S be a collection of statements. If S has a model, then S is consistent.

We leave the proof of this theorem as an exercise to the reader. The proof merely consists of showing that the inference rules of predicate logic correspond to intuitively correct deductions.

11.1 Gödel's Completeness Theorem

Is it guaranteed that the rules of propositional and first-order logic really cover "all" intuitively true statements? This is an important question to think about. Fortunately, we will prove in the following theorem below, by using the notion of models, that the answer to this question is positive. This is known as *Gödel's Completeness Theorem*. Completeness of first-order predicate logic was first proved by Gödel in his PhD dissertation [114]. Leon Henkin [143] later simplified Gödel's proof. We shall give the proof of the completeness theorem using Henkin's method to make it accesible to the reader as much as possible.[5]

---

[4]We can also call it a *system*.

[5]Completeness, in this case, is a property of logic itself. It shows that, for that particular logic, every valid statement is provable. Gödel's Completeness and Incompleteness Theorems should not be confused with each other, as the former concerns the logic itself, whereas the latter concerns formal theories.

Theorem 11.2 (Gödel's Completeness Theorem) Let S be a consistent set of statements. If S is infinite, then there exists a model of S whose cardinality does not exceed the cardinality of S, and if S is finite, then the model is countable.

The proof will use the Axiom of Choice, unless S is already well-ordered. We shall define a model of S using Henkin's construction. However, the model will be rather defined in a non-constructive manner when S is infinite. The idea of the proof is to extend the given theory to a "maximally consistent theory" for which we construct a model. More formally, a theory Γ is *maximally consistent* if whenever $\Delta \supseteq \Gamma$ is consistent, then $\Gamma = \Delta$. That is, no proper extension of Γ is consistent. We aim to show that for every given theory Γ, we can construct a maximally consistent extension of Γ for which we define a model.

Let us first consider the quantifier-free case, i.e., when statements in S do not contain any quantifier. This will give us the completeness theorem for propositional logic.

We assume that we are dealing with a collection S of sentences involving constant symbols c_α, relation symbols R_β, for $\alpha \in I$ and $\beta \in J$, where I and J are some index sets and each R_β has fixed number of variables. Let M be a non-empty set and let $c_\alpha \to \bar{c}_\alpha$ be a map from the constant symbols to elements of M. Let $R_\beta \to \overline{R}_\beta$ be a map which associates an n-ary relation symbol to a subset of M^n.

Theorem 11.3 (Completeness of propositional logic) If S is a consistent set of quantifier-free sentences, then there exists a model $\mathcal{M}$ for S such that every element of M is of the form $\bar{c}_\alpha$ and every $\bar{c}_\alpha$ corresponds to some c_α in S.

We first prove the following lemma.

Lemma 11.1 Let T be a consistent set of sentences and let A be a proposition. Then, either $T \cup \{A\}$ or $T \cup \{\neg A\}$ is consistent.

Proof. If $T \cup \{A\}$ is inconsistent, then for some $B_i \in T$ and for some proposition C,

$$A \wedge B_1 \wedge \ldots \wedge B_n \to C \wedge \neg C$$

is a tautology. Similarly, if $T \cup \{\neg A\}$ is inconsistent, then for some $B_i' \in T$,

$$\neg A \wedge B_1' \wedge \ldots \wedge B_m' \to C \wedge \neg C$$

is a tautology. We know from propositional logic that

$$B_1 \wedge \ldots \wedge B_n \wedge B_1' \wedge \ldots \wedge B_m' \to C \wedge \neg C$$

is a tautology. Then, T is inconsistent. A contradiction. $\qquad \square$

Now we prove Theorem 11.3. Suppose that S is well-ordered. Then, there is a well-ordering for every constant, function, and relation symbol. This gives us a well-ordering for all sentences of the form $c_i = c_j$ or $R_\beta(c_1, \ldots, c_n)$, where c_i is a constant symbol and R_β is a relation symbol appearing in S. Let us call each of these sentences F_α. Using induction on α, we define G_α sentences. If F_α is consistent with $S \cup \{G_\beta : \beta < \alpha\}$, then let $G_\alpha = F_\alpha$. Otherwise, let $G_\alpha = \neg F_\alpha$. By the previous lemma, and by induction on α, $S \cup \{G_\beta : \beta \leq \alpha\}$ is consistent for any α. Since any contradiction is derived from a finite number of sentences, $H = S \cup \{G_\alpha\}$ is consistent. For each c_α in S, we do the following: for smallest β, define $\bar{c}_\alpha = c_\beta$ such that the sentence $c_\alpha = c_\beta$ is in H. Now such β exists since the sentence $c_\alpha = c_\alpha$ is in H. Let M be the set of all $\bar{c}_\alpha$'s and let $\overline{R}_\beta$ be the set of all n-tuples $(\bar{c}_{\alpha_1}, \ldots, \bar{c}_{\alpha_n})$ such that $R_\beta(c_{\alpha_1}, \ldots, c_{\alpha_n}) \in H$. Then, our model is a subset of the set of symbols c_α. Any statement in H or its negation can be derived from G_α since any relation (or its negation) appears in the statements of S also appears in G_α. Note that the negation of a statement in S is not a consequence of G_α, since otherwise H would be inconsistent. We defined our model $\mathcal{M}$ so that every G_α is true in $\mathcal{M}$. So every statement in H, and furthermore every statement in S, is true in $\mathcal{M}$. Therefore, $\mathcal{M} \models S$.

Now, we prove the main theorem by including the case where we have quantified formulas. The idea is that given a theory S, we take a theory in which all sentences are quantifier-free. We will then define a model for this quantifier-free theory. But it will also happen to be a model for S. For this, we introduce new constant symbols. Now let T be a theory (whether quantifier-free or not). By Rules D and G, we can assume this without loss of generality. We define the new theory as follows: Given a theory T, if the sentence $\forall x\, A(x)$ and the constant c symbol appears in T, then add the sentence $A(c)$ into T. Similarly, for every sentence of the form $\exists x\, A(x)$, we pick a constant symbol c (called a *witness*) that was not previously defined in T. We then add the sentence $A(c)$ into T. In the end, we get a new system T^*. Clearly $T \subseteq T^*$.

Lemma 11.2 If T is consistent, then so is T^*.

Proof. It follows from Rule E that for any constant c, $\forall x\, A(x) \to A(c)$ is a tautology. $A(c)$ is the consequence of the sentence $\forall x\, A(x)$ in T. By Rule F, if the theory which contains $A(c)$ is inconsistent, then so is the theory which does not contain $A(c)$. $\qquad \square$

Using Rule D and Rule G, we may replace T with an equivalent system T^*. That is, given T, we can always define such T^*.

Let S be a consistent set of sentences given in the hypothesis of the theorem. We define S_n for any integer n in stages as follows:

We let $S_0 = S$ and let $S_{n+1} = S_n^*$. Then let $\overline{S} = \bigcup_{n \in \mathbb{N}} S_n$. Let $\overline{S}_p \subseteq \overline{S}$ be a set such that all sentences in $\overline{S}_p$ are quantifier-free. Using Theorem 11.3, let $\mathcal{M}$ be a model for $\overline{S}_p$. Suppose that A is a sentence in $\overline{S}$ which contains r

many quantifiers. We proceed by induction on r. Suppose that every sentence in $\overline{S}$ which contains less than r many quantifiers is true in $\mathcal{M}$. Theorem 11.3 proves the case for $r = 0$. If $r \neq 0$, assume that A is of the form $\forall x\, B(x)$. Every element of M must be of the form $\overline{c}_\alpha$ such that c_α occurs in S. Since $S_n \subseteq S_{n+1}$ for every n, both A and c_α must appear in some S_k. Then, $B(c_\alpha)$ must appear in S_k^* and, thus, in $\overline{S}$. Moreover, $B(c_\alpha)$ contains less than r many quantifiers. Hence, for every $\overline{c}_\alpha \in M$, $B(c_\alpha)$ is true in $\mathcal{M}$. Therefore, $\forall x\, B(x)$ must be true in $\mathcal{M}$. The proof of the case that A is of the form $\exists x\, B(x)$ is similar. We leave this to the reader as an exercise.

For any theory T, if T is infinite. then the cardinality of T^* is same as the cardinality of T. If T is finite, then so is T^*. This completes the proof of Theorem 11.2.

Corollary 11.1 If a sentence A is not derivable from a theory S, then there exists a model of S in which A is false.

Proof. Since $S \cup \{\neg A\}$ is consistent, it has a model. $\qquad\qquad\square$

Corollary 11.2 If a sentence A is true in every model, then A is provable.

Proof. Take $S = \emptyset$ in Corollary 11.1. $\qquad\qquad\square$

Completeness is a very nice property indeed. But there are certain aspects that first-order theories behave strangely. There appears to be some freedom on the choice of the cardinality of models of theories in first-order logic. This is implied in the following corollary.

Corollary 11.3 If S has an infinite model or even has arbitrarily large finite models, then S has models of arbitrarily large cardinalities.

Proof. Suppose that we are given an arbitrary set of new constant symbols c_α. We adjoin to S the sentences $c_\alpha \neq c_\beta$ for all $\alpha \neq \beta$ to form a new theory T. Now T is consistent since any contradiction must be a consequence of finitely many sentences of the form $c_\alpha \neq c_\beta$ together with the theory S. But since S has models in which we can find arbitrarily large finite sets of distinct constants, and since T puts no further restrictions on these c_α's, we see that this subset of T has a model and so it is consistent, which means that T itself is consistent. Thus, T has a model and hence S has a model whose cardinality is at least that of the set of c_α's introduced, and this proves the corollary. $\quad\square$

This results shows that unless a theory is finite, it has no unique models up to isomorphism. The following result is known as the *compactness theorem*, which follows from completeness.

Corollary 11.4 (Compactness Theorem) Let S be a theory. If every finite subset of S has a model, then S has a model.

It would be interesting to look at the cardinality behaviour of models of a theory. Set theory proves the existence of many cardinals of different sizes. Furthermore, by the Löwenheim-Skolem Theorem, it is known that if set theory has any model at all, then it also has a countable model. One interesting question is whether sets in this countable model are really "sets" in the intended meaning, in the absolute sense. What about the set-membership relation in the same model? How do we know if the set-membership relation is really the intended "membership" relation? There is no *a priori* reason to believe this so. Next is to consider the relationship between the submodels of a theory.

Definition 11.8 Let $\mathcal{M}_1$ and $\mathcal{M}_2$ be two models. If $\mathcal{M}_1$ and $\mathcal{M}_2$ satisfy the same sentences, then $\mathcal{M}_1$ and $\mathcal{M}_2$ are called *elementarily equivalent*. Given that $M_1 \subseteq M_2$ and $\vec{a} \in M_1^n$, we say that $\mathcal{M}_1$ is an *elementary submodel* of $\mathcal{M}_2$ if for all $\phi(\vec{x})$,

$$\mathcal{M}_1 \models \phi(\vec{a}) \leftrightarrow \mathcal{M}_2 \models \phi(\vec{a}).$$

Although the following definition will not be used in this section, we will require it for another chapter. We shall give it anyway since this is the right place. We say that a function $h : \mathcal{M} \to \mathcal{N}$ is called an *elementary embedding* if for every formula φ with parameters $a_1, \ldots a_n$, we have that $\mathcal{M} \models \varphi(a_1, \ldots, a_n)$ if and only if $\mathcal{N} \models \varphi(h(a_1), \ldots h(a_n))$.

11.2   Löwenheim-Skolem Theorem

We are now ready to prove the mathematical result that leads to Skolem's paradox. In fact, the Löwenheim-Skolem Theorem consists of two theorems, namely the *upward* and *downward* versions. The version we shall prove is the *Downward Löwenheim-Skolem Theorem*. We will also state the upward version later on.

Theorem 11.4 Let $\mathcal{M}$ be a model for a collection of sentences T. There is an elementary submodel of $\mathcal{M}$ whose cardinality does not exceed that of T if T is infinite and is at most countable if T is finite.

Proof. The proof uses the Axiom of Choice. If $\mathcal{M}$ is a model for a certain set of sentences, the submodel will also be a model for the same set. For any submodel $\mathcal{N}$ of $\mathcal{M}$, we show that if $\mathcal{M}$ is a model for T, then so is $\mathcal{N}$ satisfying the conditions in the theorem. Let N be a subset of M containing all constants $\bar{c}_\alpha$, and let $\phi(y, \vec{x})$ be an arbitrary formula of $n + 1$ variables. For each $\bar{x}_1, \ldots, \bar{x}_n$ in N, whenever there is some $\bar{y}$ in M such that $\phi(y, \vec{a})$ is true in $\mathcal{M}$, at $\bar{y}, \bar{x}_1, \ldots, \bar{x}_n$, choose *one* such and adjoin it to N.[6] If we do

---

[6]This is the step where we invoke the Axiom of Choice.

this for all formulas ϕ and all possible $\overline{x}_i$, we obtain a set N^* containing all elements of N. Since the number of n-tuples of elements of N is same as the cardinality of N if N is infinite and countable if N is finite, it easily follows that the cardinality of N^* is at most that of $|N| + |T| + |\mathbb{N}|$. Define N_0 as the set of constants $\overline{c}_\alpha$, and let $N_{k+1} = N_k^*$ for every $k \in \mathbb{N}$. We now claim that if $N' = \bigcup_{k \in \mathbb{N}} N_k$, then $\mathcal{N}'$ has the desired property stated in the theorem. It is clear that $\mathcal{N}'$ satisfies the cardinality requirement. We must show that $\mathcal{N}'$ is an elementary submodel of $\mathcal{M}$. Let $\phi(x_1, \ldots x_m)$ be a formula with r many quantifiers. We proceed by induction on r.

If $r = 0$, then there is nothing to prove. Assume for an induction hypothesis that for any formula $\phi(x_1, \ldots, x_m)$ with less than r quantifiers and any $\overline{x}_1, \ldots, \overline{x}_m$ in N', ϕ is true in $\mathcal{N}'$ at $\overline{x}_1, \ldots, \overline{x}_m$ if and only if it is true in $\mathcal{M}$ at $\overline{x}_1, \ldots, \overline{x}_m$. We prove the case for that ϕ is given as an existential statement, i.e., ϕ is of the form $\exists y \psi$ for some formula ψ. We leave the proof of the other cases to the reader as an exercise. Suppose ϕ is of the form $\exists y \, \psi(y, x_1, \ldots x_n)$. Let $\overline{x}_1, \ldots, \overline{x}_n$ be arbitrary elements of N'. We must show that $\phi(x_1, \ldots, x_n)$ is true in $\mathcal{M}$ at $\overline{x}_1, \ldots, \overline{x}_n$ if and only if it holds in $\mathcal{N}'$. We know that all $\overline{x}_i$ lie in N_k for some k. If there is a $\overline{y}$ in M such that $\psi(y, x_1, \ldots, x_n)$ is true in $\mathcal{M}$ at $\overline{y}, \overline{x}_1, \ldots, \overline{x}_n$, there is also such a $\overline{y}$ in N_{k+1}, thus in N'. Since ψ involves $r - 1$ many quantifiers, it follows from the induction hypothesis that ψ is true in $\mathcal{N}'$ at $\overline{y}, \overline{x}_1, \ldots, \overline{x}_n$. This means that $\phi(x_1, \ldots x_n)$ is true in $\mathcal{N}'$. If there is no $\overline{y}$ such that ψ is true in $\mathcal{M}$ at $\overline{y}, \overline{x}_1, \ldots, \overline{x}_n$, then there can be no $\overline{y}$ for which the same formula holds in $\mathcal{N}'$ since otherwise by the induction hypothesis it would hold in $\mathcal{M}$. This completes the proof. □

The upward version on the other hand states that if a theory has a countably infinite model, then it has models of arbitrary infinite cardinalities.

Discussion on Skolem's paradox.

Now that we have given the proof of the theorem, let us discuss its consequences. Applying the Löwenheim-Skolem Theorem on ZFC set theory leads to an interesting problem. The result was came to known as *Skolem's paradox*, shown by Thoralf Skolem [267] in 1922. So what is Skolem's paradox? From Cantor's theorem, we know that the set of real numbers is uncountable. In fact, from ZFC, using the Axiom of Power Set and the Axiom of Infinity, it can be proved that there exists an uncountable set. The language of set theory is countable. That is, there are countably many symbols in the language of set theory. So there can only be countably many statements written in this language. Thus, ZFC is a countable theory. Then, if we apply the Löwenheim-Skolem Theorem on ZFC, we will have the following antinomy: if ZFC has a model, then it has a countable model. To be more specific, suppose that T is a standard first-order axiomatisation of set theory, such as ZFC. If T has a model, then by the Löwenheim-Skolem Theorem, it has a countable model. Call this model $\mathcal{M}$. Thus, since T proves the statement "there exists an uncountable set x", there must be some $m_0 \in M$ such that $\mathcal{M} \models$ "m_0 is uncountable". But $\mathcal{M}$ itself is countable and so there can only be countably many $m \in M$ such that $\mathcal{M} \models m \in m_0$ holds. Although there are merely countably many sets in the underlying set of the model, it is said that there exists an uncountable set in the very same model. Of course, each element of this uncountable set is expected to be an element of the model. This is somewhat paradoxical and puzzling. On the one hand, m_0 looks uncountable. On the other hand, it is countable. In other words, it looks uncountable "inside" the model, but it is countable "outside" the model. This puzzling result tells us that the concept of cardinality of sets and, in particular, the size of models, is not absolute. In fact, the situation might be no different for other theories, as the Löwenheim-Skolem Theorem can be applied to any first-order countable theory.

How is it possible that the statement asserting that there exists an uncountable set holds in a countable model? That is, how can the statement "there are uncountably many elements", in first-order set theory, be true in a model in which there are merely countable elements? The problem is probably philosophical rather than mathematical. Prominently by Putnam [228] and Resnik [242], Skolem's paradox was thoroughly analysed on a philosophical basis. For Putnam [228], the problem in Skolem's paradox stems from the semantic indeterminacy of our language. Nonetheless, Löwenheim-Skolem Theorem does not bother the mathematician in practice. Mathematicians who are aware of this antinomy do not, in fact, consider this as a result that restrains mathematics. Perhaps one mathematical indication of Skolem's paradox is that there is no unique model of set theory, i.e., a model in which the axioms of ZFC are true. According to Stephen Cole Kleene [164], when we take a model of set theory, a *set* "inside" the model does not arbitrarily represent the concept of set in the general sense. It only represents a set that exists in the model. Now, a set is countable if there exists a bijection between that set and the set of natural numbers. Moreover, the bijection itself can be constructed as a set. So a set A may well be perceived in a model $\mathcal{M}$ as countable, yet perceived as uncountable in another model. The reason is that there might not exist, in model $\mathcal{M}$, a bijection between A and the set of natural numbers. In particular, the concept of non-enumerability, i.e., uncountability, is relative in set theory.[7] Aside from this, van Heijenoort [139] viewed Skolem Paradox as a 'novel and unexpected feature of formal systems'.[8] Kleene [165] viewed this as 'not a paradox in the sense of outright contradiction, but rather a kind of anomaly'.[9] The moral of story that the mathematicians need to know here is that the concept of uncountability in ZFC, or in similar theories for that manner, is not absolute. So contrary to popular belief, there are relative notions even in the purest branches of mathematics. The investigation of the reasons

---

[7]Kleene, pp. 426–237, 1950.

[8]van Heijenoort, p. 290, 2002.

[9]Kleene, p. 324, 1967.

behind this relativity is left for the studies of philosophers and philosophically minded mathematicians.

Review and Discussion Questions

1. Define informally an infinite set of sentences S such that any finite subset of S is consistent, yet S is inconsistent.

2. Let φ be a first-order sentence that is not contained in any complete theory. Show that $\{\varphi\}$ proves $\neg\varphi$.

3. Prove that if every proper subset of a countable theory is decidable as a set, then T is decidable.

4. Answer the following questions.

   (a) Show that the sentence $\forall x \neg \exists y P(x, y)$ is satisfiable.

   (b) Show that the sentence $\forall x R(x)$ is not a consequence of the sentence $\exists x R(x)$.

   (c) Define a structure in which the sentence $\exists x \exists y P(x, y)$ is true but $\exists x \forall y P(x, y)$ is false.

   (d) What can you say about the logical relationship between the two sentences given in (c)? That is, which one implies which?

5. Does Skolem's paradox undermine mathematical realism by any means? If so, in what sense mathematics is affected by this?

6. Is Skolem's paradox really a threat to mathematics?

12

Axiom of Choice

The *Axiom of Choice* is one of the postulates of ZFC set theory over which many debates and discussions have been made in the philosophical and mathematical community. The axiom, since the time it was first proposed, was particularly exposed to many criticisms by constructivists. Their primary concern is that the description of the function that is asserted to exist is not given explicitly. Another problem, yet more concrete, is the relationship between the Axiom of Choice and the Law of Excluded Middle, i.e., the statement that symbolically says $p \lor \neg p$. Constructivists have criticised the Axiom of Choice due to the fact that this axiom, as we shall show, logically implies the Law of Excluded Middle (see Diaconescu's theorem).

Generally, the justification that the mathematical community accepts the Axiom of Choice has been mostly relied on pragmatic intentions. The reason as to why to accept the Axiom of Choice as a "natural" postulate has been given by an argument something along the lines of "since the Axiom of Choice leads to a rich mathematical theory and proves abundance of desired properties, let us just accept it". In this chapter, we will first analyse Forster's mind-opening paper [95] and try to explain what the Axiom of Choice really does, based on *spatio-temporal* and *actual* differences such as *performing/performed, ordering/ordered, countable/counted,* etc.[1] We then criticise the pragmatic views and reasons for their justification in arguing for the "naturality" of the Axiom of Choice. We emphasise that if one wishes to argue for the "naturality" or "obviousness" of any axiom for that matter, then it seems better to argue from the viewpoint of certain propositions for which the Axiom of Choice is a necessary condition. Finally, we introduce an alternative postulate to the Axiom of Choice which was came to known as the *Axiom of Determinacy*. Axiom of Determinacy is an alternative axiom candidate which actually contradicts the Axiom of Choice, yet can prove many useful properties that the Axiom of Choice is able to prove. From this perspective, we hope that the problem of how and whether or not the Axiom of Choice should be accepted will become more clear. The argument that will be presented here can also be used for "naturality" arguments of other statements, including candidate axioms that are considered to be added in mathematics, in particular, set theory. So our

---

[1] I was lucky enough to listen Thomas Forster's talk during his visit at the University of Leeds, before then I was not completely aware of the nature of the Axiom of Choice and what it was really doing on the fundamental level. For this reason, a part of this chapter is devoted to presenting his eye opening work.

DOI: 10.1201/9781003223191-12

first purpose in this chapter is to survey Forster's analysis of the Axiom of Choice, and secondly we discuss about why and how we should accept/reject the candidate axioms based on their naturality rather than fruitfulness. We will assume in this chapter that the reader is familiar with the notions given in Chapter 2.

12.1   Applications of the Axiom of Choice

Let us recall what the Axiom of Choice is. Suppose that X is a collection of non-empty sets. Let f be a function such that for every $x \in X$, $f(x) \in x$. Then, we call f the *choice function* of X.

Axiom of Choice: Every set has a choice function.

In some cases, we do not even need to invoke the Axiom of Choice. These include the following cases.

(i) Every $x \in X$ is a singleton.

(ii) X is finite.

(iii) Every $x \in X$ contains finitely many real numbers.

In (i), since we have a single element and have no other choice, we automatically choose whatever is available. In (ii), we may define a choice function by finite induction. In (iii), $f(x)$ can be chosen to be the smallest real number in x.

There are plenty of examples that use the Axiom of Choice. It is a common practice to teach mathematics students, in their first year, some basic logic and set theory. One statement which uses a weak form of the Axiom of Choice in its proof is the statement that "countable union of countable sets is countable". It may not be very straightforward how the axiom is invoked here. In fact, it is often not told at all to the students that the Axiom of Choice is invoked in the proof. For conventional reasons, the mathematician accepts the Axiom of Choice so as to prove results that are either intuitively true or wanted to be true. From this pragmatic perspective, the line of thought for accepting this axiom is as follows: Since the Axiom of Choice is sufficient to prove the theorems that we want, we must accept the axiom.

12.1.1   Russell's Example

Bertrand Russell [251] explains in an example, in what situations one actually requires the Axiom of Choice in mathematics. Recall that if there exists a bijection between a set A and the set $\mathbb{N}$ of natural numbers, then we call A *countably infinite*. In general, given two sets A and B, the *cardinality* of A and B are equal if and only if there exists a bijection $f : A \to B$. We denote this relation by $|A| = |B|$. If two sets have the same cardinality, then this means they are equinumerous.

Suppose we have a countable infinity of pairs of shoes. We want to show that there are countably (infinite) many shoes. For this, we need to find a bijection between each shoe and natural numbers. It might be straightforward to claim that there exists such a bijection. Indeed, it is obvious that for any pair of shoes, given the nth pair, we map the left shoe to $2n$ and map the right one, say, to $2n + 1$. Thus, we conclude that there are countably many shoes. Let us now assume we have countable infinite pair of socks. This time we want to prove that there are countably many socks. Intuition tells us that there must be a one-to-one correspondence between the socks and natural numbers, same as the previous case. The existence of a bijection might prima facie come across as obvious. This means that for every sock there must be a distinct natural number. In this case, we may choose any one of the socks from a given pair, say the one whose image under the bijection corresponds to the smaller natural number in the pair. Now let us follow the inverse route. If we can choose some sock from each pair, then we can map the sock we picked, in the nth pair, to $2n$ and map the one we left out to $2n + 1$. Hence, there are countably many socks *if and only if* there is a choice function. Now in the show case, since the shoes are arranged in their natural order as the *left* shoe and the *right* shoe, it is possible to choose a specific shoe from each pair. We can define our own choice function for shoes without needing to use the Axiom of Choice. But what about the socks? Assuming that socks are idealised, they do not have a natural order like the shoes, nor do they have a property which separates the socks in each pair one from another. In order to show that there are countably many socks, give each pair we need to choose a sock and then map it to a natural number. But which sock are we supposed to *choose*? Certainly, we cannot distinguish them as left/right pairs, nor we can distinguish them with a unique property. They do not have an intrinsic property which separates one sock from another for a given pair. The situation is dramatically different in shoes as they are, by their own virtue, distinguishable. Having this capability of separating the *left* shoe from the *right* shoe is sufficient to choose *some* shoe from every pair in a uniform way. One does not need to use the Axiom of Choice in the case of shoes to show that there are countably many of them. We *know* how to choose them as we can explicitly define a choice function. However, unless we accept the Axiom of Choice, it seems that we cannot define a choice function on our own for socks. This does not necessarily mean there is absolutely no way to explicitly define a choice function for them. But for us—as human beings—given that it is not easy to find one, we can rely on the Axiom of Choice to posit that there is indeed a choice function for every set. If we want to prove that there are countably many socks—and our intuition tells us that there should be—then it seems that it is best to accept the Axiom of Choice. Without this axiom, it is unclear if we could choose a sock. The reader may ask at this point if we could just choose an *arbitrary* (or random) sock from each pair. Regarding this, Russell says:

> [...] with the socks we shall have to choose arbitrarily, with each pair, which to put first; and an infinite number of arbitrary choices is an impossibility. Unless we can find a rule for selecting, i.e. a relation which is a selector, we do not know that a selection is even theoretically possible.[2]

The choice sequence we get in the end, as a result of making arbitrary choices, may not end up being a legitimate rule or a function. A rule is a definite finite description, or at the very least it should be reducible to a finite description. A random sequence of infinite choices cannot be reduced to a simpler sequence than it is and thus, there is no guarantee that an arbitrary sequence constitutes a legitimate rule.

Suppose that for every $i \in \mathbb{N}$, X_i is countable. Let $\{X_i : i \in \mathbb{N}\}$ be a countable collection of countable sets. Is $\bigcup \{X_i : i \in \mathbb{N}\}$ countable? It might at first seem obvious that it is indeed countable. Just like we showed that the set of rational numbers are countable, using a similar method, we may show that the latter collection is countable. Let $x_{i,j}$ be the jth element of X_i. We write the elements of X_i on the ith row such that $x_{i,j}$ is the jth element of the ith row. Similar to the *zigzag* method (see Figure 9.6), we count all $x_{i,j}$'s. The usual inclination of mathematicians has been in favour of the view that if the Axiom of Choice is sufficient to prove that $\bigcup \{X_i : i \in \mathbb{N}\}$ is countable, then by convention we should accept the axiom.

12.1.2   König's lemma

We shall now look at another statement which may be regarded as "intuitively true" yet which still needs a limited version of the Axiom of Choice. Suppose that there exists a *node*, and for every node x there exist at most two *branches* pointing upwards, each containing a node, that stems from x. We call the resulting structure a (binary) *tree*. Roughly defined, a *path* is a sequence of consecutive branches.

Given the definition of tree, we can imagine it looks something similar to as in Figure 12.1. A tree is infinite if it has infinitely many nodes. *König's lemma* says that every infinite binary tree has an infinite path.[3] This might seem obvious at first. In fact, the proof of König's lemma uses a restricted form of the Axiom of Choice.

Let us shortly prove König's lemma. Suppose that T is an infinite binary tree. We will inductively define an infinite path A on T such that A will be the

---

[2] Russell, pp. 126–127, 1919.
[3] In fact, this version is known as Weak König's Lemma.

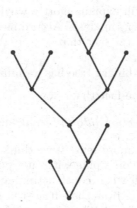

FIGURE 12.1
An example of a binary tree. Each line stemming out from nodes represents a branch.

union of all finite paths A_s, satisfying that A_{s+1} extends A_s. Then we define the infinite path $A = \bigcup A_s$. Let A_0 be the initial node of the given tree, i.e., the node for which there is no predecessor. For the induction hypothesis, suppose that a finite path A_s is given on T such that there exists an infinite path on T which extends A_s. Given A_s in T for which there are infinitely many extensions in T, let A_{s+1} be an immediate successor of A_s in T. Now such A_{s+1} exists because A_s has infinitely many extensions in T, but only finitely many immediate successors since T is finitely branching. Therefore, at least one of the immediate successors of A_s must have infinitely many extensions in T. The union of all A_s gives an infinite path on T. This completes the proof.

Most mathematicians are in favour of accepting the Axiom of Choice for the reason that it is sufficient to prove the statements that we want to be true. However, we will claim that determining whether or not the axiom is "natural" or "intuitive" on a pragmatic basis is rather inadequate, although it might comform to the naturalist philosophy (see Chapter 13).

12.2 Which statements are obvious?

Although the statements given in the examples so far look intuitively obvious, there is a fallacy of equivocation, as Forster puts it.[4] In fact, the reason that we perceive them to be obvious is due to some *spatial* and *actual* effect of

---

[4]Forster, p. 194, 2006.

the Axiom of Choice. We will explain shortly what we mean by these. First, however, let us look at Forster's analysis. According to Forster, what is obvious about König's lemma is not the fact that

"An infinite binary tree has an infinite path."   (i)

Rather what is obvious is the fact that

"The one *with an injection into the plane* has an infinite path."[5]   (ii)

The subtle difference makes more sense if we think that, given a tree without any projection or any image on a plane, it is not possible to choose the rightmost or the leftmost branch when constructing an infinite path. How do we separate the rightmost branch from the leftmost branch if there is no reference point? The "split" between the "left" and the "right" branch is not something which is provided intrinsically by the notion of a binary tree. The only way to "choose the leftmost branch" is to have a projection of the tree on some realisable plane. But we observe that as soon as we have a "spatial realisation" of a binary tree, the branches automatically become distinguishable. Thus, the Axiom of Choice "normalises" the tree, defines an order between its branches, and makes the tree "perceivable" in some sense. Forster asks if (i) and (ii) are not the same. If the given trees are isomorphic to each other, then (i) and (ii) are indeed the same. Are any given two infinite binary trees isomorphic to each other? This might at first seem quite obvious. However, according to Forster, what is obvious is not the fact that

"Any two infinite binary tree are isomorphic."   (iii)

What is in fact obvious is

"Those which are *equipped with injections into the plane* are isomorphic."
(iv)

On contrary to what most people might guess, (iii) and (iv) are not the same.[6]

12.2.1   Countable unions

Forster makes another interesting separation, relying on the oral tradition, between a *counted set* and a *countable set*. He describes the distinction in the following manner:

> [...] *counted set*, which is a structure $\langle X, f \rangle$ consisting of a set X with a bijection f onto $\mathbb{N}$, and a *countable set*, which is a naked set that just happens to be the same size as $\mathbb{N}$, but which does not come equipped with any particular bijection.[7]

---

[5]ibid, p. 194.
[6]ibid, p. 194.
[7]ibid, p. 195.

To understand the thin line between these two notions, let us consider the following case. If a set A is countable, then there exists a bijection between A and the set of natural numbers. But is it the case that the bijection must be explicitly constructed? Of course not. It is sufficient that there just exists a bijection, as no further requirement is needed for a set to be countable. On the other hand, a *defined/constructed* function is more than a function that merely exists in the metaphysical sense. We observe that Forster's distinction of the countable and the counted set stems from the difference between non-constructible and constructible way of asserting the existence of an object. According to Forster, counted union of counted sets is counted. A countable union of counted sets is countable. However, it is not very obvious what a counted union of countable sets and particularly a countable union of countable sets is. What is obvious for Forster is not that

"A union of countably many countable sets is countable." (v)

Rather what is obvious is that

"A union of countably many counted sets is countable." (vi)

Moreover, what is obvious is the statement

"A counted union of counted sets is counted."[8] (vii)

Despite that (vi) and (vii) can be proved to be true without invoking the Axiom of Choice, (v) invokes the choice axiom for its proof. The fallacy of equivocation, for Forster, is to mistake (v) for one of (vi) and (vii).

12.2.2 Countable pairs of "identical" objects

According to Forster, there are two lines of thought that lead anyone think in Russell's example of pairs of socks that there are countably many socks. It is basically the fallacious perception that all countable pairs of sets look similar. Indeed, one might think that all sets that are unions of countably many pairs must be the same size. That is, suppose that X_i is a countable collection of pairs. For every i, the size of each $\bigcup\{X_i : i \in \mathbb{N}\}$ can be thought as equal to one another. Countable pairs of shoes is no different than countable pairs of socks. All we need to do is to replace each shoe with a sock and conclude that they have the same cardinality. Countable pairs of gloves, countable pairs of dice, countable pairs of ear rings, etc., their unions are of the same size with each other. But why should we believe that, for $i \neq j$, the sets $\bigcup\{X_i : i \in \mathbb{N}\}$ and $\bigcup\{X_j : j \in \mathbb{N}\}$ are of the same size? Consider for a moment a countable pairs of white socks and a countable pairs of black socks. Let us denote the union of all pairs of white socks by W, and denote the union of all pairs of black socks by B. We ask if $|W| = |B|$. To show that they are of the same

---

[8]ibid, p. 195.

cardinality, it is sufficient that we map the elements of each nth pair of white socks with the elements of each, say, nth pair of black socks. Given no further information about the socks, we assume they are *idealised*. So we need to consider each pair of socks as an unordered pair. For if the pairs were ordered by some relation R, then we would be able to *choose*, say, the least element with respect to the given order R. In that case, it would be no problem in defining the choice function explicitly on our own. Since we take each pair as an unordered set, we can define two possible bijections, none of which can be separated from the other due to the fact that there is no order between the elements of the pair.

First possibility: $\{white, white\} \rightarrow \{black, black\}$
Second possibility: $\{white, white\} \rightarrow \{black, black\}$

Since we are taking each pair of socks as an unordered pair, the two possible bijections do not seem to look different which leaves us no choice but to select an *arbitrary* mapping. But then considering the rest of the pairs, we cannot just make infinitely many arbitrary choices, as Russell suggests that it might not constitute a legitimate function in the end. Maybe the socks *could* be distinguished but it is just that we as human beings have not figured out how to do so. Fortunately, the Axiom of Choice implicitly tells us that these two bijections, in fact, can be "distinguished" from one another and that an infinite sequence of choices indeed *defines* a function.

According to Forster, another line of thought that leads one to perceive that the set of socks is countable concerns our conception of *space*. He says:

> The very physical nature of the setting of the parable has smuggled in a lot of useful information. It cues us to set up mental pictures of infinitely many shoes (and socks) scattered through space. The shoes and socks are — all of them — extended regions of space and so they all have non-empty interior.[9] Every non-empty open set contains a rational, and the rationals are well-ordered.[10] This degree of asymmetry is enough to enable us to choose one sock from each pair.[11]

The purpose of using the Axiom of Choice is to determine the "first" of the two (or many) objects in a given set by introducing a well-ordering, after which we may determine the "next" element with respect to the $\in$ relation, and so on.

---

[9] A set of real numbers whose interval is not closed is called an *open set*. For instance, the real number interval $(0, 1)$ is an open set since it does not contain the end points 0 and 1. However, the interval $[0, 1]$ is closed, since it contains the end points as elements.

[10] For a set A, if every subset of A has a least element with respect to the relation $\in$, then A is called a *well-ordered* set. The set of rationals are well-ordered since it can be put into one-to-one correspondence with the set of naturals.

[11] Forster, p. 196.

We shall also state some of the equivalent forms of the Axiom of Choice, one of which will be left out for discussion in the next section. The *Well-Ordering Theorem* says that every set can be well ordered. This is logically equivalent to the Axiom of Choice. Another important theorem of order theory is called *Zorn's Lemma*, which states that if P is a partially ordered non-empty set and that every totally ordered subset of P has an upper bound, then P has a maximal element. Zorn's Lemma is used in the proof of many theorems, yet it is usually not mentioned that Zorn's Lemma is equivalent to the Axiom of Choice. The *Law of Trichotomy* is the statement which says that for any two sets A and B, either $|A| < |B|$ or $|A| > |B|$ or $|A| = |B|$, which also happens to be equivalent to Choice. From general topology, *Tychonoff Theorem* says that the product of compact topological spaces is compact. This is yet another equivalent form of the Axiom of Choice. One particularly interesting statement concerns linear algebra.

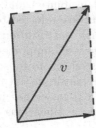 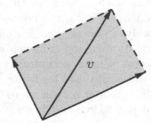

FIGURE 12.2
Two different bases, respectively, define different planes for the same vector v.

Students who take an introductory linear algebra course are often asked to define a basis for a given vector space or asked to prove certain things provided a basis. The vector basis theorem of linear algebra states that every vector space has a basis. But how do we know if there exists a basis for an arbitrary vector space in the first place? Indeed this is an assumption equivalent to the Axiom of Choice. This can be visualised as defining a *perspective*, so to speak, for a given set of vectors (see Figure 12.2). Given a vector v, there may be many bases for v each of which may define different planes.

12.3 Determining the Naturality or Otherwise

We now give examples as to why to accept or, on the contrary, reject the Axiom of Choice and to determine whether the axiom is natural or otherwise. We will rely on the following tradition: Undoubtfully, the Axiom of Choice provides many conventional features and proves many useful statements. However, we claim that this approach is inadequate in determining the naturality of an axiom. To determine if the Axiom of Choice is intrinsically natural, it is more efficient to analyse first the *sufficient conditions* of the axiom which are "intuitively true" and which conform to our mathematical intuition. In contrast with this, if we want to reject the Axiom of Choice, we can look at its "ill" *consequences*. Recall from propositional logic that given a conditional statement $P \rightarrow Q$, we say that here P is a *sufficient condition* for Q, and Q is a *necessary condition* for P. That is, if P is true then so is Q. But if Q is true, this does not necessarily mean that P is true. Similarly if P is false, then Q may be either true or false. However, if Q is false, then P must be false as well. Is P necessary for Q? It is not necessary. P is just sufficient to prove Q. That is to say, if P were false, there might well be another statement, say R, which would be sufficient to prove Q.

The justification relying on the intrinsic "plausability" and, separately, on the "practicality" of a mathematical statement are referred to in Maddy as, respectively, *intrinsic justification* and *extrinsic justification*.[12] Adoping an axiom for its own self-evidence and obviousness is supported by intrinsic justificatons. Whilst, accepting a statement as an axiom by solely relying on pragmatic reasons and its fruitfulness is an extrinsic justification for the statement. The reader may refer to Shoenfield [261] to see the intrinsic justifications of the Axiom of Choice.[13] The extrinsic justifications of the same axiom was investigated in Boolos [39].[14]

We will criticise the pragmatic based views which support extrinsic justifications for determining the "naturality" of the Axiom of Choice. Instead, we argue for that intrinsic justifications play a major role in the determination of whether the axiom is natural and plausible. The other case, i.e., the idea that the Axiom of Choice is rather "artifical" and counter-intuitive, may be supported through extrinsic justifications by looking at the paradoxical consequences of the axiom.

We begin with noting that intrinsic justifications must be necessarily taken into account in determining the naturality of an axiom. Again, by naturality we mean intuitive plausability and congruency with our mathematical "intuitions". For if intrinsic justifications did not play any role in determining the naturality of a statement, then we would be willing to accept a contradictory statement such as "0=1". After all, using the *ex falso* rule, from contradiction we derive anything.[15] From a pragmatic point of view, a contradiction is the most "fruitful" statement which technically proves so many statements. In fact, it proves everything. Of course we do not want our axiomatic system to be inconsistent. But this is due to the fact that a contradiction has no

---

[12]Maddy, p. 47, 2011.

[13]Shoenfield, pp. 335–336, 1977.

[14]Boolos, pp. 28–29, 1971. Page reference to the reprint (1998) version.

[15]*Ex falso*: For any statements p and q, $(p \wedge \neg p) \rightarrow q$.

intrinsic justification whatsoever. Thus, we must take intrinsic justifications into account whenever we want to add an axiom to our system, as intrinsic justifications play *some* role (whether major or minor) in determining the naturality of the axiom. Otherwise, from the pragmatic point of view, there would be no *a priori* reason not to add a contradiction to our system.

To decide whether the Axiom of Choice is "intuitively true", we should be able to find another intuitively true statement for which the Axiom of Choice is a *necessary* condition. The examples given in the previous sections were necessary conditions for the Axiom of Choice. However, there is another statement whose *negation* seems quite counter-intuitive. Let I be an index set and let $\{X_i : i \in I\}$ be a collection of non-empty sets. Then,

$$\prod_{i \in I} X_i \neq \emptyset.$$

Let us denote the statement above by CP, standing for Cartesian product. It tells us that the Cartesian product of an arbitrary family of non-empty sets is non-empty. In fact, many readers might find it rather absurd to claim the opposite. How could the product of arbitrarily many non-empty sets be empty? Claiming that the product is empty would be something to immediately revolt against. If CP seems intuitively true, then we must *necessarily* accept the Axiom of Choice due to the fact that CP is a sufficient condition for Choice. Such a statement being a sufficient condition for the Axiom of Choice is *sufficient* to consider the axiom as an intuitively true statement. In fact, CP is logically equivalent to the Axiom of Choice.[16] Let us prove this. First we show that the Axiom of Choice is a sufficient condition. If there exists a choice function f, then the sequence $\langle f(i) : i \in I \rangle$ is an element of the product $\prod_{i \in I} X_i$. Now we show the other direction. If $\prod_{i \in I} X_i \neq \emptyset$, then the function $f = \langle x_i : i \in I \rangle$ must be in this product, and it gives a sequence of elements x_i such that $x_i \in X_i$. Therefore, $f(i) = x_i$ defines a choice function. This completes the proof.

The fact that the Cartesian product is non-empty implies that there exists a choice function. Hence, the outcome of the Cartesian product determines whether the choice function exist or not. Let us summarise this particular intrinsic justification of the Axiom of Choice. If it is obvious for us that the Cartesian product of non-empty sets is non-empty, then the existence of a choice function for any set is *on a par* obvious as CP. If we accept CP, then we must equally accept the Axiom of Choice. The examples we gave earlier in the previous sections do not necessarily imply the existence of choice functions. For König's lemma, the full version of the Axiom of Choice is a sufficient condition, not a necessary condition. The full version of Choice is too strong to be a necessary condition for König's lemma. If, however, we restrict the Axiom of Choice to countable sets, then the countable version of the Axiom of Choice becomes a necessary condition for König's lemma. On the other

---

[16]Even if this equivalency did not hold, the idea that the Axiom of Choice is intuitively true would still follow from the relation that CP is a sufficient condition.

hand, CP is logically equivalent to the full version of Choice. In particular, since CP is a suffcient condition for Choice, it may be best to rely on this very fact in order to establish the claim that the Axiom of Choice is intuitively plausible and natural.

Some might think that CP does not constitute a valid intrinsic justification for the Axiom of Choice to be regarded as an "intuitively true" mathematical statement. They might point out that a possible vicious circle appearing in the relationship between CP and the Axiom of Choice, as they logically imply each other. For this reason, although we suggest solely taking into account the fact that CP is a sufficient condition for Choice, we may also find propositions for which the Axiom of Choice is a necessary condition but not vice versa. Due to this biconditional relationship between CP and the Choice, it is possible to modify our argument to use another statement as an intrinsic justification by eliminating the sufficient condition case to ensure that the Axiom of Choice is strictly a necessary condition. So we shall say:

"The Axiom of Choice is an intuitively true statement if and only if there exists an intuitively true statement φ such that the Axiom of Choice is (strictly) a necessary condition for φ."

Then, besides CP, we may also consider other statements to argue for the naturality of the Axiom of Choice. As a matter of fact, as a personal view, we may even consider the (General) Continuum Hypothesis as an intrinsic justification of the Axiom of Choice. Sierpinski [265] showed that the General Continuum Hypothesis is a sufficient condition for Choice. However, the other way around does not hold. Thus, the Continuum Hypothesis and the Axiom of Choice are not logically equivalent. Despite that the Axiom of Choice, CP, and the Continuum Hypothesis are independent of the Zermelo-Fraenkel set theory, any "intuitively true" statement which implies the Axiom of Choice can be used to establish that Choice is accordingly an intuitively true axiom. It is still being argued today whether there are intuitively true statements which imply Choice. It is more convenient to use such statements to determine the naturality of the Axiom of Choice. The position we criticise, in determining the naturality of the Axiom of Choice or an axiom candidate to be added into mathematics (such as the Continuum Hypothesis or the Axiom of Constructibility, i.e., $V = L$), is the pragmatic views which solely suggest considering the consequences and the fruitfulness of the statements.[17] Considering that the Continuum Hypothesis implies Choice, we can as well claim that the Axiom of Choice is naturally true if we believe that the Continuum Hypothesis exhibits an intrinsically plausible feature. CP can be considered just as an example to show that the Axiom of Choice is intuitively true. Since CP seems intrinsically plausible and CP implies Choice, then Choice is at least

---

[17]$V = L$ (Axiom of Constructibility), roughly says that if x is any set, then there exists a formula ϕ such that ϕ "defines" x. This axiom, in fact, tells us that every set is "definable" in the first-order language of set theory. Many philosophers of mathematics, particularly Penelope Maddy [186], questioned if the set theoretic universe V could really be equal to L.

as intrinsically plausible as CP. Nevertheless, we should not restrict ourselves to CP, as there may be other "intuitively true" statements for which the Axiom of Choice is a necessary condition. One may regard, including myself, the Continuum Hypothesis as an intuitively true statement like CP. Moreover, unlike CP, the Continuum Hypothesis is not logically equivalent to the Axiom of Choice.

The argument for the naturality of CP, or the Continuum Hypothesis, relies on a weak form of Law of Excluding Middle. It seems counter-intuitive to me that the Cartesian product of non-empty sets is empty. Hence, intuition tells that it must be non-empty. Kunen [175] says that 'it is not generally considered to be intuitively true any more than CH is generally considered to be false'.[18] So it may not be feasible to apply the same "excluded middle" argument for the Continuum Hypothesis. If we had to follow a similar approach, though, it would be to claim that it is hard to imagine a tree with uncountably many nodes while keeping the splittings non-dense. That is, in terms of cardinalities, it is counter-intuitive to me that the smallest infinite cardinal greater than $\aleph_0$ is strictly less than $2^{\aleph_0}$. Thus, from this point of view, the Continuum Hypothesis seems intuitively true and natural. There are many mathematicians, especially set theorists, who believe that the Continuum Hypothesis is false for the reason that rejecting the hypothesis leads to rich set-theoretical universes. While those who favour a neat and ordered universe usually accept the Continuum Hypothesis. Both Gödel and Cohen were profound believers in the falsity of the Continuum Hypothesis. Gödel says:

> I believe that adding up all that has been said one has good reason for suspecting that the role of the continuum problem in set theory will be lead to the discovery of new axioms which will make it possible to disprove Cantor's conjecture.[19]

Paul Cohen [63] also was on the side of the belief that CH is false. In the conclusion section of his book, he says:

> A point of view which the author feels may eventually come to be accepted is that CH is *obviously* false. The main reason one accepts the Axiom of Infinity is probably that we feel it absurd to think that the process of adding only one set at a time can exhaust the entire universe. Similarly with the higher axioms of infinity. Now $\aleph_1$ is the set of countable ordinals and this is merely a special and the simplest way of generating higher cardinal. The set $\mathcal{C}$ (the set of reals), in contrast, generated by a totally new and more powerful principle, merely the Power Set Axiom. It is unreasonable to expect that any description of a larger cardinal which attempts to build up that cardinal from ideas deriving

---

[18]Kunen, p. 171, 1980.
[19]Gödel, p. 480, 1964.

from the Replacement Axiom can ever reach $\mathcal{C}$. Thus $\mathcal{C}$ is greater than $\aleph_1, \aleph_\omega, \aleph_\alpha$ where $\alpha = \aleph_\omega$ etc.[20]

We shall elaborate more on why CP and the Continuum Hypothesis seem as intuitively true statements. Let us abbreviate the Law of Excluded Middle by LEM, Axiom of Choice by AC, and the Continuum Hypothesis by CH. By Diaconescu's theorem (see Theorem 4.2), we know that AC $\rightarrow$ LEM. By Sierpiński [265], it is known that the General Continuum Hypothesis (GCH) implies AC. Hence, GCH $\rightarrow$ LEM. The effect of LEM in determining the naturality of AC and CH is evidently clear. As a matter of fact, since CP implies LEM in a more explicit manner, where LEM is manifested in CP in the form of a dichotomy of the product being non-empty or not, the naturality of CP seems stronger and more intuitive than other propositions equivalent to AC such as Zorn's Lemma and Tychonoff Theorem. Due to these implications, if one believes that CH is a natural self-evident axiom, then she should also believe the same for AC.

The naturality of the Axiom of Choice is a relative view, as the mathematical community was divided into two camps in the early 20th century regarding the status of this axiom. Whether the Axiom of Choice is a natural axiom depends on what kind of mathematics we want to develop. Do we want mathematics to be fruitful regardless of the intrinsic plausibility of statements we accept, or do we want mathematics to mostly contain intuitively true statements regardless of how fruitful and practical they are? On contrary to popular view that mathematics goes hand in hand with science and that mathematics should serve the scientific practice, my personal view on this matter is that mathematical truth is attained by adopting the maxim "more obviousness, less fruitfulness", or in other words "obviousness before fruitfulness". Mathematical knowledge, as soon as we define mathematical objects, is *a priori*. Due to this reason, we must take intrinsic justifications into account majority of the time when determining the naturality of a statement. Of course, we also require some practical implication, as solely relying on intrinsic justifications does not extend our mathematical knowledge. In fact, logical tautologies are of this kind. All tautologies are, by virtue of themselves, intuitively true. But they do not extend our mathematical knowledge and they do not entail too many statements. More specifically, tautologies only entail other tautologies. Thus, we should neither solely rely on intrinsic justifications nor on extrinsic justifications of a statement to see if it is a natural axiom candidate. Omitting the extrinsic justifications would be, in fact, an anti-naturalist approach. Such approaches were known to be criticised by Quine and Maddy. Since naturalism abandons first philosophy, our position in this chapter is not a naturalistic one, but we rather take the intrinsic qualities of mathematical statements prior to their practical implications. If, however, one happens to decide to develop a mathematical theory which serves our best scientific and other

---

[20]Cohen, §13, p. 151, 2008. Parenthesis added

mathematical theories, then the definition of naturality would be changed in favour of fruitfulness of the statements and their extrinsic justifications.

The other side of the coin concerns on what grounds the Axiom of Choice should be rejected. For this it is reasonable to consider either (i) to look at the intrinsic plausibility of statements that contradict the Axiom of Choice, or (ii) to look for counter-intuitive statements for which the Axiom of Choice is a *sufficient* condition. In fact, the second scenario is analysed though extrinsic justifications. The most popular example of this type of argument involves the *Banach-Tarski Theorem* [22], also known in the mathematical literature as the *Banach-Tarski Paradox*. The paradox is roughly described as follows. Consider a three-dimensional ball in the Euclidean space, i.e., in $\mathbb{R}^3$. We can decompose this ball into finitely many pieces (apparently five pieces would suffice) such that after a series of applications of rotations and moving the pieces (no bending or tearing), when we reassemble the pieces we get two new balls of the exact size as the original ball. Since the proof of the Banach-Tarski Theorem uses the Axiom of Choice, we conclude that the axiom causes such anomalies. However, the discovery of the Banach-Tarski Theorem did not affect ordinary mathematical practice. Mathematicians usually have no problem getting away with paradoxes and anomalies in their theories. The reason behind the anomaly is that the partitions of the ball must be *unmeasurable*, in the sense that the volume of each piece is undefinable. So the Axiom of Choice allows some paradoxical decompositions. Nevertheless, a major part of mathematics uses the Axiom of Choice. It is generally agreed that the Axiom of Choice should not be rejected just because it produces paradoxical decompositions. Although the axiom produces results that contradicts our basic geometric intuition, it also proves things that are intuitive. Maddy argues against the theses which suggest rejecting the Axiom of Choice based on the Banach-Tarski Theorem. In [190], she says:

> [...] if physical regions aren't literally modelled by sets of ordered triples (coordinates) of real numbers, then we can't assume that all consequences of our mathematical theory of those sets will hold for those (physical) regions; therefore, we can't conclusively draw our false empirical conclusion.[21]

The point that Maddy objects is that the refutation of the axiom is based on a physical observation. It is not that the Axiom of Choice is counter-intuitive, but we understand from the passage below that rather the physical space might not be completely modelled by $\mathbb{R}^3$. Maddy says:

> But isn't it at least as reasonable to conclude that the full power set of $\mathbb{R}^3$ was a poor choice as a model for physical regions?[22]

Can it be that the Axiom of Choice makes non-conceivable and unordered

---

[21]Maddy, p. 35, 2011. Parentheses added.
[22]ibid, p. 35.

objects conceivable and ordered? From an Aristotelian viewpoint, objects with
no property are unconceivable. Consider, for example, the notion of *circle*.
Although we may think of different sorts of circles of different sizes, colours,
etc., if we omit all these physical features, we end up having a unique property
of all circles: *A circle is a collection of points that are equidistant from a
centre point.* All circles must have this feature. Objects that do not bear any
property are chaotic. Thus, no chaotic object can be said to have a structure,
and even if they exist, they cannot be conceived by any means. When given
a tree, projected on the two-dimensional plane, we immediately tend to think
that König's Lemma is intuitively true. Nevertheless, trees, in general, can
exist as raw non-chaotic objects with no natural order between its branches,
where we use the term *non-chaotic* to refer to objects with some intrinsic
feature, yet not concretely conceivable. No mathematical object is chaotic
unless contradictory, but they may be non-chaotic. A non-projected raw tree
or a collection of pairs of socks are some examples of what we call non-chaotic
objects. However, these are still unconceivable objects. A raw tree, given in its
pure form, is non-chaotic for the fact that it can be described as earlier: There
exists a *node*, and for every node x, there exist at most two *branches*, each
containing a node, that stems from x. This defines a raw tree. However, for
the tree object to be conceivable, at least more concretely, it must contain a
property that distinguishes the branches from one another. Projecting the tree
on the plane *is* the result of automatically assuming that there is a natural
order between the branches of the tree. As we *draw* the tree, we define an order,
and not only does the tree remains as a non-chaotic object, but furthermore
it transforms into a perceivable entity in a more concrete sense. Imagining in
our mind the *projection* of the tree on a spatial dimension is similar to how the
choice function acts on a raw tree. The Axiom of Choice, then, transforms raw
objects to conceivable objects. This is quite congruent with the Kantian view
that we tend to project raw non-chaotic mathematical objects on a spatial
and actual basis in order to "normalise" it. As a result, the raw object gets
normalised and it transforms into a perceivable entity. The Kantian "intuition"
seems to be the act of normalising mathematical objects, and this is what the
Axiom of Choice does, particularly when there is no immediate natural order
in the given set. As in Russell's example of pairs of socks, it may not be always
possible for the mathematican to find a choice function for the given collection
of non-empty sets.

12.3.1   Axiom of determinacy

Besides statements that imply the Axiom of Choice, there are also statements
which contradict this axiom. These statements play a significant role in rejecting Choice. The set of axioms of set theory including the Axiom of Choice

is known today as ZFC.[23] The set theory in which the Axiom of Choice is not included is called ZF. We know that the Axiom of Choice is *independent* from ZF. That is, if ZF is consistent, then so is ZFC, but so is the system we obtain when we add the negation of the Axiom of Choice to ZF. Thus, ZF cannot prove or disprove the Axiom of Choice.

Set theorists have proposed various statements to be added in set theory as an axiom, some of which contradict the Axiom of Choice. One of them, introduced by Mycielski and Jan [202], is called the *Axiom of Determinacy*. We need to look at two-player *infinite games* to understand it. We summarise the essential part we need from Jech [155], to which the reader may refer for a detailed account to learn the consequences of the Axiom of Determinacy and other properties. Let us denote the set of natural numbers by ω. Consider the Baire space ω^ω and let us denote this space by $\mathcal{N}$. The Baire space contains all infinite sequences of natural numbers. That is, every element of the Baire space is an infinite sequence of naturals. For every $A \subseteq \omega^\omega$, we define an infinite game G_A played by two players. In this game, Player I begins by choosing a natural number a_0, and then Player II chooses some natural number b_0. Then, Player I takes turn and chooses another number a_1, and later Player II chooses some number b_1, and so on. The game is of length ω, so it does not end in finite steps. If $\langle a_0, b_0, a_1, b_1, \ldots \rangle \in A$, then we say that Player I *wins*. Otherwise, Player II *wins*. A *strategy* (for either of the players) is a rule that tells the player what move to make depending on the previous moves of both players. A strategy is called a *winning strategy* if the player who follows it always wins. If any one of the players has a winning strategy, then say that the game G_A is *determined*.

The Axiom of Determinacy: For every $A \subseteq \omega^\omega$, G_A is determined.

For a given $A \subseteq \omega^\omega$, let G_A denote the corresponding game. For every $n \in \omega$, suppose a_n denotes the nth move of Player I, and let b_n denote the nth move of Player II. Then, a *play* is a sequence of moves $\langle a_0, b_0, a_1, b_1, \ldots \rangle \in \omega^\omega$. A strategy for Player I is a function σ with values in ω whose domain consists of finite sequences (if Player I makes the first move in a play, then the length of the sequence must be even). Let σ be a strategy. If $a_0 = \sigma(\emptyset)$, $a_1 = \sigma(\langle a_0, b_0 \rangle)$, $a_2 = \sigma(\langle a_0, b_0, a_1, b_1 \rangle)$, and so on, then Player I plays $\langle a_0, b_0, a_1, b_1, \ldots \rangle$ *by the strategy* σ. If Player I plays by σ, then the play is determined by σ and the sequence $b = \langle b_n : n \in \omega \rangle$. Let us denote the play by $\sigma * b$. For a strategy σ, if $\{\sigma * b : b \in \mathcal{N}\} \subseteq A$, then we say that σ is a *winning strategy* for Player I. That is, all plays that Player I plays by σ must be in A. Strategy for Player II is defined similarly. A strategy for Player II is a function τ with values in ω, defined on finite sequences of odd length. If $a \in \mathcal{N}$ and τ is a strategy for Player II, then let us denote the play $a * \tau$ in which the first

---

player plays a and the second player plays by τ. If for some τ, we have that $\{a * \tau : a \in \mathcal{N}\} \subseteq \mathcal{N} - A$, then the strategy τ is a winning strategy for Player II.

We know that the cardinality of the set of natural numbers is $\aleph_0$. Consider the set of all subsets of the set of natural numbers, which is of cardinality $2^{\aleph_0}$. Hence, there are $2^{\aleph_0}$ many possible strategies.

The fact that the Axiom of Determinacy contradicts the Axiom of Choice is followed by the theorem given below. The proof of the theorem, in fact, uses the diagonalisation method.

Theorem 12.1 (Mycielski and Jan [202], 1962) Suppose that the Axiom of Choice holds. Then there exists some $A \subseteq \omega^\omega$ such that G_A is indetermined.

Proof. Let $\{\sigma_\alpha : \alpha < 2^{\aleph_0}\}$ and $\{\tau_\alpha : \alpha < 2^{\aleph_0}\}$ denote, respectively, sets of all strategies for Players I and II. We construct sets $X = \{x_\alpha : \alpha < 2^{\aleph_0}\}$ and $Y = \{y_\alpha : \alpha < 2^{\aleph_0}\}$ such that $X \subseteq \mathcal{N}$ and $Y \subseteq \mathcal{N}$. Suppose that we are given $\{x_\xi : \xi < \alpha\}$ and $\{y_\xi : \xi < \alpha\}$. Choose some y_α such that $y_\alpha = \sigma_\alpha * b$ for some b and $y_\alpha \notin \{x_\xi : \xi < \alpha\}$. Now such y_α exists since the cardinality of $\{\sigma_\alpha * b : b \in \mathcal{N}\}$ is $2^{\aleph_0}$. Similarly, choose some x_α such that $x_\alpha = a * \tau_\alpha$ for some a and $x_\alpha \notin \{y_\xi : \xi \leq \alpha\}$. Clearly $X \cap Y = \emptyset$. That is, for every α there exists some b such that $\sigma_\alpha * b \notin X$ and that there is some a satisfying that $a * \tau_\alpha \in X$. Therefore, neither Player I nor Player II has a winning strategy in the game G_X. Hence, G_X is indetermined.   □

In fact, we may also think of the game G_A as a sequence of 0's and 1's which furthermore can be seen as a coded real number. Then, the Axiom of Determinacy tells us that every real number is determined. Despite that the Axiom of Determinacy contradicts the Axiom of Choice, a weak form of Choice called the *Axiom of Countable Choice* is still consistent with the Axiom of Determinacy. The Axiom of Countable Choice is the statement which says that every countable collection of non-empty sets has a choice function. Note that the full version of Choice does not put any restriction on the size of the collection.

The Axiom of Determinacy is, of course, not an axiom of ZFC. However, just as the Axiom of Choice produces a rich theory, determinacy too has fruitful consequences. On the one hand, the Axiom of Determinacy proves a specific form of the Continuum Hypothesis, that is the statement

"There is no set S such that $|\mathbb{N}| < |S| < |\mathbb{R}|$".

On the other hand, it cannot prove that $2^{\aleph_0} = \aleph_1$. Under the Axiom of Choice, these two forms are, in fact, logically equivalent. If we do not assume Choice, we get different unequivalent forms of the Continuum Hypothesis. We said that a paradoxical consequence of Choice was the Banach-Tarski Theorem. Nevertheless, the Axiom of Determinacy produces certain anomalies regarding cardinalities as well. Let us say that a *partition* of a set A is a collection $\mathcal{C}$

of non-empty sets C_i such that $C_k \cap C_l = \emptyset$ for all distinct k and j and that $\bigcup \mathcal{C} = A$. As a matter of fact, it is a consequence of the Axiom of Determinacy that a set A with cardinality $2^{\aleph_0}$ can be partitioned into its disjoint subsets such that the number of elements in the partition is greater than the number of elements of A. This is due to the fact that when Choice fails, there is no injection from $\dot{\omega}_1$ to 2^ω.[24]

12.4 Concluding Remarks

We summarised how the Axiom of Choice is used in the proof of some fundamental statements in mathematics and what the axiom really does. We also criticised the pragmatic (naturalistic) views that are in favour of accepting the Axiom of Choice for its *fruitfulness* and convenience rather than for its *intrinsic justifications*. The naturalistic account of taking "fruitfulness" in deciding the status of an axiom is mostly based on what Maddy refers to as the *extrinsic justifications*. We argue for that if the Axiom of Choice is wanted to be accepted as a natural axiom, it should be adopted for the most part by checking whether or not intrinsically justifiable (or intuitively true) statements imply the axiom, instead of looking its consequences and service to other statements that we wish to prove. This approach would serve better for a development of a "first philosophy" oriented mathematics as opposed to a "second philosophy" orientation, which is more suitable for a naturalistic treatment of mathematics. In order to know if the Axiom of Choice is intrinsically plausible, we also need to take into account counter-intuitive statements implied by the Choice, such as the Banach-Tarski Theorem and maybe similar anomalies. We may then compare both extremes to draw a conclusion on the status of the Axiom of Choice. Is the Continuum Hypothesis or the Axiom of Determinacy more "intuitive"? What about the other extreme? Is the fact that the Cartesian product of an arbitrary collection of non-empty sets being empty more catastrophic than the Banach-Tarski Theorem? Prima facie, mathematical intuition says that ¬CP would be more problematic than the Banach-Tarski Theorem. Solely based on this intrinsic justification, we may say that the Axiom of Choice is a natural axiom. Nevertheless, one should make similar comparisons so as to determine whether or not an axiom can be regarded as "naturally true" after assessing each extreme and weighing their values to see which case could be preferred to another. The arguments concerning the naturality and plausibility of the Axiom of Choice, thus, may serve better for creating a self-evident mathematics if they focus on claims

---

[24]See Halbeisen and Shelah [126] (2001), p. 258, for more details. For a fuller discussion on the paradoxical consequences of Choice, we refer the reader to Hamkins [132] (2021), §8.8.

about the intuitiveness of sufficient conditions of the Axiom of Choice and the counter-intuitiveness of the necessary conditions.

Discussion Questions

1. We know that invoking LEM in existence proofs involves metaphysics to a certain extent. But does this metaphysical reference bring more advantages than the disadvantages or not?

2. Recall in Russell's example that a pair of socks is indistinguishable. Can we say that though it is *absolutely* indistinguishable? Or is it merely a defect of our mind that we do not have the apparatus to see the difference in them?

3. The fact that every vector space has a basis is logically equivalent to the Axiom of Choice. Discuss what a choiceless linear algebra would look like. Is it even possible to work with vector spaces with no guarantee of having a basis?

4. What justifies the Axiom of Choice? Should we prioritise intrinsic justifications over extrinsic justifications, or vice versa?

5. Can it be said that the Banach-Tarski paradox should not be seen as an antinomy? Do you agree with Maddy's claim that the physical space may not be correctly conceptualised in $\mathbb{R}^3$?

13

Naturalism

Ontological realism, or Platonism, which we studied in Chapter 3, is the philosophical position that mathematical objects exist independent of the language, mind, and senses. There are many variations and modifications of mathematical realism, one of which can be classified as the view that science and mathematics go hand in hand. In this chapter, we will look at Gödel's [118] [120] realism, and then study a contemporary position in the philosophy of mathematics called *mathematical naturalism*, led by Quine [232] and Maddy [183], [188]. We will first discuss Gödel's and Quine's versions of realism, and then move on to Maddy's set-theoretical realism and naturalism.

According to Quine's [237] definition, naturalism is the recognition that it is within science itself, and not in some prior philosophy, that reality is to be identified and described.[1] Quine claims that naturalism sees science as an inquiry into reality.[2] Naturalism abandons the goal of first philosophy and the view that philosophy (or mathematics) comes prior to its applications in science.[3]

Before discussing Quine and Maddy, we shall first look at Gödel's realism, from whom Maddy herself was also influenced.

13.1 Gödel's Realism

Gödel's views on the philosophy of mathematics can be found for the most part in his 1944 [118] and 1947 (revised in 1964) [123] papers.[4] His remarks on the Continuum Hypothesis is a strong indication that he can be regarded as an epistemological realist. Gödel says:

> It is to be noted, however, that on the basis of the point of view
> here adopted, a proof of the undecidability of Cantor's conjecture

---

[1] Quine, p. 21, 1981.

[2] ibid, p. 72.

[3] Naturalism is known to be against *first philosophy* and to the idea that philosophy determines natural sciences. On the contrary, naturalism supports what Maddy calls *second philosophy*, the view that science is prior to any discourse on first philosophy.

[4] Page references of [118] and [123] are to the reprint version in Benacerraf and Putnam (1983).

DOI: 10.1201/9781003223191-13

from the accepted axioms of set theory (in contradistinction, e.g., to the proof of the transcendency of π) would by no means solve the problem. For if the meanings of the primitive terms of set theory [...] are accepted as sound, it follows that the set-theoretical concepts and theorems describe some well-determined reality, in which Cantor's conjecture must be either true or false. Hence its undecidability from the axioms being assumed today can only mean that these axioms do not contain a complete description of that reality. Such a belief is by no means chimerical, since it is possible to point out ways in which the decision of a question, which is undecidable from the usual axioms, might nevertheless be obtained.[5]

In fact, Gödel's realism concerns the relationship between the philosophy of mathematics and natural sciences. He claims that classes and concepts may be conceived as real objects used in natural sciences, namely classes as "plurality of things", existing independently of our definitions and constructions.[6] On the epistemological side, Gödel believed that we have a perception of objects of set theory and set theoretical truths, as the axioms of set theory force themselves upon us as being true.[7] We need to explain what this set-theoretical perception is more in detail. Gödel explains this in the following manner:

It should be noted that mathematical intuition need not be conceived of as a faculty giving an *immediate* knowledge of the objects concerned. Rather it seems that, as in the case of physical experience, we *form* our ideas also of those objects on the basis of something else which *is* immediately given. Only this something else here is *not*, or not primarily, the sensations. That something besides the sensations actually is immediately follows (independently of mathematics) from the fact that even our ideas referring to physical objects contain constituents qualitatively different from sensations [...].[8]

It can be said that Gödel implies that there is a thin line between *perceiving* physical objects through senses and *conceiving* what they represent through intuition. Sensing a physical object is one thing, conceiving what it constitutes in our mind is another thing. On the one hand there is the reality of object. On the other hand, there is the conception it forms in our idea.

According to Maddy, what bridges the gap between retinal stimulation and perception is the neural cell-assembly.[9] For Gödel, the conceptual elements of sensory experience may represent an aspect of objective reality, but, as

---

[5] Gödel, p. 476, 1964. References to the reprint version in Benacerraf and Putnam (1983).
[6] Gödel, p. 456, 1944. References to the reprint version in Benacerraf and Putnam (1983).
[7] Gödel, pp. 483-484, 1964.
[8] ibid, p. 484.
[9] Maddy, p. 76, 1990.

opposed to the sensations, their presence in us may be due to another kind of relationship between ourselves and reality.[10] Due to Maddy, this 'other kind of relationship' is the complex causal process that produces the cell-assembly, i.e., the evolutionary pressures on our ancestors and our childhood experiences with objects.[11]

Although Maddy admits that intuitive beliefs (or intrinsic justifications) underlie the simplest set theoretic axioms, both Gödel and Maddy agree that not all axioms can be justified on intuitive grounds.[12] Gödel explains this as in the following manner:

> [B]esides mathematical intuition, there exists another (though only probable) criterion of the truth of mathematical axioms, namely their fruitfulness in mathematics and, one may add, possibly also in physics.[13]

In another passage, Gödel writes:

> [...] even disregarding the intrinsic necessity of some new axioms, and in even in case it has no intrinsic necessity at all, there might exist axioms so abundant in their verifiable consequences, shedding so much light upon a whole field, and yielding such powerful methods for solving problems [...] that, no matter whether or not they are intrinsically necessary, they would have to be accepted at least in the same sense as any well-established physical theory.[14]

It is understood from the following passage that Gödel's version of realism firmly connects mathematical and natural sciences together, as we mentioned earlier. He states:

> It seems to me that the assumption of such (abstract) objects is quite as legitimate as the assumption of physical bodies and there is quite as much reason to believe in their existence. They are in the same sense necessary to obtain a satisfactory system of mathematics as physical bodies are necessary for a satisfactory theory of our sense perceptions [...].[15]

It is worth noting that the naturalistic idea of justifying mathematical statements by their consequences was also somewhat supported by Russell himself, though not entirely, for the purpose of practicality and "maximising" our mathematical theories. His motivation was to discover what are (logically) the simplest axioms for mathematics, which eventually led him to abandon foundationalism.[16] In his work [249], Russell says:

---

[10]Gödel, p. 484, 1964.
[11]Maddy, pp. 76–77, 1990.
[12]ibid, p. 77.
[13]Gödel, p. 485, 1964.
[14]ibid, p. 477.
[15]Gödel, p. 456, 1944. Parenthesis added.
[16]See Irvine [154] (1989) for a detailed analysis of Russell's regressive method.

Thus in mathematics, except in the earliest parts, the propositions from which a given proposition is deduced generally give
the reason why we believe the given proposition. But in dealing with the principles of mathematics, this relation is reversed.
Our propositions are too simple to be easy, and thus their consequences are generally easier than they are. Hence we tend to
believe the premises because we can see that their consequences
are true, instead of believing the consequences because we know
the premises to be true.[17]

Another prominent adherent of the naturalist philosophy is Willard Van
Orman Quine, about whom we shall discuss next.

13.2   Quine

Quine stated in his work with Goodman [113] that he did not believe in the
existence of abstract objects and that he was not a proponent of realism.[18]
However, he later changed his views after his acquaintance with Rudolf Carnap.[19] We said in section 5.3 of Chapter 5 that for Carnap, metaphysical questions such as "Does there exist a set?" or "Does there exist a number?" were
non-sensical as they should instead be asked within a *language framework*.
Of course, *within* ZFC or Peano arithmetic, respectively, we can positively
answer these questions, since the axioms necessarily imply their existence.
However, according to Carnap, such questions concerning general metaphysical concepts of "existence" and "numbers", when asked independently of any
framework reference, are meaningless. After all, logical positivism back then
was passionately campaigning against metaphysics. According to Carnap [55],
unless we supply a clear cognitive interpretation for metaphysical objects or a
language framework for metaphysical terms, these external questions remain
as "pseudo-questions".[20]

Should we then adopt a language framework or not? For Carnap, this
depends on how the language framework would serve our purposes. He claims
that the acceptance of new linguistic forms can only be judged as being more
or less expedient, fruitful, conductive to the aim for which the language is
intended.[21] Quine argues for that the mathematical language must be adopted
on a pragmatic basis. Moreover, since the internal and external questions both
are based on pragmatic grounds, Quine claims that these two types cannot be
differentiated from each other. He writes:

---

[17] Russell, p. 573, 1907.
[18] Goodman and Quine, p. 105, 1947.
[19] See Quine's work [233] for details.
[20] Carnap, p. 245, 1950. .
[21] ibid, p. 250.

Ontological questions, under this view, are on a par with questions of natural science [...] Carnap has maintained that this is
a question not of matters of fact but of choosing a convenient
language form, a convenient conceptual scheme or framework for
science. With this I agree, but only on the proviso that the same
can be conceded regarding scientific hypotheses generally.[22]

The scientific language contains both mathematical language which includes mathematical terms, variables, logical symbols, and concepts of our
physical theories. Carnap separates the mathematical portion of the scientific language from the physical portion. Whilst he classifies mathematical
knowledge as analytic, he classifies statements of physics as synthetic. The
analytic/synthetic *indistinction* was famously argued by Quine in his celebrated paper *Two Dogmas of Empiricism*, which is widely regarded as one of
the most important works of the 20th century philosophy. From the viewpoint
of Quine's naturalistic philosophy, it is understood that philosophy is not prior
to science but they rather go hand in hand with each other, as 'the naturalistic
philosopher begins his reasoning within the inherited world theory as a going
concern'.[23] From this it follows that epistemology is investigated based on the
facts of natural sciences and so it is ultimately based on physics. As a matter
of fact, this line of thought characterises Quine's and Mill's views in a similar
kind. Both believe that *a priori* knowledge in the absolute independent sense
is non-existent.

Contrary to Carnap, Quine disagrees with the idea that analytic and synthetic judgments are distinct. He argues for that the language factors and
empirical factors are directly correlated with each other in the following way:

It is obvious that truth in general depends on both language
and extralinguistic fact. The statement 'Brutus killed Caesar'
would be false if the world had been different in certain ways,
but it would also be false if the word 'killed' happened rather
to have the sense of 'begat'. Hence the temptation to suppose
in general that the truth of a statement is somehow analysable
into a linguistic component and a factual component. Given this
supposition, it next seems reasonable that in some statements
the factual component should be null; and these are the analytic
statements. But, for all its a priori reasonableness, a boundary
between analytic and synthetic statements simply has not been
drawn. That there is such a distinction to be drawn at all is
an unempirical dogma of empiricists, a metaphysical article of
faith.[24]

The second dogma of empiricism that Quine rejects is *reductionism*, the

---

[22]Quine, p. 43, 1951.
[23]Quine, p. 72, 1981.
[24]Quine, p. 34, 1951.

belief that each meaningful statement is equivalent to some logical construct upon (simpler) terms which refer to immediate experience. Moreover, these terms are expected to be empirically verifiable.

In place of these dogmas, Quine [235] proposes that our scientific observations corresponds to some kind of *web of belief* where each node of the web represents a portion of knowledge in our system of scientific observations and that they may be connected to arbitrary number of other nodes. The connections between the nodes may be logical, some may be linguistic and some may be empirical. Empirical nodes are placed at the edge of the web. Whenever we observe a new information, it might affect the inner web and the connections through which the nodes are related might change until the web is settled. For Quine, 'science is a tool for predicting future experience in the light of past experience'.[25] Ultimately, what is relevant to our scientific theories is empirical evidence. However, for Quine, no scientific proposition is verified by itself. All propositions are subject to revision in the web, including mathematical ones. Quine writes:

> [...] our statements about the external world face the tribunal of sense experience not individually, but only as a corporate body. The totality of our so-called knowledge or beliefs [...] is a man-made fabric which impinges on experience only along the edges [...] Any statement can be held true come what may, if we make drastic enough adjustments elsewhere in the system [...] Conversely [...] no statement is immune to revision.[26]

The view what is known today as *holism* is the idea that no hypothesis can be justified on its own without considering other beliefs in the web. It is easy to see that holism is a natural consequence of the abandonment of first philosophy and as well as foundationalism.

The indistinction of analytic and synthetic statements was also studied in Quine's *Word and Object* [234]. His *indeterminacy of translation* was covered thoroughly in his book. For Quine, the justification behind the analytic and synthetic indistinction is the indirect reference to synthetic propositions when one grounds analyticity through interchangability of synonyms. Thus, in his theory, he strengthens the claim that there is no cleavage between analytic and synthetic propositions due to this circularity.

What can then we say about Quine's philosophy of mathematics? Clearly, Quine's views show similarity to Mill's philosophy of mathematics. On the one hand, as Mill didn't supply a sufficient epistemological background, his theory was somehow incomplete. On the other hand, Quine's idea of *web of belief* really offers an alternative to *a priori* knowledge.

For Quine, mathematics has a central place in the web of belief. Without mathematics, it is hard to imagine having a firm scientific theory. Quine

---

[25]ibid, p. 41.
[26]ibid, pp. 39-40.

accepts the existence of mathematical entities for the same reason that he accepts physical entities, which is to serve the web of belief or, in other terms, scientific knowledge. Not only does Quine believe that this criterion holds for mathematical entities, but to what extent we adopt a discipline is measured by how much it takes place in our knowledge web. Disciplines such as physics, chemistry, biology, and so on take a large part in the web of belief. For this very reason, we tend to believe in the existence of objects of natural sciences. However, mythology or mystical objects do not take any part in our scientific knowledge; thus, there is no reason to accept the existence of, say, mythological creatures like unicorns, dragons, or things like fairy tales and fictions.

The disciplines in the web of belief are not entirely co-independent. An adherent of holism would accept the hypotheses that these disciplines posit for their mutual relationship in the web. The idea that entities in our knowledge web are all codependent gives rise to ontological and epistemological realism in the *relative* sense. Objects of science exist due to their practical benefits in science. Since mathematical statements are for the most part heavily relied on the concepts of point, number, and set, they are assumed to exist independent of the language, mind and the sensory experience of the mathematician.

Recall that *nominalism* is the view which argues for that abstract objects do not exist. A nominalistic language, thus, is one that does not refer to abstract object whatsoever and that which does not quantify over such entities. Prominent supporters of nominalism include Hartry Field [91], Charles Chihara [59], and Mark Balaguer [17]. For a nominalist, everything that exists is concrete. We will not discuss in this chapter if science can be nominalised, as we shall leave this to Chapter 18. But we shall say a few things about Putnam's *Philosophy of Logic* [226] for reasons that will become clear in a moment. Putnam, although was in favour of adopting nominalism in the beginning, expressed that it would be impossible to "do" physics in nominalistic language.[27] By a *nominalistic language*, we mean a formalized language whose variables range over individual things, in some suitable sense, and whose predicate letters stand for adjectives and verbs applied to individual things, yet these individuals do not need to correspond to observable things. For Putnam, it is inevitable to use real numbers in the modelling of physical magnitudes like volume, force, mass, temperature, pressure, etc. According to Putnam, if the numericalisation of physical magnitudes is to make sense, we must accept such notions as function and real number.[28]

Let us also note that sometimes physical phenomena involve mathematical facts. Shapiro gives an interesting case regarding this matter which concerns the construction of a rectangular area from a given number of tiles where the number of tiles is prime.[29] If we are given n many tiles such that n is a prime number, we cannot form a rectangular area with the condition of using all the given tiles.

---

[27]Putnam, p. 37, 1971.
[28]ibid, p. 43.
[29]Shapiro, p. 217, 2000.

Acceptance of mathematical objects in Putnam's account relies on an argument which is used to support realism. The argument is called the *Quine-Putnam indispensability argument*, which goes as follows.

Premise 1: We should be committed to the existence of objects which are indispensable to our best scientific theories.
Premise 2: Mathematical objects are indispensable to our best scientific theories.
Conclusion: We should be committed to the existence of mathematical objects.

Thus, asserting the existence of mathematical objects is no different than asserting the existence of physical objects. They merely exist for the sole purpose of describing our best scientific theories.

Quine, as an empiricist, argues for that mathematics, on contrary to the popular belief, is not *a priori*. In fact, he does not believe in the existence of "pure" *a priori* knowledge.[30] Furthermore, he is a profound adherent of the dynamic nature of mathematics along with science. For Quine, the idea that mathematical knowledge is static and eternal is not credible. In his *Two Dogmas*, he writes:

> Revision even of the logical law of the excluded middle has been proposed as a means of simplifying quantum mechanics; and what difference is there in principle between such a shift and the shift whereby Kepler superseded Ptolemy, or Einstein Newton, or Darwin Aristotle?[31]

Needless to say that the Law of Excluded Middle has also been proposed so as to reform the mathematical method, i.e., to make logic more compatible with the constructivist ideology. From this point of view, mathematics is subject to reformation even for the sake of mathematics. To what extent it should be revised is a controversy. Although mathematics is not purely *a priori* for Quine and his supporters, it is more likely to be *a priori* in the relative sense. That is to say, once we fix the language and the entities, as long as no revision occurs in the knowledge web, we may say that mathematical knowledge remains "locally" *a priori*.[32]

As long as science requires a branch of mathematics or it requires to use a type of mathematical entity, Quine is willing to accept their existence. To state it inversely, if certain mathematical objects are dispensable for what he calls the *web of belief*, then there is no *a priori* reason to accept their existence.

---

[30]In fact, as a consequence of indispensability, mathematical knowledge was also regarded as *a posteriori*. See Colyvan [64] (2001), ch. 6.

[31]Quine, p. 40, 1951.

[32]I find Quine's views to be quite compatible with the pluralist conception of mathematics. As I will argue in Chapter 16, naturalistic philosophers would find pluralism rather a tempting philosophical position.

13.3 Maddy

Penelope Maddy is a leading figure among the contemporary philosophers of mathematics, who is known for her adaptation of the naturalist philosophy to the mathematical methodology. We should start with her earlier 1990 work *Realism in Mathematics* [183], and then move on to discuss her later works.

The earlier view that Maddy attempts to establish defines some kind of *compromised Platonism* which reconciles Quine's and Gödel's realism.[33] According to Maddy, there are two tiers in Gödel's epistemological theory, namely the *lower level* and the *higher level*.[34] Propositions that belong to the lower level are supposed to be justified based on mathematical intuition and intrinsic properties. The lower level statements are expected to be obvious and that they are intuited from experience due to their obviousness. These types of statements contain basic principles of logic and mathematics. Statements that belong to the upper level are those which are extrinsically justified via their usefulness in mathematics and natural sciences (see Figure 13.1). Maddy writes:

> [T]he higher, less intuitive, levels are justified by their consequences at lower, more intuitive, levels, just as physical unobservables are justified by their ability to systematise our experience of observables. At its more theoretical reaches, then, Gödel's mathematical realism is analogous to scientific realism. Thus Godel's Platonistic epistemology is two-tiered: the simpler concepts and axioms are justified intrinsically by their intuitiveness; more theoretical hypotheses are justified extrinsically, by their consequences.[35]

It has been a long standing epistemological problem of Platonism that how humans, as physical beings, get in contact with the knowledge of abstract mathematical entities. How do we, as human beings, acquire set theoretic or number theoretic knowledge? In Gödel's Platonistic epistemology, mathematical intuition plays a major role. Maddy, in her early work [183] attempted to replace the Kantian influenced Gödelian "intuition", for items located at the lower level, with a naturalistic epistemology by providing a scientific ground for the lower level intuition. Maddy concerns sets as objects to be scientifically justified since *set* is a primitive object, that all mathematical objects can be seen as collections of things. Although in her later works [188] and [189], she departed from realism—but not completely forsaken it—her earlier view was called *set theoretic realism*. Maddy claims that the mathematician, in fact, perceives macro scale entities as *sets of objects*. Essentially, she fills the gap

---

[33]Maddy, p. 35, 1990.
[34]ibid, p. 33.
[35]ibid, p. 33.

between "perception" and "intuition" with methods congruent with naturalism so that Gödel's the lower level "intuition" can be justified on scientific grounds. The connection is strongly tied to the concept of set containment. When we discussed Gödel's realism in the first section, we talked about neurophysiological cell-assemblies in humans that allow us to perceive physical objects. For Maddy, these assemblies are *object detectors* that 'bridge the gap between what is interacted with and what is perceived'.[36] Furthermore, she claims that we perceive sets in the same way we perceive physical objects via *set detectors*. She gives the following scenario:

> Steve needs two eggs for a certain recipe. The egg carton he takes from the refrigerator feels ominously light. He opens the carton and sees, to his relief, three eggs there. My claim is that Steve has perceived a set of three eggs. By the account of perception just canvassed, this requires that there be a set of three eggs in the carton, that Steve acquire perceptual beliefs about it, and that the set of eggs participate in the generation of these perceptual beliefs in the same way that my hand participates in the generation of my belief that there is a hand before me when I look at it in good light.[37]

For further elaboration, consider four crowns shown in Figure 13.2. Call this perceptual schema A. Consider now each crown in A perceived as set singletons. Call this case B. Although physically speaking, objects of A and B are the same, for Steve—in Maddy's hypothetical scenario—A and B are indeed different. In fact, Steve has the ability to perceive primitive objects of mathematics relying on his "set detectors", an assembly of functions in the human brain.

Any physical object can be perceived as a *set* containing that object. Thus, for Maddy, our set detectors allow us to perceive crowns as sets of crowns (or a set of crowns when taken all together). Moreover, we may arbitrarily iterate the concept of *set-containment* as set of crowns, set of set of crowns, set of set of set of crowns, etc., which leads us to perceive natural numbers in the set-theoretic domain.

One problem with this account arises when introducing infinite sets in our theory. The *cumulative hierarchy* of sets, i.e., the von Neumann hierarchy, is defined by transfinite induction as follows:

$$V_0 = \emptyset,$$
$$V_{\alpha+1} = \mathcal{P}(V_\alpha),$$
$$V_\alpha = \bigcup_{\beta < \alpha} V_\beta, \text{ if } \alpha \text{ is a limit ordinal.}$$
For all ordinals α, $V \doteq \bigcup_\alpha V_\alpha$.

---

[36] ibid, p. 58.
[37] ibid, p. 58.

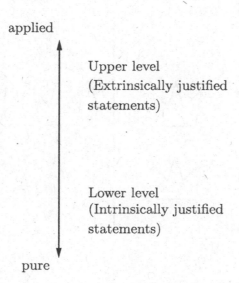

applied

Upper level
(Extrinsically justified
statements)

Lower level
(Intrinsically justified
statements)

pure

FIGURE 13.1
Depiction of mathematical knowledge in Gödel's and Maddy's two-tiered epistemological theory.

The set-theoretical universe V is then seen as a cumulative hierarchy of sets as depicted in Figure 13.3. The work domain of the modern mathematician is usually taken as V, although there has been an on-going debate on what V could actually be.[38] One possible problem in Maddy's epistemology concerns the question of how we perceive infinite sets. In fact, Maddy does not claim that we perceive infinite sets the same way we perceive *sets* of physical objects. The set theorist works with highly abstract objects including sets of size of uncountable cardinality. For Maddy, 'in fact, pure sets aren't really needed', and that the 'set-theoretic realist can locate all these sets she needs in space and time'.[39] In other words, pure sets are dispensable as they can be replaced by physically grounded objects, by which we mean sets that are ultimately based on *urelements*.[40] Recall that each level of Maddy's epistemological two-tiered knowledge supports one another. The existence of infinite sets is supported by the upper level knowledge in her epistemology. Thus, the justification behind their existence, possibly including the set of natural

---

[38]The structure $(V_\omega; \in)$ is, in fact, mathematically identical to $(\mathbb{N}; +, \cdot)$, i.e., the structure of arithmetic. V_ω is the set of all *hereditarily finite* sets. Number theory lies in V_ω.

[39]Maddy, pp. 156–157, 1990.

[40]This was motivated in the view what she calls *physicalistic Platonism*. See Maddy [184] (1990) for details.

A B

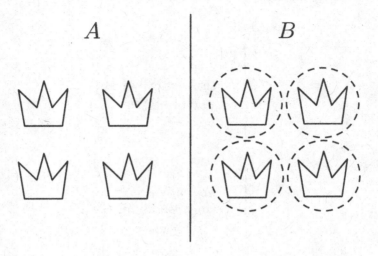

FIGURE 13.2
Distinction between crowns and sets of crowns.

numbers, is merely pragmatic within mathematics.[41] From a naturalistic point of view, infinite sets are adopted in mathematics for pragmatic and conventional reasons, e.g., applications in mathematics or in modelling physical phenomena, for instance *fluid dynamics* to analyse the behaviour of water by introducing divisible parts within so as to allow *arbitrarily* small amounts of fluxions.

We said that the Gödelian interpretation of mathematical intuition for the lower level epistemology was based on perceptions. Recall that the Gödelian account of mathematical knowledge takes mathematical statements as analytic, hence *a priori*. According to Maddy, one needs experience to form (mathematical) concepts, but once they are in place, no further experience is

---

[41] In a personal discussion with Penelope Maddy in April 2016 at UC Irvine, she asked me why I accept the existence of ω, i.e., the set of natural numbers. I think the reason she asked me this question was that she wanted to know if I accept ω on pragmatic grounds or solely based on its intrinsic properties. My answer was rather straightforward. I do not see ω as a completed totality. In fact, I do not need to consider ω as a totality to use it in mathematical theorems. For me, ω is a *name* that refers to a potentially infinite sequence of objects each of which has a unique successor. Moreover, the successor is an object of the same type as its predecessor, i.e., a natural number. Since the successor function is *closed* in natural numbers, the very fact that every number has a successor, defines a potentially infinite sequence. I can call this potentiality ω. Now the reason that ω is potentially infinite is entirely based on my perception that the structure of natural numbers has no greatest element due to that every number has a successor, which I think can be regarded as an intrinsic justification for the existence of infinite collections as *potentials*, but not as completed objects. As a matter of fact, the distinction is not so relevant when it comes to ordinary mathematical practice.

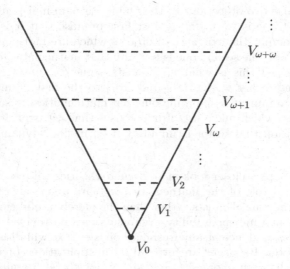

FIGURE 13.3

The set-theoretical universe, also known as the *von Neumann hierarchy*.

needed to produce intuitive beliefs; for Maddy, the support for the intuitive statements are called "impurely" *a priori*.[42]

Quine is widespread considered to be one of the most important philosophers in the American continent and that Maddy is indeed influenced by him. Just like Quine, Maddy too claims that the role of intuition in mathematics is weak and that intrinsic properties do not provide much justification for the naturalist. On the contrary, the naturalist is more inclined to take extrinsic justifications as reference. This doesn't mean naturalism omits intrinsic justifications of mathematical claims. But for Maddy, pure *a priori* knowledge does not exist. What leads us to impurely *a priori* knowledge is the presumption that we know by sensory experience that intuition is reliable.

13.3.1   The naturalist second philosopher

Maddy, just like Gödel, is an epistemological realist, which is the view that every mathematical statement has an absolute truth value independent of us. So from this perspective, adding axiom candidates to ZFC would settle the independent problems of set theory, in particular, the Continuum Hypothesis. Maddy later takes a slight turn, though not completely, from her set-theoretic realism towards mathematical naturalism. Her turn is due to the reliance on the Quine-Putnam indispensability argument in her earlier view.[43]

---

[42]Maddy, s. 74, 1990.

[43]See the preface of Maddy (2011), p. ix.

One can raise two objections to the Quine-Putnam indispensability argument. The first one is to object against first premise that it is, in fact, not necessary to commit the existence of objects which are indispensable for our scientific theories. The second one is to claim that mathematics is dispensable for science. Indeed this is what nominalists argue for. On the other hand, Maddy opposed in her works [185] and [187] to the first premise of the indispensability argument by pointing out that mathematics, to some extent, is used in theories which make use of hypotheses that are explicitly false, such as the assumption that water is infinitely deep in fluid dynamics.[44] Maddy says:

> But perhaps a closer look at particular theories will reveal that the actual role of the mathematics we care about always falls within the true elements rather than the merely useful elements; perhaps the indispensability arguments can be revived in this way. Alas, a glance at any freshman physics text will disappoint this notion. Its pages are littered with applications of mathematics that are expressly understood not to be literally true, e.g., the analysis of water waves by assuming the water to be infinitely deep or the treatment of matter as continuous in fluid dynamics or the representation of energy as a continuously varying quantity. Notice that this merely useful mathematics is still indispensable; without these (false) assumptions, the theory becomes unworkable.[45]

In her [187], she goes about to argue against a similar point. She writes:

> It would be foolish, however, to argue for the reality of the infinite simply because it appears in our best theory of water waves.[46]

Then, for Maddy, the Quine-Putnam indispensability argument does not apply to pure mathematics since it is has nothing to do with the applications in physical sciences. The assumption that the water is infinitely deep received a criticism from Colyvan [65]. He says:

> [...] there is nothing troubling here. The sonar theory is surely correct and the assumption of infinitely deep oceans is a mere idealisation.[47]

In *Naturalism in Mathematics* [188], Maddy extensively studies one of the open problems for the set-theoretic realist. She asks the question "How are set theoretic axioms to be judged?". As a matter of fact, the question that what

---

[44] Apparently the assumption that fluid dynamics is continuous is also called "the continuum hypothesis". See Tritton [286], pp. 58–51, 2008.

[45] Maddy, p. 281, 1992.

[46] Maddy, p. 254, 1995.

[47] Colyvan, p. 117, 2008.

justifies the axioms of set theory has been a long lasting theme in Maddy's works.

Maddy starts her investigation with what she calls the Wittgensteinian anti-philosophy.[48] By anti-philosophy, Maddy means 'a work done by professional philosophers that lacks the traditional theories, controversies, and the recommendations for reform'.[49] Wittgenstein's late anti-philosophy seems to have hints of naturalism in the primitive sense. In his *Philosophical Investigations* [297], he says that 'philosophy may in no way interfere with the actual use of language [...] It leaves everything as it is. It also leaves mathematics as it is'.[50] From Wittgenstein's anti-philosophy Maddy claims the victory of practice over philosophy by saying that 'in a conflict between philosophy and practive, practice wins—by the extreme measure of eliminating philosophy altogether [...] practice will grow in stronger, healthier directions'.[51]

How does one then picture the "practice within itself" which sets philosophy aside? Quine favoured Neurath's image of science, which is like 'a boat which, if we are to rebuild it, we must rebuild plank by plank while staying afloat in it. The philosopher and the scientist are in the same boat'.[52] Quine, in his [236], summarises this analogy as follows:

> The naturalistic philosopher begins his reasoning within the inherited world theory as a going concern. He tentatively believes all of it, but believes also that some unidentified portions are wrong. He tries to improve, clarify and understand the system from within. He is the busy sailor adrift on Neurath's boat.[53]

A major difference between the Quinean naturalism and Maddy's naturalism is Maddy's rejection of holism. For Quine, science—including mathematics—is not subject to supra-scientific tribunal. Maddy proposes a mathematical naturalism which extends to mathematical practice. She writes:

> To judge mathematical methods from any vantage-point outside mathematics, say from the vantage-point of physics, seems to me to run counter to the fundamental spirit that underlies all naturalism: the conviction that a successful enterprise, be it science

---

[48]The reader might have come to notice that the Wittgensteinian theme hasn't been covered much in our book. The first reason for this is that, I think many philosophers would appreciate the fact that it is not easy to interpret Wittgenstein's views. The investigation, thus, is left to the professionals who study his philosophy more seriously. Secondly, a central theme in Wittgenstein is the philosophy of language, a subject which if studied here would deviate from the main goal of this book. As Wittgenstein said in the Proposition 7 of his Tractatus [296], 'What we cannot speak about we must pass over in silence'. Hence, I should leave the experts who are more familiar with Wittgenstein to discuss the matter in more detail despite that, ironically, he is my academic great grandfather, i.e., the doctoral advisor of the advisor of my advisor.

[49]Maddy, pp. 163–164, 1997.

[50]Wittgenstein, §124, 1953.

[51]Maddy, p. 169, 1997.

[52]Quine, p. 3, 1960

[53]Quine, p. 72, 1975.

or mathematics, should be understood and evaluated on its own terms, that such an enterprise should not be subject to criticism from, and does not stand in need of support from, some external, supposedly higher point of view [...] Where Quine holds that science is 'not answerable to any supra-scientific tribunal, and not in need of any justification beyond observation and the hypothetico-deductive method', the mathematical naturalist adds that mathematics is not answerable to any extra-mathematical tribunal and not in need of any justification beyond proof and the axiomatic method. Where Quine takes science to be independent of first philosophy, my naturalist takes mathematics to be independent of both first philosophy and natural science [...].[54]

We said that naturalism rejects first philosophy, at least placing first philosophical principles before empirical data. Methodologically, Maddy's naturalistic philosopher is referred to as a 'second philosopher'. The second philosopher is in conflict with Descartes' meditator. Maddy, in her [189], says:

Where Descartes's meditator begins by rejecting science and common sense in the hope of founding them more firmly by philosophical means, our inquirer proceeds scientifically and attempts to answer even philosophical questions by appeal to its resources. For Descartes's meditator, philosophy comes first; for our inquirer, it comes second—hence 'Second Philosophy' as opposed to 'First'. Our Character now has a name: she is the Second Philosopher.[55]

The methodology of mathematics and science before the 19th century was not largely separated from each other. The developments in the mid-19th century, however, divided mathematics and science into two distinct disciplines. Pure mathematics became a study on its own for its own sake. One problem though the naturalist second philosopher faces is whether other pure studies should be treated in the same manner. Astrology, for instance, can be considered as a study for its own sake, pursuing its own goals. From the second philosopher's viewpoint, is pure astrology on a par with pure mathematics then? Comparing two disciplines, pure mathematics is obviously used for modelling scientific descriptions of the physical world. Whereas pure astrology has no connection to natural sciences. The second philosopher, as Maddy puts it, 'should confine her pursuit of pure mathematics to those portions most directly connected to natural science.'[56]

Maddy also discusses how her second philosopher would pursue to settle independent problems of set theory. In the case of CH, since ZFC has a model in which CH is true and a model in which ¬CH holds, it was mentioned that

---

[54]Maddy, p. 184, 1997.
[55]Maddy, pp. 18-19, 2007.
[56]ibid, p. 347.

it is no different than deciding whether a group is commutative or not. As a matter of fact, there are models for which groups are commutative and models for which they are not. For the methodology of pursuing the question about CH, Maddy says:

> [...] the Second Philosopher sees the appropriateness or inappropriateness of a given mathematical method as determined by its effectiveness toward the goals of the practice for which it is proposed [...].[57]

The second philosopher then transforms the problem into a problem of practical usefulness. Maddy continues to argue that there are axiom candidates that decides the truth value of CH. But we shall leave this discussion to Chapter 17, as it greatly concerns the problem whether we should confine mathematics to a fixed number of axioms or continually search for new axioms to be added in our formal system.

Discussion Questions

1. Do you think there is any way to eliminate intrinsic jusitifations in accepting the axioms without compromising on mathematical realism?

2. Can there be an intrinsic motivation in working with non-standard analysis, non-standard arithmetic, or ill-founded set theory? Do you think these subjects were invented for purely experimental reasons within the domain of pure mathematics, or do they contribute to other sciences as well?

3. For Quine, the existence of mathematical objects are on a par with the existence of, say, theoretically posited entities of physics, such as strings and quarks. It may be the case in the future that we will be able to experimentally observe these physical objects. But the same might not hold for mathematical objects. How can we explain this asymmetry?

4. In Quine's web of belief, does one really need to put mathematical objects in the centre of the web as far as ontological relativity goes? What do you think gives priority to mathematical objects when they are really ontologically on a par with objects of physical theories? How would the web be affected if we switched the priority between the two?

5. By which lines of argument would someone who does not believe in the existence of anything counter the indispensability argument?

6. Argue whether or not the indispensability argument is consistent with the idea that mathematical knowledge is purely *a priori*?

7. Compare the approach of a "first philosophy" mathematician and a "second philosopher" mathematician in terms of their mathematical practice?

---

[57]ibid, p. 352.

14

Structuralism

In this chapter, we will study another contemporary philosophical position called *structuralism*. Structuralism was partly emerged from the studies in the foundations of mathematics of the 20th century, predominantly concerning set theory and model theory. Proponents of structuralism include Paul Benacerraf [28], Michael Resnik [244], and Stewart Shapiro [259], all of whose views slightly differ from each other. On structuralism, mathematics is considered as a science that studies abstract structures. Readers who are familiar with abstract algebra or model theory will be able to associate with this philosophical position with such subjects.

14.1 Characteristic Properties

Majority of the structuralists are epistemological realists, in the sense that they believe mathematical statements have an absolute truth value independent of the mind, language, and sensory experience. However, they usually disagree on the ontology mathematical objects.

Recall that ontological realism (or Platonism in the ontological domain) asserts that mathematical objects exist independently of sensory experience, the mind, etc. On realism, numbers, for instance, just like physical objects, exist independently and irregardless of their relationship to each other. That is, every number is *ontologically independent* from any entity of mathematics. Each mathematical entity is an independently existing abstract object, as every object has its own "essence". The structuralist rejects all kinds of ontological independence. Numbers exist in so far as to the *relationship* with other numbers. Arithmetic by its own nature concerns the natural number structure. This structure bears a "natural" ordering by which any two entities in the structure are related. For example, the natural number structure contains a least element, that every element has a successor element in the structure, and *ipso facto*, it has no greatest element. The number "2", for instance, has no real essence on its own. What matters to the structuralist is the *characteristic property* of "2". In fact, the number "2" is the predecessor of the number "3" and the successor of the number "1". Thus, "2" merely denotes a *position* in structures that are isomorphic to the structure of

DOI: 10.1201/9781003223191-14

natural numbers. No number is ontologically independent from the others. Since "2" and "3" have this successor/predecessor relationship with one another, according to structuralism, they do not earn their own independent existence but they exist merely by their relationship to other entities of the structure.

In any axiomatic set theory course, it is taught that ordered pairs are defined in terms of unordered pairs satisfying a particular condition. More specifically, an *ordered pair* (x, y) is defined as an unordered pair $\{\{x\}, \{x, y\}\}$ satisfying the following condition:

$$(x, y) = (u, v) \text{ if and only if } x = u \text{ and } y = v.$$

In fact, the *characteristic property* written above is sufficient to define ordered pairs. As long as the ordered pair (x, y) satisfies the characteristic property, it is irrelevant what type of object the pair (x, y) amounts to.

What about of epistemology? Are numbers *epistemologically* independent? Shapiro [260] writes:

> To be sure, a child can learn much about the number 2 while knowing next to nothing about other numbers like 6 or 6,000,000. But this *epistemic* independence does not preclude an ontological link between the natural numbers. By analogy, one can know a great deal about a physical object, like a baseball, while knowing next to nothing about molecules and atoms. It does not follow that the baseball is ontologically independent of its molecules and atoms.[1]

We may find many exemplifications of the natural number structure. For instance, instead of using the numerals

$$1, 2, 3, \ldots$$

one could also describe the same structure with symbols

$$*, **, ***, \ldots.$$

The relation $1 + 2 = 3$ could have been described, for example, something along the lines of $\Delta(*, **) = ***$. While one representation uses numerals, the other could denote the objects of the *same* structure by $*$ or in terms of entirely different entities. The concept of *isomorphism* is a happily endorsed notion by structuralists. One could define a binary function Δ preserving the rules of addition and the characteristic property of each number so as to ensure that Δ is isomorphic to the addition function.

In abstract algebra, for instance, group theory does not concern a single system. It is about *any* "structure" $(G, \circ)$, where G is a set and $\circ : G \times G \to G$ is a binary function, which satisfies the following axioms:

---

[1]Shapiro, p. 258, 2000.

(i) For all $a, b \in G$, $a \circ b \in G$. (Closure)

(ii) For all $a, b, c \in G$, $a \circ (b \circ c) = (a \circ b) \circ c$. (Associativity)

(iii) There exists some $e \in G$ such that for all $a \in G$, $e \circ a = a = a \circ e$. (Identity)

(iv) For every $a \in G$, there exists some $b \in G$ such that $a \circ b = e = b \circ a$. (Inverse)

Since there may be multiple systems that satisfy the group axioms, a structure can be described by many systems. As we demonstrated, there are many systems that exemplify the natural number structure. The system we know from childhood using numbers and standard arithmetical operations is one and perhaps the foremost exemplification of the structure of naturals. Aside from the standard numerical system, the same structure can also be captured by a system of objects under some relation Δ which satisfies the characteristic properties, i.e., axioms of arithmetic. Each exemplification constitutes a separate system, yet describes the same structure. In Platonic terms, a structure can be thought of as the "Form" of all of its exemplifications. A structure is determined by its characteristic properties, such as the ordering of the elements, closureness, neighbourhood, and so on. A system, on the other hand, is an instance of the structure in a fixed language and with fixed entities for which the characteristic properties of the structure are satisfied.

A straightforward way of conceiving structures is by *abstraction*, that is, omitting all the attributes (or unit-properties). In a given system, we tend to comprehend the relationship between the entities by abstracting some of their unit-properties, subject to the condition of preserving the characteristic properties of the objects. For example, when we write on a piece of paper that "0 has no predecessor", we omit the irrelevant exclusive attributes such as the colour or the size of 0, as these properties does not change our understanding of the role of number 0 and its relation to other numbers in the natural number structure. In any system of natural number arithmetic though, we still keep in mind that 0 is the "least element" of the natural number structure with respect to the natural order. However, if we were to withdraw the characteristic property of 0 being the "first" element and the axiom of arithmetic $a \times 0 = 0$ for any given element a in the structure, then this would yield a completely different mathematical structure which is not isomorphic to the structure of naturals. By a similar line of though, if the characteristic property that the structure of naturals has no greatest element is changed in a way that a "greatest" element is now defined, then the outcome would be a finite structure. Numbers exists by their characteristic properties and their relation to other numbers. In fact, according to Resnik [244], numbers can be seen as "positions" in the structure. He writes:

> In mathematics, I claim, we do not have objects with an 'internal' composition arranged in structures, we only have structures. The

objects of mathematics, that is, the entities which our mathematical constants and quantifiers denote, are structureless points and
positions in structures. As positions in structures, they have no
identity or features outside a structure.[2]

Imagine a government bureau, for example, the police department. Suppose that each officer has a distinct duty. They all have junior and senior
officers related to them in some way. They are recognised in the department
by their duty and position, in relation to the others. Once they are retired
from the department, however, they are no longer recognised in the same manner as an officer, as they used to be. If the person X, say, is a cashier and
he retires, another person will take over his duty as a cashier. The department needs a cashier no matter which person is in charge in that position.
The cashier *position*, in this scenario, represents a "position" in the structure,
whereas the *cashier person* represents the object assigned for that position.
From this perspective, on structuralism, the number '2' cannot exist independent of the other numbers and thus, it cannot be conceived outside the natural
number structure as an independent entity. This, in turn, leads to ontological dependency. Of course, the ontological dependency of numbers does not
guarantee their existence, as nominalists would claim that they simply do not
exist.[3] The ontological dependency just ensures that *if* numbers exist, then
their epistemology would need to be dependent on each other and they would
be conceived via to their characteristic properties.

Although we study nominalism in Chapter 18 more thoroughly, it is worth
noting one point that the structuralist is really not concerned about, that
is, the nominalist distinction between *practice* and *theory*. Hartry Field [91]
attempted to nominalistically reconstruct Newton's theory of gravity. Despite
the fact that Field does not believe in the existence of mathematical entities,
he maintains that the space contains infinitely many points. Furthermore, for
Field, points in space are *concrete*. One objection Field considers is that there
doesn't seem to be a significant difference between postulating such a rich
physical space and postulating the irrational numbers. Field writes:

> The nominalistic objection to using real numbers was not on the
> grounds of their uncountability or of the structural assumptions
> (e.g. Cauchy completeness) typically made about them. Rather,
> the objection was to their abstractness: even postulating one real
> number would have been a violation of nominalism as I'm con-
> ceiving it. Conversely, postulating uncountably many physical
> entities [...] is not an objection to nominalism; nor does it become
> any more objectionable when one postulates that these physical

---

[2]Resnik, p. 530, 1981.
[3]Recall the logical fallacy of circularity that we discussed in section 2.2.5 of Chapter 2.

entities obey structural assumptions analogous to the ones that platonists postulate for the real numbers.[4]

Concerning this distinction, Shapiro [260] replies as follows:

> The structuralist demurs from this distinction. For her, a real number *is* a position in the real number structure. It makes no sense to 'postulate one real number' since each real number is a part of a large structure.[5]

We can ask two interrelated questions concerning the ontology of structuralism. The first question is whether structures really exist as entities. The second question on the other hand concerns the existence of mathematical objects and their position in the structures. Since there may be many systems that describe a structure, as Shapiro puts it, structures are 'one-over-many' entities.[6] A natural number structure, as we said earlier, can be exemplified by many systems. Shapiro immediately makes a platonic reference to *forms*. Platonic forms are 'one-over-many' beings in the same way as structures. The *universal* "tree Form" may have many *particular* tree instances. Though, there is a unique property that all particulars bear, which is the property of *treeness*. A similar notion arose in section 6.1 of Chapter 6 when we introduced *types* and *tokens* in type formalism. While tokens are physically inscribed symbols, types are universal properties that every instantiation possesses. The string 10001, for instance, contains five tokens but two types. On ontological realism, even if we discarded physical tokens, types would still remain to exist in the Platonic universe independently on their own.

A variation of Platonism, regarding the metaphysics of structures, is known as *ante rem realism*. On this view, structures exist independent of the mathematician and even independent of whether or not the structure is exemplified in the mathematical realm. On this view, structures are prior to their instantiations. For example, the natural number structure exists prior to—and independent of—any of its exemplifications. The semantics is rather straightforward: the variables range over the "places" in the structure. For example, in the equation $2 + 2 = 4$, the numerals "2" and "4" are merely singular terms (proper names) that denote certain places in the natural number structure. The equation then tells us a certain fact about that structure. On ante rem structuralism though, any structure is prior to singular terms. The natural number structure is ontologically prior to any number.

Unlike Plato's realism, Aristotelian realism claims that Platonic forms are dependent on physical objects. Forms manifest themselves *within* physical objects. The Aristotelian realism can be as well interpreted for the metaphysics of structures. On *in re realism*, structures are assumed to exist dependent on

---

[4]Field, p. 31, 1980. In the second edition of *Science Without Numbers*, Field notes that postulating just one real number so as to violate nominalism was intended to be a joke.

[5]Shapiro, p. 260, 2000.

[6]ibid, p. 261.

the instantiations and, in fact, systems are prior to structures. On this view, if we destroyed all natural number systems, we would have destroyed the natural number structure.[7] Since particulars are ontologically prior to universals on in re realism, systems can exist independent of the structures. Apart from the two realist views, one can also have an idealistic metaphysical interpretation of structures, claiming that structures are merely the product of human mind. One can even go one step further, as we shall discuss shortly, to claim that structures do not exist. Whether structures exist independent of the systems that exemplify them, though, depends on our preference between *ante rem* and *in re* realisms.

14.2   Identification Problem

For an ante rem structuralist, a natural number, say, is a proper name which denotes a specific place on the natural number structure. We said that there might be many systems that exemplify the natural number structure. On ante rem structuralism though, the natural number structure is unique and it exists independent of any of their instances.[8] In the game of chess, for example, every piece has its own movement rule. The piece which only moves horizontally or vertically, on any number of unoccupied squares is called the *rook*. The rook may also participate with the king in a special action called short or long *castling*, a strategic move to protect the king. Now there may be many chess-like games, having the exact same rules as chess except played on a different platform with different objects. One may invent a game played with leaves or sea shells. While one person plays the game of chess on the chessboard, the other person may play the same game with leaves on the branches of the tree. The only requirement is to preserve the rules of the game and the neighbourhoodness of squares of the chessboard. Despite that the instance of the game of chess, when played on a chessboard and with chess pieces, seem to perfectly exemplify the game of chess, the other exemplification can be another instance another instantiation of chess, where the leaves take the place of chess pieces and the tree branches take the place of the squares in the chessboard. From the ante rem point of view, the game of chess is ontologically prior to the chess pieces.

As tempting as it may seem to adopt ante rem realism, we are faced with an *identification problem* concerning the characterisation of objects. By

---

[7]We encourage the reader to argue whether or not Aristotelian realism is time independent. If a table exists at time t, then we simultaneously have the "table form". If we destroy all the tables at a later time, does that mean the table form ceases to exist? Can we have a time-independent version of Aristotelian realism?

[8]It is worth noting that it is the *structures* that are claimed to exist independently, not numbers, functions, or sets. The only abstract objects that exist are the natural number structure, the real number structure, the Euclidean geometry, etc.

indentification problem, we mean the following concern. Although Platonists can easily define natural numbers, they are not obliged to describe what they really are. For a Platonist, a number is basically what it *is* on its own. It exists independent of any sensory experience, the mind, the language, and so on. As we mentioned in Chapter 5, Frege derived in his *Grundlagen*, before even Basic Law V, an abstraction principle what is known as *Hume's principle*, as a formal statement of second-order logic. He used things like "the number of the concept F is x". Hume's principle fails to provide an exact characterisation of what a number is or is not. A similar problem in set theory occurs in the definition of equipollency of two sets. We say that A and B have the same *cardinality* if and only if there exists a one-to-one correspondence between them. Setting aside the "cardinality" of a set, this does not determine or characterise what a "cardinal number" is. It cannot be determined from Hume's principle, for instance, what the number "2" is equal to or what it is not and *ipso facto*, it is undetermined whether the number "2" is same as or different from any other object. As a matter of fact, with Hume's principle alone, we cannot maintain that numbers are independent "objects". How do we ensure, for instance, that the number "2" is same as or different from Julius Caesar? Frege's earlier account does not determine the truth value of such inequalities. This is called the *Julius Caesar problem*. Concerning this issue, Frege writes:

> [...] we can never —to take a crude example— decide by means of our definitions whether any concept has the number Julius Caesar belonging to it, or whether that same familiar conqueror of Gaul is a number or is not. Moreover we cannot by the aid of our suggested definitions prove that, if the number a belongs to the concept F and the number b belongs to the same concept, then necessarily $a = b$. Thus we should be unable to justify the expression "*the* number which belongs to the concept F", and therefore should find it impossible in general to prove a numerical identity, since we should be quite unable to achieve a determinate number. It is only an illusion that we have defined 0 and 1; in reality we have only fixed the sense of the phrases
>
> "the number 0 belongs to"
> "the number 1 belongs to";
>
> but we have no authority to pick out the 0 and 1 here as self-subsistent objects that can be recognised as the same again.[9]

Paul Benacerraf is considered as one of the earliest proponents of structuralism for his classic article titled *What numbers could not be* [28]. Benacerraf

---

[9]Frege, §56, 1884.

presented an argument against set-theoretic Platonism to point out a similar problem to that of Frege's. Benacerraf's identification problem stems from a fundamental dilemma concerning the reduction of natural numbers to sets. It is known that all mathematical objects can be in principle reduced to pure sets. Recall from Chapter 2, the identification of natural numbers in terms of ordinals. As a matter of fact, in set theory, we can represent natural numbers in at least two different ordinal notations, namely the Zermelo notation, introduced by Ernst Zermelo, and the notation introduced by John von Neumann. Benacerraf first considers the Zermelo notation of natural numbers in the set-theoretic domain.

$$0 = \emptyset$$
$$1 = \{\emptyset\}$$
$$2 = \{\{\emptyset\}\}$$
$$3 = \{\{\{\emptyset\}\}\}$$
$$\vdots$$
$$n + 1 = \{n\},$$
$$\vdots$$

Based on the Zermelo notation, the set-theoretic representation of each number, except 0, is a singleton. The representation of the n, in this case, is obtained by iterating the set containment, beginning with the empty set, n number of times. On the other hand, the von Neumann representation of natural numbers is as follows:

$$0 = \emptyset$$
$$1 = \{\emptyset\}$$
$$2 = \{\emptyset, \{\emptyset\}\}$$
$$3 = \{\emptyset, \{\emptyset\}, \{\emptyset, \{\emptyset\}\}\}$$
$$\vdots$$
$$n + 1 = n \cup \{n\}$$
$$\vdots$$

In the von Neumann notation, every number n has n many elements. The problem is to identify which one of these gives the correct account of numbers? On Platonism, since objects exist independent of anything, there must be a correct identification for each number. As we can identify numbers with sets, for a set-theoretic Platonist, both accounts cannot be correct simultaneously. What is then the number 2? Is it $\{\{\emptyset\}\}$ or $\{\emptyset, \{\emptyset\}\}$? The former set-identification of 2 contains a single element, whilst the latter representation has two elements. This is the dilemma that Benacerraf presents.[10] In

---

[10]See Balaguer [16] (1998) for an argument that this multiple reduction problem does not actually cause any problem for the Platonist.

the last part of his paper, he argues for structuralism as a way out from this dilemma. He writes:

> [...] numbers are not objects at all, because in giving the properties (that is, necessary and sufficient) of numbers you merely characterise an *abstract structure* – and the distinction lies in the fact that the "elements" of the structure have no properties other than those relating them to other "elements" of the same structure [...] To *be* the number 3 is no more and no less than to be preceded by 2, 1, and possibly 0 [...] *Any* object can *play the role of* 3; that is, any object can be the third element in some progression.[11]

Such epistemological questions are not relevant to the structuralist. The structuralist is rather interested in the progression that the objects exhibit. Instead of asking what a particular object *is*, regardless of how it is represented in the set-theoretic language, the structuralist asks how the object *relates* to the others in the "pattern" they all take place. What a number could or could not be only makes sense in the structure of arithmetic. Thus, any arithmetical statement is internal to the structure of natural numbers. The question whether or not 2 is less than 5 is something to be answered within the aforementioned structure for the fact that the binary relation "less than" is thought to be an intrinsic relation between any two numbers in the structure of naturals. On the other hand, a statement such as "2 is darker than 4" is absurd and has no answer in the natural number structure.[12] The structuralist would accept any of the set-theoretic identification of "2", whether it is the Zermelo notation or the von Neumann notation. Replacing "2" with another object which preserves the characteristic properties of it would still suffice to preserve the natural number structure.

It may naturally be thought that there is a distinction between an *object* and its *place* in the structure. What is essentially a "place" in a structure? Is it an "object" or is it merely a "characterisation" for an object to be *placed* in the structure? Let us go back to our example in the previous subsection about the police department. In this case, there are offices and office holders. According to Shapiro, to maintain that numbers, sets, and points are *objects*, the ante rem structuralist invokes a distinction in linguistic practice which is explained by two different orientations. Shapiro says:

> Sometimes the places of a structure are treated in the context of one or more *systems* that exemplify the structure [...] we treat each position of a structure in terms of the objects or people that occupy the positions. Call this the *places-are-offices* perspective.

---

[11] Benacerraf, p. 70, 1965. Also in Benacerraf and Putnam (1983), p. 291.

[12] Even so that numbers are colourless entities, we encourage the reader to argue about whether it would be possible to define a natural number system based on the *degrees of shades* of a fixed colour, in place of *quantities*.

> So construed, the positions of a structure are more like properties than objects.[13]

He continues to claim that this orientation presupposes a background ontology that supplies objects to be filled in places of the structure. For example, if the natural number structure is defined in terms of sets, then sets will form the background ontology of arithmetic. In contrast to the first orientation, Shapiro says that 'there are contexts in which the places of a given structure are treated as *objects* in their own right', and he calls this *places-are-objects* perspective.[14] In this sense, every "place" to which the number 2, 3 or, say, the number 1000 corresponds is regarded as an object. The term *place* needs a little explanation for clarity. The natural number structure can be thought as a collection of infinitely many boxes each of which can contain an object that preserves the characteristic properties of the usual arithmetical progression. Each box is a place that admits the characteristic property of the specific number it corresponds to. In fact, we can view each box as a variable for number-like objects. Each *place*, from the perspective of the second orientation, is considered to be an object. From this point of view, arithmetic is a mathematical discipline which treats the natural number structure in the way that the places in the structures are objects and the domain of discourse consists of these "places". For instance, a statement like $7 + 5 = 12$ is treated so as to ensure that

"any object which takes places of the number 7 and any object which takes place of the number 5, when applied on an operator which takes place of the addition, is equal to the object which takes place of the number 12".(*)

So the numbers in the equation are merely variables for the objects that would take the role of those numbers. In this sense, the entire natural number structure consists of *places* in which any two of them are related by the notion of being a "successor" or "predecessor".

14.3　Eliminative Structuralism

Ante rem and in re views of structuralism correspond to, respectively, the Platonistic and Aristotelian accounts of structuralist ontology. On ante rem structuralism, from the *places-are-objects* perspective, it is presupposed that each term in a given equation like $7 + 5 = 12$ is a proper name. According to Shapiro, however, adherents of in re structuralism 'might hold that places-are-objects statements are no more than a convenient rephrasing of corresponding

---

[13]Shapiro, p. 268, 2000.
[14]ibid, p. 268.

generalisations over systems that exemplify the structure in question'.[15] Anyone who favours an in re account of structures will hold that if S is a system that exemplifies the natural number structure, then the equation $7 + 5 = 12$ is interpreted in S as (*). That is, suppose $+_S$ corresponds to the addition function in S, and 7_S, 5_S, 12_S correspond in S, respectively, to places of the numbers 7, 5, 12 of the natural number structure. Then, the equation $7 + 5 = 12$ would amount to $7_S +_S 5_S = 12_S$. The consequence of interpreting the equation in this manner is explained by Shapiro as follows:

> When paraphrased like this, seemingly bold ontological claims lose their teeth. For example, the sentence '3 exists' comes to 'every natural number system has an object in its 3-place', and 'numbers exist' comes to 'every natural number system has objects in its places'.[16]

In fact, phrasing the statements as generalisation in the form of "If S is a system that exemplifies the structure of natural numbers, then S has property φ" leads us to another form of structuralism introduced by Charles Parsons, in his work [208], called *eliminative structuralism*. He characterises this view as which 'avoids singling out any one simply infinite system (i.e., inductive set) as the natural numbers [...]'.[17] He continues as follows:

> [...] (eliminative structuralism) is a very natural response to the considerations on which a structuralist view is based, to see statements about a kind of mathematical objects as general statements about structures of a certain type and to look for a way of eliminating reference to mathematical objects of the kind in question by means of this idea.[18]

A version of eliminative structuralism is defended by Paul Benacerraf. He claims that 'number theory is the elaboration of the properties of all structures of the order type of the numbers'.[19] Benacerraf is, in fact, on the nominalistic side of structuralism, as the closing statement in his paper writes:

> They think that numbers are really sets of sets while, if the truth be known, there are no such things as numbers; which is not to say that there are not at least two prime numbers between 15 and 20.[20]

One challenge we encounter in eliminative structuralism is the requirement

---

[15]ibid, p. 270.
[16]ibid, p. 271.
[17]Parsons, p. 307, 1990. Parenthesis added.
[18]ibid, p. 307. Parenthesis added.
[19]Benacerraf, p. 291, 1965.
[20]ibid, p. 294.

for having a background ontology. The nature of objects in the ontological universe does not matter though as long as there are sufficiently many objects. Given a statement φ of arithmetic, eliminative structuralism treats it as "if S is a system that exemplifies the natural number structure, then φ_S", where φ_S is the statement in S that corresponds to φ in terms of the objects of S. The successor function and the constant symbols in the natural number structure must all have correspondences in S. The result of replacing the "standard" arithmetical terms in φ with the terms in S will yield φ_S. Now any system that exemplifies the structure of naturals must contain infinitely many objects. For if it contained finitely many entities, it would not be able to exemplify the natural number structure for the fact that the finite system will satisfy the statement "There exists a maximal element", which is essentially false in the natural number structure. If our background ontology contained only finitely objects, then no system constructed from this ontology would be able to exemplify the natural number system. By the truth table definition of implication ($\rightarrow$), thus, regardless of what φ tells us, φ_S would always be true since the hypothesis that "S is a system that exemplifies the natural number structure" is false. Hence, to have a reasonable account of eliminative structuralism, the background ontology should contain infinitely objects. Not only that but the relationship between a sufficient amount of objects must suit the "places" of the natural number structure. Alas, arithmetic is just a small part of mathematics. The number of objects in the background ontology needs to be uncountable for the real number structure. Similarly for the Euclidean geometry. For set theory, one even needs a larger background ontology considering the fact that ZFC proves the existence of cardinalities larger than that of the set of real numbers. In fact, we cannot have a system of set theory and describe the $\in$-relationship between the ordinals relying on a background ontology containing less objects than the size of the class of all ordinals, assuming that we accept all axioms of ZFC. Another question, then, for an eliminative structuralist to ask herself is how large exactly our ontological background should be. This, of course, depends on what structures of mathematics we want to exemplify. We at least know for certain that the background ontology should not be finite. A simple answer would be that a background ontology of the size of the class of all ordinals actually suffices. Since ZFC formalises all ordinary mathematics, it is sufficient for an eliminative structuralist to adopt the aforementioned background ontology of objects. As a matter of fact, due to the Löweheim-Skolem Theorem, it suffices to have a countable universe of objects; since if ZFC has a transitive model, then it has a countable model. There is already no point in having more than countable infinitely many objects in the background ontology to exemplify the structure of natural numbers alone. The existence of ordinals greater than ω does not affect systems of arithmetic. In Shapiro [260], for each legitimate field, the assumption that 'there are enough objects to keep that field from being vacuous' is called *ontological choice*, and the structuralist view accordingly is called *ontological eliminative structuralism.*[21] The effect of Skolem's paradox is particularly puzzling in this case. If ZFC is a foundational theory of mathematics and that the Löwenheim-Skolem Theorem allows us to have countable models of set theory—even so that there are uncountably many "places" in the real number structure—what is the sufficient amount of objects we need to exemplify the real number structure? It turns out that this depends on whether we wish to have a system *within* ZFC or not.

On the ontological eliminative structuralism, the background ontology should not be understood in structuralist terms, as Shapiro puts it. He says:

> If the set-theoretic hierarchy is the background, then set theory is not [...] the theory of a particular structure. [...] Perhaps from a different point of view, set theory can be thought of as the study of a particular structure U, but this would require another background ontology to fill the places of U.[22]

Thus, the never-ending need for having a background ontology leads to an infinite regress. Ultimately, the ontological eliminative structuralist is bound to accept the lack of structure for the beginning ontological background.

We said that Benacerraf was more on the nominalistic side of structuralism. Indeed, Platonistic or Aristotelian realisms are not the only positions for the interpretation of the background ontology. A structuralist who supports nominalism is required to abandon the idea of the necessity of having a background ontology. From this point of view, statements asserted by a nominalistic structuralist are merely hypothetical, that the assertions are grounded on the *possibility* of the existence of structures rather than their *actual* existence. The difference between the two standpoints is critical. Even so that it does not quite affect the mathematician in practice, the philosopher usually separates the actual and potential existence of objects. Given a statement φ about natural numbers, a nominalistic structuralist then treats φ as

"If S is a system of arithmetic that could potentially exist, whenever S exemplifies the natural number structure, then φ_S".

The challenge for the nominalists is, of course, to make sense of the progressions and ensure the soundness of the assertions made about arithmetic, geometry, or set theory without assuming the existence of such structures.

Hellman [141] introduced this type of a structuralist philosophy called *modal structuralism* in which he invoked the *possibility* and *necessity* operators of modal logic. Modal logic is a formal logic system which includes operators expressing modalities, namely *possibly* and *necessary*. For example, the statement

---

[21] Shapiro, p. 272, 2000.
[22] ibid, p. 273.

"For every natural number n, there exists a number m with property P"

can be interpreted in two ways in modal logic. It is one thing to say

"For every natural number n, there necessarily exists a number m with
property P".

It is another thing to say

"For every natural number n, there possibly exists a number m with
property P".

Statements in modal structuralism consist of assertions not referring to structures that necessarily exist, but to structures that possibly exist. Shapiro argues that the model operators are not to be understood as metaphysical possibility, as for him, necessary and possible existence are equivalent. He writes:

> Most proponents and opponents of the existence of mathematical objects agree that 'the natural numbers exist' is equivalent to both 'possibly, the natural numbers exist' and 'necessarily, the natural numbers exist' [...] Thus, the existence and the possible existence of the items in the background ontology are equivalent.[23]

One fundamental question we can ask about modal structuralism is how we interpret taking *arbitrary* subsets of infinite sets. Hellman says that both modal structuralism and set-theoretic platonism accept as "meaningful" the highly non-constructive notion of "all subsets of an infinite set" in one form or another.[24]

Discussion Questions

1. On structuralism, the number "1" has the characteristic property of being the unique element between "0" and "2" in the structure of natural number arithmetic. In the structure of reals, it has additional properties such as being the limit of the partial sums

$$\sum_{n=1}^{\infty} \frac{1}{2^n}.$$

How does the structuralist make sense of the relationship between the number "1" in the natural number structure, and the number "1" in the real number structure? Does the same "place" belong to many structures?

---

[23] ibid, p. 274.
[24] Hellman, p. 32, 1989.

2. In relation to the first question, we know that there can be two structures in which the same object satisfies the characteristic property φ in one structure and satisfies $\neg\varphi$ in the other structure. Does this mean that there can be no such thing as "absolute characteristic property" and that properties are merely relative conditions which depend on the structure in consideration?

3. Due to Löwenheim-Skolem Theorem, if set theory has a model, then it has a countable model. Does this mean that our ontological choice for the ontological background of set theory suffices to be of size countable infinite? In other words, how might we settle Skolem's paradox in ontological eliminative structuralism?

4. Consider the following sentences.

    (a) If there is a natural number structure that could potentially exist, then 0 is necessarily the least element.

    (b) For any structure that exemplifies the natural number structure, 0 is potentially the least element.

    Compare (a) and (b) from ontological and epistemological viewpoints.

15

Yablo's Paradox

Many paradoxes in pure logic are believed to be caused by certain kind of circularity. Stephen Yablo [306], however, proposed a very interesting paradox which is not self-referential or circular in any way. In this chapter, we give arguments for strengthening Hardy's [136] and Ketland's [163] claims surrounding the self-referential nature of Yablo's paradox. We first begin by pointing out that Priest's [223] construction of the binary satisfaction relation in revealing a fixed point relies on impredicative definitions. We then show that Yablo's paradox is "ω-circular", based on ω-inconsistent theories. We argue that Yablo's paradox is not self-referential in the classical sense but it rather admits circularity at the least transfinite countable ordinal. In this way, we demonstrate that Yablo's paradox is circular in the *limit* sense. Hence, we aim to improve Hardy's and Ketland's theses for the ω-inconsistency of Yablo's paradox and present a compromise solution of the problem emerging from Yablo's and Priest's conflicting claims. Throughout our discussion, the term "Yablo's paradox" may occasionally be used, depending on the context, to refer to the set of sentences or the contradiction that arises from these set of sentences.

15.1 Self-reference and Impredicative Definitions

As stated in Yablo's original paper, since Russell, the reason behind all paradoxes in logic has been traditionally believed to be caused by self-reference. Clearly, in the liar paradox for instance, the sentence "This statement is false" is self-referential and hence, it is circular. This first type of circularity involves only a single sentence. The barber paradox is another famous example. There is a barber in the town who shaves every people who don't shave themselves. But then who will shave the barber? One may also consider the following: "The next statement is true. The previous statement is false". This is yet another example of a liar-like paradox. The last example gives us the second type of circularity involving more than one sentence. Many of the paradoxes that we know belong to one of these two types classes and self-reference has

This chapter is based on my paper [72], which was published in *Logic and Logical Philosophy*. I would like to thank the editors for kind permission to reprint.

usually been taken as the main reason behind them. Nevertheless, not every self-referential statement leads to a paradox. For example, there is nothing wrong with the proposition "This statement is true" for the obvious reason that the statement may just be taken as a true proposition.

It is usually *impredicativity* that allows one to construct self-referential propositions. Although we discussed impredicativity in earlier chapters, it is worth to remind the reader what it is. Roughly speaking, a definition is called *impredicative* if the object being defined is referred in a totality containing the object. Paul Cohen [63] gives a nice example regarding this.[1] Let us suppose that we accept integers and the operations for arithmetic. We can consider a set of integers defined by a property $P(n)$ which has all its bound variables ranging over integers. Such set is said to be *predicatively* defined in terms of the integers. Consider now the following property. Let us say that an integer n satisfies $P(n)$ if there is a partition of the set of integers into n disjoint sets none of which contains arithmetic progressions of arbitrary length. Let S be the set of all n such that $P(n)$ holds. Now in order to determine whether any integer, say 5, belongs to S, we have to consider *all* partitions of the set of integers into 5 sets possibly including partitions in which S occurs. Therefore, in order to define S we are forced to assume a meaningful totality of the set of all sets of integers, i.e., the power set of the set of integers. This is an example to an impredicative definition.[2]

Impredicativity was after all the main problem behind Russell's paradox when Frege's abstraction rule called *Basic Law V* was used in the language of naive set theory. It was Whitehead and Russell who later proposed *type theory* on a predicative basis in *Principia Mathematica* to overcome these problems caused by impredicativity.

Frege's *Basic Laws of Arithmetic* formally allows one to construct impredicative definitions when working with naive set theory. Let us revisit Basic Law V solely by using *concepts* and *extensions*. We may think of *concepts* as unsaturated predicates, i.e., functions. The *extension* of a concept is the collection of all objects falling under that concept. For instance, if we let K be the concept "x is an integer which satsfies $x^2 = 4$", then the objects which fall under K are 2 and -2. So the extension of K is the collection $\{2, -2\}$. Let us denote an object a falling under a concept F by $F(a)$, and let $Ext(F)$ denote the extension of F. Basic Law V can be summarized as follows: Let F and G be two concepts. The extension of F is equivalent (or equinumerous) to the extension of G if and only if $F(a)$ is true if and only if $G(a)$ is true. That is,

$$Ext(F) = Ext(G) \text{ if and only if } \forall a(F(a) \leftrightarrow G(a)).$$

It is a known fact that Basic Law V, when applied to the Axiom Schema of

---

[1] Cohen, p. 86, 1966.

[2] For another example, see the definition of greatest lower bound we gave in section 4.2 of Chapter 4.

(unrestricted) Comprehension in naive set theory, leads to Russell's paradox. The reason why we have a paradoxical class $R = \{x : x \notin x\}$ is due to the assumption that the extension of R *includes* the class R. Any predicative definition, however, bans the use of such assumptions. We leave our discussion of predicativity here and will refer to it later when needed. The reader may refer to Feferman [86] for a rich discussion on predicativism.

15.2 What is Yablo's Paradox?

According to Stephen Yablo, in defiance of the traditional view, paradoxes without self-reference do exist. If a proposition A is true, let us denote this by $T(A)$. *Yablo's paradox*, and it would perhaps be fairer to refer to it as the *Visser-Yablo paradox* due to that a very similar antimony occurs in Visser [289], is an infinite conjuction of statements $S_1, S_2, S_3, \ldots$ defined as follows:

$$S_1 : \forall k > 1 \ \neg T(S_k),$$
$$S_2 : \forall k > 2 \ \neg T(S_k),$$
$$S_3 : \forall k > 3 \ \neg T(S_k),$$
$$\vdots$$

Of course, we may view these statements as an infinite sequence $\{S_i\}$. We shall use this notation later on. Now let us assume that $T(S_n)$ holds for some n. Then $\forall k > n \ \neg T(S_k)$. Therefore, in particular, $\neg T(S_{n+1})$ holds. But if $\forall k > n \ \neg T(S_k)$, then

$$\forall k > n + 1 \ \neg T(S_k) \tag{*}$$

holds. But then (*) really says the same thing as S_{n+1}. Then $T(S_{n+1})$ must hold. A contradiction. So our assumption that $T(S_n)$ holds cannot be true. That is, $\neg T(S_n)$ must be true. Moreover, since n is arbitrary, S_n must be false for all n. Then, every statement that proceeds S_n must be false as well. Therefore, it must be the case that $T(S_n)$—since S_n actually says that any statement that proceeds itself is false. Again, a contradiction. Therefore, we cannot assign any truth value to any of S_n. At the end of his paper, Yablo states:

> I conclude that self-reference is neither necessary nor sufficient for Liar-like paradox.[3]

---

[3]Yablo, p. 252, 1993.

15.3　Priest's Inclosure Schema and ω-inconsistency

We shall now look at Priest's criticism of Yablo's claim that his paradox does not quite rely on self-reference. Priest claims that all liar-like paradoxes contain a *fixed point*. He writes:

> To put the discussion into context, think, first, of the standard Liar paradox, 'This sentence is not true'. Writing T as the truth predicate, then the Liar sentence is one, t, such that $t = \text{'}\neg Tt\text{'}$. The fact that '$t$' occurs on both sides of the equation, makes it a fixed point of a certain kind, and, in this context, codes the self-reference.[4]

Priest objects, in Yablo's proof, to the derivation

$$T(S_n) \rightarrow \forall k > n \; \neg T(S_k) \tag{†}$$

He writes:

> What is their justification? It is natural to suppose that this is the T-schema, but it is not. The n involved in each step of the reductio argument is a free variable, since we apply universal generalization to it a little later; and the T-schema applies only to sentences, not to things, with free variables in. It is nonsense to say, for example, T'x is white' iff x is white. What is necessary is, of course, the generalization of the T-schema formulas containing free variables. (For the purpose of this paper, I will call such things 'predicates'.) This involves the notion of satisfaction. For the line marked (†) to work, it should therefore read:
>
> $$S(n, \dot{s}) \rightarrow \forall k > n \; \neg T_{S_k}$$
>
> where S is the two-place satisfaction relation between numbers and predicates, and $\dot{s}$ is the predicate $\forall k > x, \neg T_{S_k}$.[5]

Essentially, Priest suggests using satisfiability in place of the truth schema. In this case, we use the symbol $\dot{s}$ to denote the formula $\forall k > x \; \neg S(k, \dot{s})$.

We observe here that, though, using $\dot{s}$ as a predicate object defines the binary satisfaction relation S impredicatively as both objects refer to the totality of the same class of entities, namely the class of *all n-ary relations*. The application of impredicative definitions, which is a general form of circularity, is therefore, critical and required in revealing a fixed point. Even though an

---

[4]Priest, p. 236, 1997.
[5]ibid, p. 237.

explicit circularity is not present, the existence of a fixed point is shown under the assumption of an existing totality of all n-ary relations possibly containing the fixed point itself as a satisfaction relation. The passage from Priest below verifies how the fixed point is being used as a predicate and its connection to the satisfaction relation.

Priest continues as follows:

> [...] it focuses attention on the fact that the paradox concerns a predicate, $\dot{s}$, of the form $\forall k > x \; \neg S(k, \dot{s})$; and the fact that $\dot{s} = `\forall k > x, \neg S(k, \dot{s})'$ shows that we have a fixed point, $\dot{s}$, here, of exactly the same self-referential kind as in the liar paradox. In a nutshell, $\dot{s}$ is the predicate 'no number greater than x satisfies this predicate'. The circularity is now manifest.[6]

Priest also claims that all liar-like paradoxes, including Yablo's paradox, are bound to have a structure which he calls the *inclosure schema*. An *inclosure* is a triplet $\langle \delta, \Omega, \theta \rangle$, where Ω is a set of objects, θ is a property defined on subsets of Ω, such that $\theta(\Omega)$, and δ is a partial function from subsets of Ω to Ω, defined on the sets of which θ is true, and such that if $X \subseteq \Omega$:

$$\delta(X) \notin X \text{ (Transcendence)},$$
$$\delta(X) \in \Omega \text{ (Closure)}$$

must hold.[7]

Priest showed that Yablo's paradox satisfies the transcendence and closure conditions in the inclosure schema after which he concludes the following:

> As we can see, then, Yablo's paradox does involve circularity of a self-referential kind. However one formulates it, it has the characteristic fixed-point structure.[8]

An interesting question to ask here is that whether or not it would be possible to assert the existence of a fixed point without defining the satisfaction relation *impredicatively*, i.e., by restricting S and $\dot{s}$ to have different types so as to not use one in place for another. A positive solution would genuinely prove the existence of circularity in the paradox since the construction in that case does not rely on an impredicative definition.

Hardy [136] points out two basic differences between the liar paradox and Yablo's paradox.[9] The first obvious difference is that in the liar paradox, the contradiction arises from a single statement, whereas in Yablo's paradox, the contradiction is obtained from the totality of infinitely many sentences. The second difference, an apparent consequence of the first, is that there

---

[6] ibid, p. 238.
[7] ibid, p. 240.
[8] ibid, p. 242.
[9] Hardy, p. 198, 1995.

are infinitely many T-schema conditions in Yablo's paradox. Nevertheless, Hardy argues that no contradiction arises from finitely many Yablo sentences $S_1, \ldots, S_n$ for any $n > 1$.

Similarly, Ketland [163] also suggests that Yablo's paradox forms an "ω-inconsistent" theory. We strengthen Hardy's and Ketland's arguments by showing that Yablo's paradox admits circularity at the least transfinite ordinal.

Let $\mathbb{N} = \{1, 2, 3, \ldots\}$ be defined as the set of positive natural numbers.[10] Let $P_1 \wedge P_2 \wedge P_3 \wedge \ldots$ be an infinite conjunction of propositions.[11] We use the notation $\bigwedge_{i=1}^{n} P_i$ to denote $P_1 \wedge P_2 \wedge \ldots \wedge P_n$, and use $\bigwedge_{i \in \mathbb{N}} P_i$ to denote the infinite conjunction of all P_i's. Taking each P_i as an individual statement, we call a set of statements a *theory*. Let $\mathcal{F} = \{P_i\}_{i \in \mathbb{N}}$ be a theory. If no contradiction is obtained from any finite subset of $\mathcal{F}$, then $\mathcal{F}$ is called *consistent*. If for each $n \in \mathbb{N}$, we are able to show that $\bigwedge_{i=1}^{n} P_i$ is consistent and that $\exists j \neg P_j$ holds, then we say that $\mathcal{F}$ is ω-*inconsistent*. If $\mathcal{F}$ is not ω-inconsistent, then it is called ω-*consistent*. An n-*circular* theory is a finite set of statements $\{P_i\}_{i \in \mathbb{N}}$ such that for some $n \in \mathbb{N}$, $\bigwedge_{i=1}^{n} P_i$ implies some $\neg P_e$, where $e \leq n$. We remind the reader that the classical liar paradox forms an n-circular theory, in particular, for $n = 1$ by definition. Define $P_\omega = \bigwedge_{n \in \mathbb{N}} P_n$. A theory $\{P_i\}_{i \in \mathbb{N}}$ is called ω-*circular* if P_ω implies $\neg P_e$ for some $e \in \mathbb{N}$. Now P_ω is not a finitary statement. Using transfinite induction, it is safe to assume that logical implication can be generalised to indices of transfinite ordinals since we are concerned with infinite theories and that Yablo's paradox makes use of infinitary logic. In particular, the paradox involves an infinite conjunction of sentences. This is a straightforward application of infinitary logic since no finite conjunction of any of these sentences causes any contradiction.

An ω-circular theory has to be necessarily ω-inconsistent, but not necessarily vice versa. We can easily define an ω-inconsistent theory that is not ω-circular in the following way. Since an ω-inconsistent theory proves $\exists j \neg P_j$, it must be the case that for some $k > j$, the theory also proves that $\bigwedge_{i=1}^{k} P_i$ is consistent. Since for every j there exists an index $k > j$, the theory cannot be ω-circular.

We argue that not only is Yablo's paradox ω-inconsistent, but it also forms an ω-circular theory.

In fact, any set of Yabloesque sentences is ω-circular. This way we hope to establish a compromise solution to the debate between the supporters who argue for the circularity of the paradox, like Sorensen [273] and of course Yablo himself, and those who argue for the opposite view, like Priest, Beall [24], and

---

[10]We did not include 0 in $\mathbb{N}$ so as to be consistent with the indices used in the original definition of the paradox in Yablo's paper. We merely use $\mathbb{N}$ for the purpose of indexing without loss of generality.

[11]We can also denote them by an infinite sequence $\{P_i\}_{i \in \mathbb{N}}$, but for the purpose of emphasising the infinitary logical nature of Yablo's paradox, we are explicitly writing the conjunctions.

Cook [67]. Consequently, we will be holding the position that Yablo's paradox is only circular in the limit sense by emphasizing the connection between ω-inconsistent theories and ω-circularity.

Now Hardy has already shown that any finite set of Yablo sentences is consistent. The paradox arises when there are infinitely many sentences $\{S_n\}_{n \in \mathbb{N}}$. Since no contradiction is entailed by $\bigwedge_{i=1}^{n} S_i$, this finite theory is consistent.

In fact, an ω-circular theory is sufficient to produce a Yabloesque type paradox. It follows by definition that if $\mathcal{T}$ is an ω-circular theory, then $\mathcal{T}$ cannot be finite, i.e., the cardinality of $\mathcal{T}$ cannot be equal to any natural number. So an ω-circular theory cannot be circular in the classical finite sense and this means it is possible to define a set of Yabloesque sentences. Let us now argue that Yablo's paradox forms an ω-circular theory. Now consider, for every $n \in \mathbb{N}$, the sentences

$$S_n : \forall k > n \ \neg S_k.$$

Assume for a contradiction that the set $\{S_n\}_{n \in \mathbb{N}}$ is not ω-circular. In this case, for $m \in \mathbb{N}$, no truth value can be assigned to any S_i in $\bigwedge_{i=1}^{m} S_i$. However, we know that Yablo's paradox cannot be derived from a finite number of sentences. Moreover, since m is arbitrary, using induction on the size of the set $\{S_i\}_{i<j}$, since $\bigwedge_{i=1}^{j-1} S_i$ is consistent, then so is $\bigwedge_{i=1}^{j} S_i$, and so $\bigwedge_{n \in \mathbb{N}} S_n$ must be consistent. However, the latter totality of infinite conjunction of sentences, i.e., $\bigwedge_{n \in \mathbb{N}} S_n$, actually gives us Yablo's paradox. Then $\bigwedge_{n \in \mathbb{N}} S_n$ must be ω-inconsistent. The fact that it is also ω-circular follows from the fact that if it were not ω-circular, then it would not be ω-inconsistent since $\bigwedge_{n \in \mathbb{N}} S_n$ would imply $\neg P_e$ for no $e \in \mathbb{N}$ and so that there would be no contradiction.

We conclude that Yablo's paradox is not circular in the classical sense but admits circularity in the limit, i.e., the contradiction arises at the first transfinite countable ordinal ω. The reason why we have used ω-inconsistent theories is because Yablo's paradox cannot be described in terms of proper inconsistent theories. Expressing Yablo's paradox in the form of an ω-circular theory allows us to find a compromise solution to the debate on the self-referential nature of the paradox. The controversy, of course, stems from the fact that Yablo's paradox uses infinitary logic. A rather interesting question to ask, therefore, is whether there exists a finite set of paradoxical sentences without self-reference. To judge by a recent paper by Beringer and Schindler [31], the answer to this interesting question appears to be negative.

Discussion Questions

1. Can impredicativity cause an issue in finite universes?

2. To what degree do you think impredicativity or self-reference plays role in geometric paradoxes and optical illusions?

3. In what other forms might Yablo's paradox be written, e.g., geometric, set-theoretic, algebraic?

16

Mathematical Pluralism

Euclidean geometry consistently served science for more than two millennia. It was generally believed to model the physical space that we inhabit in. However, the discovery of non-Euclidean geometries in the early 19th century undermined the belief that the Euclidean perception of geometry was the unique model associated with the notion of "space". Furthermore, it brought doubt into the fundamental assumptions in Kant's interpretation, in his *Critique of Pure Reason*, of the pure space form that is present *a priori* in the human mind.

The key notion behind *pluralism* is the change of "context". Our first experience of mathematical pluralism perhaps stems from the interpretation of Euclid's fifth postulate of geometry in different contexts of "space". Let us recall the axioms of Euclidean geometry.

(i) A straight line segment can be drawn joining any two points.

(ii) Any straight line segment can be extended indefinitely in a straight line.

(iii) Given any straight line segment, a circle can be drawn having the segment as radius and one endpoint as center.

(iv) All right angles are congruent.

(v) If two lines are drawn which intersect a third in such a way that the sum of the inner angles on one side is less than two right angles, then the two lines inevitably must intersect each other on that side if extended far enough (see Figure 16.1).

The fifth postulate has a special place in Euclid's *Elements*. Euclid himself used only the first four axioms in proving the first 28 propositions in the *Elements*, but could not avoid provoking the fifth postulate in the demonstration of the 29th proposition. The fifth postulate is also known today as the *parallel postulate*. The name originates from the work of the Scottish mathematician John Playfair [219], introduced in 1846. *Playfair's axiom* is the assertion that given a line L and a point a outside the line L, there exists a unique line which passes from point a and is parallel to L.

Playfair's axiom (see Figure 16.2) was shown to be equivalent to the fifth postulate. Despite the great efforts of mathematicians, no one was able to prove the fifth postulate as a theorem using the first four postulates.

DOI: 10.1201/9781003223191-16

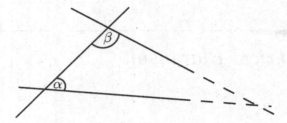

FIGURE 16.1
Euclid's fifth postulate.

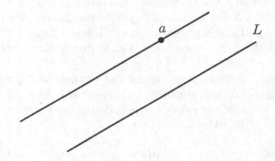

FIGURE 16.2
Playfair's axiom.

The early 19th century was a major turn in geometry for understanding the mathematical conception of space. It was shown independently by Janos Bolyai, Nikolai Lobachevsky, and Carl Friedrich Gauss that self-consistent non-Euclidean spaces could be defined in which the parallel postulate does not hold. These geometries, which can be seen in Figure 16.3, are also known as hyperbolic and elliptic (or Riemann) geometries. The surface of the Earth and the top of a horse saddle would be good examples of an elliptic geometry and a hyperbolic geometry, respectively.

Although it may look quite tempting to use the three-dimensional Euclidean geometry for modelling the physical space in our scientific theories, general relativity suggests we should use non-Euclidean geometries to model the spacetime due to the fact that gravity curves the spacetime structure. Indeed, physical space(time) is non-Euclidean. Even so that the parallel postulate is true in the Euclidean geometry, it certainly does not hold in non-Euclidean geometries that we mentioned above. This shows that the parallel postulate is true in some models, yet false in other models. In other words, the parallel postulate is *independent* from the other four axioms.

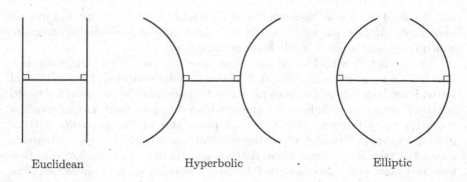

Euclidean · Hyperbolic · Elliptic

FIGURE 16.3
Comparison of three different geometries.

The existence of non-Euclidean geometries supports the *pluralist* conception of mathematics, whereas the *monist* conception, the view that only a single absolute model should be adopted, argues for that the truth or falsity of mathematical propositions are justified within the frame of the absolute model. But if we confine ourselves to geometry, which geometry is actually true? As Poincaré says, 'one geometry cannot be more true than another; it can only be more convenient'[1] In fact, the pluralist view that there exist different interpretations of concepts is not restricted to just geometry, as we can also generalise the pluralist conception to sets. Results in set theory give rise to pluralism in the set-theoretic arena. In Linnebo writes:

> Monists hold that the structure of the cumulative hierarchy has been singled out uniquely up to isomorphism. Pluralists deny this. The fact that CH is independent of ZFC is often thought to support set-theoretic pluralism.[2]

A careful investigation of pluralism then becomes a essential considering the fact that set theory is generally regarded as the foundation of ordinary mathematics.

Knowing that there are many different geometries, which geometry should be adopted? Should we choose, in the first place, an absolute model of geometry? Negative answer to this question finds its roots in naturalism. Naturalism suggests that the universe we pick, as a basis for our studies, is determined by pragmatic reasons.[3] These reasons are expected to be based on extrinsic justifications. If we were to refer to our intrinsic justifications in order to decide

---

[1]Poincaré, p. 50, 1902.
[2]Linnebo, p. 176, 2017.
[3]There is a strong connection between pluralism and the question whether we should add more axioms in mathematics. As we discussed in Chapter 11, a theory may have more than one models. A pluralist is not obliged to choose a particular model.

which model to choose, then we would have had to endorse a kind of monist view that our justifications actually rely on a unique meaningful structure which complies with our mathematical intuition.

The monist view and foundationalism have a strong connection with each other. To a great extent, a pluralist is anti-foundationalist in the traditional sense. Foundationalism demands an absolute model, call it the *standard model* (or the *intended model*), in which the postulated axioms hold. Let us consider geometry for a moment. The first four postulates of the *Elements*, for the most part, are sufficient for proving most of our intuitively true statements about the two-dimensional plane that we are familiar with. In fact, the first four postulates are also modelled by many non-Euclidean geometries. The fifth postulate is what separates the Euclidean interpretation of the notion of space from the others. By adding the fifth postulate to our set of axioms, we are faced with the problem of which one of the possible structures to choose so that the axioms come out as true statements.

16.1   Plurality of Models

Mathematical pluralism is being studied today, for the most part, under the *multiverse* conception of sets. The emergence of the multiverse conception of sets goes as far back as to the independence of the Continuum Hypothesis (CH) from ZFC. The standard interpretation of the set theoretic language and the set concept is based on the iterative conception of sets where every set is ranked in the so-called *cumulative hierarchy* with respect to its order type. The iterative conception describes a "well-behaved" universe of sets. The question whether or not the iterative conception promises a unique interpretation of the set theoretical language, at least up to isomorphism, is an interesting one. Recall from the work of Gödel [116] and Cohen [62] that CH is independent of ZFC. It is implicitly implied in Gödel's Axiom of Constructibility that the universe of sets complies with our iterative conception in which CH turns out to be true. Gödel himself did not fully believe in the multiverse conception of set theory as he separated the situation we had with non-Euclidean geometries from multiverse of sets. He says:

> Very interesting new results about the axioms of infinity have been obtained in recent years [...] I therefore would like to point out that the situation in set theory is different than that of geometry, both from the mathematical and from the epistemological point of view.[4]

Gödel justifies his claim above relying on the apparent motivation that

---

[4]Gödel, pp. 482–483, 1964.

geometry concerns mathematical modelling of the physical space, whereas the interpretation of the set theoretical language is purely based on our mathematical intuition. He writes:

> In geometry, e.g. the question as to whether Euclid's fifth postulate is true retains its meaning if the primitive terms are taken in a definite sense, i.e., as referring to the behavior of rigid bodies, rays of light, etc. The situation in set theory is similar, the difference is only that, in geometry, the meaning usually adopted today refers to physics rather than to mathematical [...].[5]

Recent pluralist studies also include works by Friend [107], Davies [75], Hellman [142], and Balaguer [20]. Davies defends pluralism under Bishop's [35] constructive approach of mathematical analysis. Hellman and Bell argue for the necessity of pluralism at the foundational level in mathematics. Balaguer argues that both Platonism and anti-Platonism entail a certain version of pluralism. Hence, he argues that this kind of mathematical pluralism is actually true. Considering the fact that both CH and ¬CH are consistent with ZFC, Balaguer asks what Platonists should say about the truth value of CH. In the beginning of his paper [20], he mentions two types of Platonisms, namely *Silly Platonism* and *Better Platonism*. According to Balaguer, mathematical truth in Silly Platonism is determined as follows:

> A mathematical sentence or theory is true just in case it accurately characterizes some collection of mathematical objects. Thus, since the mathematical realm is plenitudinous, it follows that all consistent mathematical theories are true. And so it follows that CH and ¬CH are both true, because CH is true of some parts of the mathematical realm, and ¬CH is true of others.[6]

Note that Silly Platonism is a loose version of realism in which the theory could be realized in non-standard models and even if so, the sentence (or the theory) is still said to be true. The role of intended (standard) models becomes more critical in Better Platonism. He describes mathematical truth accordingly as below:

> There is a difference between being true in some particular structure and being true simpliciter. To be true simpliciter, a pure mathematical sentence needs to be true in the intended structure, or the intended part of the mathematical realm—i.e., the part of the mathematical realm that we have in mind in the given branch of mathematics.[7]

---

[5] ibid, p. 483.
[6] Balaguer, p. 383, 2017.
[7] ibid, p. 383.

According to this view of Platonism, an arithmetical sentence is true iff it is true in the standard model of arithmetic. In addition, Balaguer claims that the intended structure may not be unique in every branch of mathematics. For this, he gives an example of two structures H_1 and H_2 in which, say, respectively, ZFC + CH and ZFC + ¬CH hold. He also assumes that both H_1 and H_2 could be intended models of set theory as they could be both fully consistent with our full conception about sets.

An intriguing question that can be asked here is, though, whether H_1 and H_2 can really be *equally* consistent with our full conception of set theory, and hence whether both at the same time can be taken as intended models *on a par* with each other, assuming that they satisfy mutually contradicting statements. If these structures satisfy mutually contradictory statements, then to what degree a structure is said to capture our full conception of sets? It seems that one of them must be "less" standard than the other. Intentionality in models is a varying concept. That is to say, some structures may serve as a "better" intended model than others, but no two structures have the same level of intentionality, up to isomorphism. Assuming this kind of *equivalency in intentionality*, otherwise, would make the line even between standard and non-standard models rather blurred.

The question that what determines a structure to be counted as an intended model was in fact considered by Balaguer himself, and he gives a criterion for this.[8] Now although we do not have a clear answer to the latter question, let us consider the negation of the same question: What determines a structure *not* to be counted as an intended model? Apart from the obvious non-standard interpretation of objects and the language, the argument we gave in the previous paragraph might give an answer to the negated version of the question. This is not to determine which structure actually *is* or *is not* an intended model. It settles a rather weaker question. So we can say, if H_1 and H_2 are two structures such that $H_1 \models \varphi$ and $H_2 \models \neg\varphi$ for some statement φ, then at least one of them cannot be an intended model to the degree that the other can. This does not mean that neither is an intended model. It just says that both structures cannot be simultaneously counted as intended (standard) models to the same extent as the other. The reason is that if φ is a statement which is modelled by some standard structure of a theory and if $\neg\varphi$ is modelled by another standard structure, then there should be no reason to believe in existence of the *full conception*, in which case this would go against Platonism. For mathematical pluralism and Platonism to fit together, it appears that every model should have a "degree of intentionality", so that some models are meant to be "more intended" or "more preferred" than the others. We will not be concerned here with how to select preferred models. Nevertheless, this can be done either pragmatically or by looking at the intrinsic properties of the model.[9] We should admit that this entails some form

---

[8]ibid, p. 384.

[9]See Magidor [192] (2012) for a discussion that some set theories can be selected as more preferable than others.

of weak pluralism which we may call *hierarchical pluralism*. So hierarchical pluralists might believe in the existence of plural standard models, but they also believe that those models can be (linearly) ordered with respect to their degree of intentionality. Determining the degree of intentionality of a theory or a model is a separate subject to discuss though. The important point here is that if one advocates for epistemological realism, then there must exist such an ordering.

The kind of Platonism that Balaguer suggests for Platonists to endorse is what he calls *intention-based Platonism* (IBP) in which he defines truth in the following manner:

> A pure mathematical sentence S is *true* iff it is true in *all* the parts of the mathematical realm that count as intended in the given branch of mathematics (and there is at least one such part of the mathematical realm); and S is *false* iff it is false in all such parts of the mathematical realm (or there is no such part of the mathematical realm); and if S is true in some intended parts of the mathematical realm and false in others, then there is no fact of the matter whether it is true or false.[10]

It is, in fact, this type of Platonism which Balaguer claims is consistent with mathematical pluralism. Regarding IBP, he concludes the following:

> Given that IBP is a *plenitudinous* version of platonism, we get the following result: which (consistent) mathematical sentences are true is wholly determined by our intentions. This is because (a) mathematical truth comes down to truth in intended structures, and (b) which mathematical structures count as intended is wholly determined by our intentions.[11]

The justification used in part (b) seemingly aims for a naturalistic philosophy if what is meant by "intentions" is "fruitfulness". However, if it the word "intention" refers to "mathematical intuition", then we need to have in hand a standard interpretation of the theory in consideration. On the IBP interpretation, however, the status of CH is left undetermined since set theorists have experiences in CH and ¬CH universes, both of which can be claimed as intended models of set theory. But on hierarchical pluralism, either CH or ¬CH universe has a higher degree of intendedness. Thus, one of them is preferred over the other. For readers who are interested in looking at the arguments for the compatibility of Platonism (as well as anti-Platonism) with Balaguer's form of pluralism, we refer the reader to his paper [20] for a detailed account.

Another well-known proponent of pluralism is Michèle Friend, who published a comprehensive book [107] on mathematical pluralism. We shall mention here several points which are discussed in more detail in her work. Pluralism is a very broad term and, in fact, not very easy to define. Pluralism has

---

[10]ibid, p. 388.
[11]ibid, p. 393.

plural definitions so to speak. In fact, pluralism concerns not just one aspect of mathematics but it concerns many aspects such as the methodology, ontology, epistemology, foundational, etc. The first few sections of her book define these types of pluralisms. She also discusses how Maddy's naturalism inspires the pluralist.[12] However, we should also note that Maddy's methodological maxim UNIFY does not endorse a multiverse perspective for set theory.[13] Maddy says:

> If set-theorists were not motivated by a maxim of this sort, there would be no pressure to settle CH, to decide the questions of descriptive set theory, or to choose between alternative axiom candidates; it would be enough to consider a multitude of alternative set theories.[14]

Friend also discusses the relationship between pluralism and Shapiro's structuralism. She writes:

> The pluralist is inspired by Shapiro's position, especially by his anti-foundationalism and by his self-avowed pluralism.[15]

Indeed, it is quite clear that the pluralist is an anti-foundationalist. That being said, however, it may not be wrong to claim that Friend takes anti-foundationalism in its radical form as in opposing to the idea of reducing of mathematics to *any* of the simplified theories. This is due to the fact that she does not consider *foundational pluralism* as a consistent and stable position.[16] At this point, we should also draw a firm line between *foundationalism* and *reductionism*. In our definition, reductionism is the idea that mathematical objects can be reduced to objects of *one of possibly many* simplified theories. On the other hand, mathematical foundationalism is the view which argues for the existence of a *unique* foundational theory for all (or most) of mathematics. I think it is reasonable to say that reductionism does not presume the uniqueness requirement. Hence, we may think of foundationalism as a rigid form of reductionism.

Friend gives four anti-foundational arguments one of which is based on the idea that the foundational theory is a continually *growing* theory. She says:

> [...] the foundationalist begins with the technical result that most of mathematics can be reduced to The Foundation. This is a twofold mis-description. First, the reduction is sometimes too contrived, and therefore, not successful. Second, any proposed foundation is only that: a foundation. That is, we can add more

---

[12]Friend, p. 31, 2014.
[13]For Maddy's account on the multiverse, see her [191]. Also see Ternullo [282] (2019).
[14]Maddy p. 210, 1997.
[15]Friend, p. 51, 2014.
[16]ibid, pp. 24–25.

to the foundation. Whatever the founding theory is, 'it' grows. As it grows we understand the founding parts in a new light. So it is not a fixed foundation.[17]

Friend's *growing foundation* looks quite compatible with our idea, which was mentioned earlier, that some standard interpretations have a higher degree of intentionality than the others. The standard model has a growing feature. In our case, a growing foundation is always improving its degree of intentionality.

The fourth anti-foundational argument, given by Friend, is what she calls "On truth in a theory". The argument is that truth only makes sense in a structure or on an interpretation. She writes:

> Together, the structuralist and the pluralist do not think that there are absolute truths in mathematics of the form: "$2 + 8 = 10$". Instead, what is true is: "In Peano Arithmetic, $2+8 = 10$".[18]

Mathematicians would be more than willing to accept this view as a sort of *local realism*.[19] Though at the same time, they would also expect the following statement to be true: *The number obtained by the concept of adding any equivalent of the number 2 to any equivalent of the number 8 is equivalent to any of the equivalents of the number 10.*[20] For otherwise, there would be a counter-structure in which the concept of *adding* 2 and 8 does not yield 10. Note that the counter-structure would necessarily have to use a different "meaning" of the concept of *addition* rather than just some non-standard "interpretation" of the + (plus) symbol.

16.2   Multiverse Conception of Sets

The *multiverse* conception of sets—a relatively new trend in the foundations of mathematics and philosophy of set theory—is a central theme in the pluralist philosophy of mathematics. In fact, most of pluralist ideas about mathematics revolve around the studies, especially by researchers who are more on the mathematical side, on the mutiverse conception of sets. Prominent supporters of this view include Joel David Hamkins [129], Geoffrey Hellman [142], and Sy-David Friedman [5].

---

[17]ibid, p. 60.

[18]ibid, p. 67.

[19]By *local realism* we mean that mathematical statements are independently true or false in structures and nowhere else.

[20]By *equivalent of the number n* we refer to any object that could potentially take the role of n in the natural number structure.

In general, mostly among mathematicians, the notion of set is usually understood as a single set concept belonging to a single set-theoretical universe, that is, the cumulative hierarchy V. However, results in set theory indicate that there might be other universes of sets just as there are other geometries. This view, that there are multiple universes of sets, is known today as the *multiverse* conception. The idea that there is only one absolute universe of sets is called the *universe* conception. Hamkins distinguishes the universe and the multiverse conceptions by first defining the universe conception of sets in the following manner:

> The *universe view* is the commonly held philosophical position that there is a unique absolute background concept of set, instantiated in the corresponding absolute set-theoretic universe, the cumulative universe of all sets, in which every set-theoretic assertion has a definite truth value.[21]

The notion of *independence* is generally regarded as a distracting issue for the adherents of the universe view. The multiverse conception of sets can avoid the infamous independence phenomenon more easily. Hamkins also describes the multiverse view. He writes:

> [...] the *multiverse view*, which holds that there are diverse distinct concepts of set, each instantiated in a corresponding set-theoretic universe, which exhibit diverse set-theoretic truths. Each such universe exists independently in the same Platonic sense that proponents of the universe view regard their universe to exist.[22]

Apart from the cumulative hierarchy V, other well-known examples of the set-theoretic domain include the constructible hierarchy L, the class of hereditarily ordinal definable sets HOD, etc.

The idea of degrees of intentionality we talked about earlier was also somewhat implied in Hamkins, at least in the sense that we are not obliged to assume that every universe in the multiverse is expected to have equal status. He says:

> The multiverse view is one of higher-order realism—Platonism about universes—and I defend it as a realist position asserting actual existence of the alternative set-theoretic universes into which our mathematical tools have allowed us to glimpse. The multiverse view, therefore, does not reduce via proof to a brand of formalism. In particular, we may prefer some of the universes in the multiverse to others, and there is no obligation to consider them all as somehow equal.[23]

---

[21] Hamkins, p. 416, 2012.
[22] ibid, p. 416.
[23] ibid, p. 417.

The independence of a given statement from ZFC is usually proved by *forcing* method which was introduced by Cohen [62]. If φ is a statement and we want to prove its independence, we should construct models for ZFC $+ \varphi$ and ZFC $+ \neg\varphi$. This is usually done by taking a countable transitive model of ZFC and adding into the model certain elements, called *generic sets*, and obtain a forcing extension of the model in which the desired statement holds along with the usual axioms of ZFC.[24]

One puzzling problem, regarding the ontology of sets, is the question whether or not there are universes, and in particular, forcing extensions of V, *outside* of the cumulative hierarchy V. With respect to the universe view, the forcing extension of V is rather regarded as an illusory object. Hamkins explains this as follows:

> On the universe view, of course, forcing extensions of V are deemed illusory, for V is already everything, while the multiverse perspective regards V as a relative concept, referring to whichever universe is currently under consideration, without there being any absolute background universe. On the multiverse view, the use of the symbol V to mean "the universe" is something like an introduced constant that might refer to any of the universes in the multiverse, and for each of these the corresponding forcing extensions are fully real.[25]

Another way to think about forcing extensions is imagining, for example, the case where we take a leap from the set of reals to complex numbers. Does the square root of $\sqrt{-1}$ exist? Certainly not in $\mathbb{R}$. So we need to consider the field extension and obtain the set of complex numbers to find the square root of $\sqrt{-1}$.

The problem about the illusory existence of forcing extensions of the set theoretical universe V, to some extent, can be settled down under the naturalist philosophy and, in a better way, the multiverse view. Hamkins says:

> Of course, one might on the universe view simply use the naturalist account of forcing as the means of explaining the illusion: The forcing extensions don't really exist, but the naturalist account merely makes it seem as though they do. The multiverse view, however, takes this use of forcing at face value, as evidence that there actually are V-generic filters and the corresponding universes $V[G]$ (forcing extension of G with respect to the filter G) to which they give rise, existing outside the universe.[26]

[24]Forcing is an exciting subject in set theory and it requires a rigorous study to fully understand the method. Due to the limited scope of this book, we will not go into the details. However, we encourage the reader to refer to Kunen [175] or Jech [155].

[25]ibid, p. 419.

[26]ibid, p. 425. Parentheses added.

Despite that the universe view is perfectly self-consistent, according to Hamkins, it limits future insights that may arise from these possible mathematical realms. He says:

> The history of mathematics provides numerous examples where initially puzzling imaginary objects become accepted as real. Irrational numbers, such as $\sqrt{2}$, became accepted; zero became a number; negative numbers and then imaginary and complex numbers were accepted; and then non-Euclidean geometries. Now is the time for V-generic filters.[27]

How does the multiverse view settle the most fundamental questions of set theory which are known to be independent of ZFC? Let us take CH as an example. The problem with the universe view is that CH remains as an undecidable proposition. Nonetheless, it has to be true or false in some absolute universe that we are yet to understand. On the other hand, CH is a settled question under the multiverse interpretation, perhaps in a non-deterministic manner. CH is true in some universes and equally false in others. Supporters of the universe view might suggest that CH would be solved using the following schema, which Hamkins refers to as the *dream solution*.[28]

Step (i). Produce an "obviously true" statement φ of set theory.
Step (ii). Prove that φ implies CH (or $\neg$CH).

Now, Koellner claims that there has never been such a dream in this sense.[29] Even if there is so, however, according to Hamkins, such a dream solution is impossible to achieve. This is due to the fact that mathematicians have had rich experiences in many various set theoretical universes. Set theorists are now deeply informed about the status of CH and they continue their investigations both in the CH universes and the $\neg$CH universes. For Hamkins, if we are to introduce an obviously true statement φ which implies $\neg$CH, then we can no longer see φ as an obviously true statement since doing so, would negate the experiences in the universes in which CH holds. Similarly, if we are to introduce some φ which implies CH, then the same argument would hold for the otherwise.

We may observe that Hamkins' argument is again based on the assumption that the CH/$\neg$CH universes have the same degree of intentionality. An independent position in the multiverse view, as we suggested, can be that some of these universes may have a higher degree of intentionality up to isomorphism. In other words, some intended structures could be more plausible or make intuitively more sense than the others. Truth definition can then be changed accordingly. Now classifying the universes with respect to their degree

---

[27]ibid, p. 426.
[28]ibid, p. 430.
[29]Koellner, p. 24, 2013.

of intentionality can be done in two ways: either by a naturalistic approach, in which case by looking at the fruitfulness of the model (that is, relying on the extrinsic justifications), or by purely based on the intrinsic justifications. Following Hamkins' own example, if someone proposes an obviously true statement S implying that Manhattan does not exist, then we might have to decide whether the existence of Manhattan or the truth of S is more obvious. In the universe view, claiming that their degrees of intentionality are on a par with each other does not help us prove or disprove a statement. It is by some sort of classification of degrees of intentionality of models, we will be able to pick the best possible model (in this case, the one with the "highest" degree of intentionality) and regard that model as the absolute background for our theory. It is expected that the ultimate intended model can only extend itself in the intentionality hierarchy in a consistent manner similar to as in Friend's *growing foundations*. For instance, if the Axiom of Choice is true in the intended model M and if M' extends M in the intentionality hierarchy, then the negation of the Axiom of Choice should not be true in M'.

Hamkins also presents axioms for the multiverse vision. It is argued that the multiverse conception allows the set-theorist to study inter-universe relationships on a relativistic basis. In Chapter 11, when we presented the Löwenheim-Skolem Theorem, we said that countability is a relative concept which depends on the set-theoretic universe and how the model *perceives* cardinalities. Any set in a given model of set theory can be made countable in the forcing extension of the model. The relativity of the notion of countability gives rise to what Hamkins calls the *Countability Principle* as one of the multiverse axioms.

Countability Principle. Every universe V is countable from the perspective of another, better universe W.[30]

A more provocative axiom was proposed regarding the well-foundedness of models due to pluralistic account of set-theoretical backgrounds. For Hamkins, some models can be seen to be ill-founded from the perspective of another (better) model. The next axiom says that there is no absolute notion of well-foundedness.

Well-foundedness Mirage. Every universe V is ill-founded from the perspective of another, better universe.[31]

Another axiom Hamkins states concerns the absorption of a model into the constructible universe L.[32]

---

[30]Hamkins, p. 438, 2012.

[31]ibid, p. 439.

[32]See our section 17.1.1 of Chapter 17 for the precise definition of L.

Absorption into L. Every universe V is a countable transitive model in another universe W satisfying $V = L$.[33]

Ternullo [282] describes Hamkins' multiverse vision of allowing one to inspect set-theoretical universes from the point of view of the others as *Multiverse Perspectivism*. According to Ternullo, given V, the set theorist is interested in knowing what universes one may have access to from V and how we access them, and what V looks like from the perspective of the other universes.[34]

We shall end our discussion on Hamkins' paper, however, we encourage the reader to refer to his work [129] for a detailed account. For a criticism of his work, we refer the reader to Koellner [170] who also published a paper [168] in which he criticises Carnap's work [54], *The Logical Syntax of Language*, and his radical pluralism. Koellner's [168] is more technical, however, and it includes some set theoretical results which may require, for full understanding, some background knowledge in large cardinals.

The paper by Antos, Friedman, *et al* [5] has also received attention in the multiverse view community. The authors present their own theory of the multiverse.[35] They define the universe and multiverse views similarly as in [129]. However, they introduce an additional criterion of differentiation between the conceptions. They call this criterion *commitment to realism* which measures how strongly each conception holds that the universe or the multiverse exist "objectively".[36] This is relatively close to what we have expressed earlier about degrees of intentionality of models. It was also mentioned in [5] that the universe and the multiverse views would each split into two further categories according to whether one is a realist or a non-realist, yielding four positions in total: universe view realism, universe view non-realism, multiverse view realism, and multiverse view non-realism.[37] The authors describe the universe view realism alternatively as Gödel's realism. On the other hand, a universe view non-realist uses the set theoretical universe merely for practical purposes without actually believing in its independent existence. The two other positions are the multiverse view realism and the multiverse view non-realism. The adherents of the former view actually coincides with Balaguer's *full-blooded Platonism* (FBP) that any mathematical object that could exist actually do exist.[38] Finally, a multiverse view non-realist does not believe in the existence of any universe, nevertheless she takes the multiverse phenomenon as a mathematical practice. The authors discuss a new multiverse conception where·

---

[33] ibid, p. 439.

[34] Ternullo, p. 56, 2019.

[35] We will refer to this paper in the next chapter as well since it addresses some important points about the quest for new axioms in set theory.

[36] At least commitment to some form of realism if not to an all-around realism.

[37] Antos, Friedman, *et al*, p. 3, 2015.

[38] See our section 18.2 of Chapter 18. Also see Balaguer's [17] and [19] for details.

they claim that Zermelo's [310] account of the universe of sets can be further extended to what they call the *vertical multiverse* conception in which they allow the cumulative hierarchy V to be heightened while keeping the width fixed. They also explain the distinction between what they call *actualism* and *potentialism*, two positions regarding the modifiability of V. They describe actualism as the belief that the cumulative hierarchy V is a fixed object and that it is impossible to "stretch" V via model theoretic constructions.[39] Any forcing extension of V, according to actualism, is, in fact, a model in V.[40] The opposite extreme is called *potentialism* which they define it in the following manner.

> A potentialist, on the other hand, sees V as an indefinite object, which can never be thought of as a 'fixed' entity. The potentialist may well believe that there are some fixed features of V, but she believes that these are not sufficient to fully make sense of an 'unmodifiable' V: the potentialist believes that V is indeed 'modifiable' in some sense.[41]

Perhaps a major section in their work is about the explanation of how the vertical multiverse view is used in their *hyperuniverse programme*. The hyperuniverse programme, originally proposed by Arrigoni and Friedman [10], can be primarily associated with an on-going project of searching new axioms for set theory. For this reason, we shall leave this discussion to Chapter 17.

The universe view non-realism is also close to Maddy's [189] *thin realism*, which can be described as follows. The subject of set theory is basically *sets*. Every set-theoretic statement has a definite truth value which is determined by the set theoretical methods. Thin realism argues for that truth about sets is exactly, no less and no more, what set theory tells us. Maddy says that 'the Thin Realist holds the set-theoretic methods are the reliable avenue to the facts about sets, that no external guarantee is necessary or possible'.[42] On thin realism, set-theoretical methods suffice to establish facts about sets. A robust realist does not believe in the reliability of set-theoretical methods and would require further account, possibly knowledge that goes beyond the epistemic limits of set theory, to establish the set-theoretical facts. Maddy compares the thin realist perspective and the Carnapian position of embracing set theory as an adoption of a linguistic framework. For Maddy, these two perspectives differ from each other as she writes:

---

[39] Antos, Friedman, *et al.*, p. 19, 2015.

[40] The "forcing extension" of V may sound rather paradoxical prima facie since V is *all* there is. But it actually depends on how we interpret forcing and whether we defend the universist or the multiversist position. For more details on this matter we refer the reader to Barton [23] (2020).

[41] ibid, p. 19.

[42] Maddy, p. 63, 2011.

> [T]he Thin Realist doesn't regard the embrace of 'the concept
> of set', the adoption of the set-theoretic framework, as a purely
> pragmatic matter of linguistic convention; she takes Cantor and
> Dedekind to have discovered the existence of these mathematical
> objects, not to have come to a convenient decision.[43]

For Maddy, there is no gap between methods of set theory and sets, as truth
about sets are known by applying these methods.[44] Although thin realism is
somewhat compatible with the traditional realist idea that set theory really
does form a body of truth, Maddy's *arealism* on the other hand rejects this
presupposition.[45] According to Maddy, arealism argues for that set theory is
just the activity of developing a theory of sets that effectively serves a concrete
and ever-evolving range of mathematical purposes.[46] In contrast with thin
realism, on arealism, set theory does not necessarily have a subject matter.
Furthermore, on arealism, statements of set theory do not have determinate
truth values.

One interesting question asked by Roland [245], regarding thin realism, is
the question why set-theoretical methods are reliable indicators of truth.[47] Of
course, we also need to ask why another foundational subject of mathematics
would not be a realiable indicator of truth. Should thin realism be merely
restricted to set theory? Some questions in mathematics can often be resolved
by transforming the question into a question about another subject domain.
Although set theory serves as a foundation and that all questions about mathematics, in principle, can be transformed into questions about sets, due to its
low-level language, it is usually inefficient to reduce the theorems of mathematics in other fields or their problems into its set-theoretical form.[48] The
aim of *category theory*, in a way, is to establish this high-level interaction and
communication between different domains of mathematics in case when one
needs to transform a question about a particular domain into a question about
another.

16.3   Liberating the Mathematical Ontology or Blurring the Mathematical Truth?

Does pluralism liberate the mathematical ontology or blur mathematical
truth? It appears to me that there is a trade-off, and that we end up having to
choose between the richness of mathematical ontology and the objectivity of

---

[43]ibid, p. 68.

[44]Maddy, p. 370, 2007.

[45]Maddy, p. 88, 2011.

[46]ibid, p. 89.

[47]Roland, p. 301, 2016.

[48]An appropriate analogy for this would be writing an algorithm in the binary machine
language.

mathematical epistemology. I shall call this *pluralist's dilemma*. According to this, it is not possible to fully gain both without compromising the other. On the one hand, we cannot deny the plurality of set-theoretical universes. This is not just due to the experiences we had and results we obtained in these different universes, but it is also that many of them are intrinsically plausible, and that they cannot be easily singled out to a unique absolute model. On the other hand, radical forms of pluralism are rather too liberal about the epistemology of mathematics. The main concern is that, as tempting as it may sound to liberate mathematical truth, we depart from objectivity in determining intrinsically plausible statements of mathematics. Radical pluralism must reject mathematical intuition. According to this type of pluralism, there are, in fact, many mathematical intuitions incompatible with each other.[49] If two contradictory statements can be conceived in two different universes respectively, then what determines a sensical mathematical intuition? For a pluralist, there is 'actually' no common mathematical intuition. A radical pluralist has no motivation in finding what is intrinsically appealing, nor can she have any justification in her objection to "counter-intuitive" statements being called "true" statements. On what grounds do radical pluralists decide to eliminate non-plausible statements? For a radical pluralist, there is a good reason not to believe in the Axiom of Extensionality, equally as to believe in the existence of a large cardinal which contradicts ZFC. Furthermore, these beliefs can be considered as *equally* plausible by the radical pluralist. The general consensus holds that the structure of natural numbers *is* the intended model of arithmetic. But how do we decide if another structure for arithmetic is not as plausible as the natural numbers? Even for a modest realist, it seems not right to put two different conceptions on the same level of intentionality.

In [224], Priest lists a number of objections raised by anti-pluralists and replies to them. One particular objection concerns the universality of the inference rules in proving mathematical statements. Priest says:

> [...] when someone breaks the rules of a mathematical game, and they are not simply making a mistake, they are, *ipso facto*, no longer playing that game. The new set of rules constitutes a new game.[50]

The liberation that radical pluralism offers the mathematician comes from assuming that pluralism also applies to the methodology. The set of rules of inference is one thing that a non-formalist mathematician would want to be certain about the semantics. This type of methodological pluralism, i.e., pluralism applied on the rules of the mathematical game, allows one to have non-standard interpretations of the logical language. Since some primitive rules such as the laws of thought or Peano axioms, etc., are more universally

---

[49] A radical realist will likely to think that liberating mathematical truth and ontology leads to blurred mathematics with no basis for a credible assessment by any means.

[50] Priest, p. 9, 2013.

accepted than the otherwise, there is a good reason these axioms or rules deserve to be called "standard". It is not because they have been taught to us that way, but it is because of how primitive and finitary facts about quantities, collections, space, and so on, are perceived by the human mind in a certain empirical manner.

We end our discussion with a quote by Pedeferri and Friend who advocated in their work [209] the idea that mathematicians are better described as pluralists rather than as formalists. Pluralism can be seen as a nice unification of the syntax (formalism) and the semantics. Formalisation, after all, failed to capture naive mathematical reasoning due to Gödel's incompleteness theorems. It did, however, serve well for the establishment of formal computing machineries and programming. Better call a mathematician a pluralist than a formalist. Nevertheless, better call a programmer a formalists than a pluralist.

Discussion Questions

1. Discuss to what extent the discovery of non-Euclidean geometries undermines Kant's philosophy of mathematics?

2. Is there a correlation and congruency between a "convenient geometry" and "intuitive geometry"? In other words, is intuition determined by practicality?

3. What might be the consequences of having non-standard models as "intended" models? Are standard models of a theory always intended?

4. Consider a structuralist position of set theory. Knowing that we have plenty of experience in universes, respectively, consistent with CH and ¬CH, how might a structuralist respond to the multiverse theory of sets? If the CH and ¬CH universes can both be intended models of set theory, does the structuralist in set theory contradict herself for endorsing plurality of set theoretic structures?

5. Argue for/against Hamkins' *dream solution* to CH (see footnote 28).

6. Do you think ontological/epistemological pluralism in mathematics entails any form of methodological pluralism? Why/why not?

7. Discuss about the merits and defects of the multiverse theory of sets.

8. Argue whether or not Maddy's thin realism for sets is compatible with the multiverse theory.

17

Does Mathematics Need More Axioms?

It has been a long standing problem of philosophy and science whether reality can be captured by a theory that explains *all* principles of the universe. Thales of Miletus (623/524–548/545 BC) thought that the primary principle behind everything was water. Anaximander claimed that the principle of all things was the *apeiron*, that which has no limit. Pythagoreans, on the other hand, suggested that everything in the universe could be ultimately explained by natural numbers. Ontologically, the Pythagoreans thought that the only mathematical objects that ever existed were natural numbers, since all geometric lengths were assumed to be measured by rational numbers. As fate would have it, throughout the history of science, we have always managed to find things beyond our conceptions and theories. In the case for mathematics, the Pythagoreans later discovered the existence of irrationals and faced a problem of whether one should add irrational magnitudes as a part of our realm. This followed by the acceptance of negative numbers, imaginary numbers, non-Euclidean geometries, generic forcing extensions, and so on. We tend to seek further objects to be added as a part of our mathematical realm. Mathematics, like many other sciences, advances in accordance with our intentions and justifications, whether they are intrinsic or extrinsic. In this chapter, we present the attempts in settling one of the most central open problems of set theory, the Continuum Hypothesis, and discuss about the quest for new axioms.[1]

17.1 Status of the Continuum Hypothesis

The Continuum Hypothesis (CH) is an old problem emerged with the discovery of set theory by Georg Cantor (see section 9.3 of Chapter 9).[2] It is the statement that every subset of real numbers is either countable or has cardinality $2^{\aleph_0}$. We demonstrated in section 9.3 of Chapter 9 that cardinal

---

[1] The problem of whether or not mathematics needs new axioms was discussed by Feferman *et al* [89] (2000) in a panel at the annual meeting of Association for Symbolic Logic, held in Urbana-Champaign at the beginning of the millennium. We encourage the reader to refer to this panel discussion for an initial motivation.

[2] See Maddy (1997), Part I, for a historical background of CH.

DOI: 10.1201/9781003223191-17

exponentiation generally admits rather more interesting properties than addition and multiplication do. The problem there was to determine exactly which aleph number $2^{\aleph_0}$ corresponds to. For most mathematicians, CH appears to be a definite statement of set theory.[3] Despite Cantor's great efforts, he wasn't able to prove his conjecture. It became clear in the mid-20th century that CH turned out to be independent from the axioms of ZFC. By Gödel [116], we know that CH is consistent with ZFC, and by Cohen [62] we know that ¬CH is consistent with the same system. The *forcing* method became an extremely poweful tool for proving independence results by constructing *outer models* of a given model of set theory. Cohen showed that if M is a model of ZFC, then M contains "blueprints" for artificial universes that extend M—in other words, *forcing extensions* of M— in which CH is false, but also contains extensions in which CH is true. Furthermore, such universes can be completely constructed and analysed within M.

As a first attempt, one may try to settle CH by adding higher axioms of infinity, also called *large cardinal* axioms. These are sets whose size exceeds all cardinalities that exist within the boundaries of ZFC.

One of the simplest types of large cardinal is called an inaccessible cardinal, which is defined as follows.

Definition 17.1 A cardinal κ is called (strongly) *inaccessible* if it is uncountable, and it is not a sum of fewer than κ cardinals that are less than κ, and $\lambda < \kappa$ implies $2^\lambda < \kappa$.

In fact, $\aleph_0$ satisfies these properties except the uncountability condition. The cardinal $\aleph_0$ cannot be reached by taking unions of finite sets, nor by taking power sets of finite sets. Of course, we are interested in uncountable cardinals as given in the definition. It is worth noting that the existence of inaccessible cardinals cannot be proved from ZFC. Because if κ is an inaccessible cardinal, then V_κ is a model of ZFC. So if such a cardinal κ exists, then ZFC is consistent. Particularly, if κ is an inaccessible cardinal, then for least such κ, $V_\kappa \models$ "there is no inaccessible cardinal". Then, there is a model of ZFC + "there is no inaccessible cardinal". Therefore, within ZFC, we cannot prove the existence of inaccessible cardinals. Furthermore, it cannot be shown in ZFC that the existence of inaccessible cardinals is consistent with ZFC. Suppose otherwise for reductio. That is, assume that if ZFC is consistent, then so is the theory ZFC + "there exists an inaccessible cardinal". Let us abbreviate the statement "there exists an inaccessible cardinal" by I. Suppose that ZFC is consistent. Since by assumption that ZFC is consistent with I, ZFC + I is consistent. Within ZFC + I, we can prove that there is a model of ZFC. Then, the sentence "ZFC is consistent" is provable in ZFC + I. But we assumed that ZFC is consistent with I and so the sentence "ZFC + I is consistent" can be proved in ZFC + I. But this contradicts Gödel's Second Incompleteness Theorem.

---

[3] For a counter argument, see Feferman [88] (2011).

Despite the fact that inaccessible cardinals cannot be shown to exist within ZFC, it is not really forbidden to work with them in our mathematical proofs. As a matter of fact, the original proof of Andrew Wiles [294] for Fermat's Last Theorem relies on the existence of Grothendieck universes, named after Alexander Grothendieck.[4] The existence of Grothendieck universes is equivalent to the assumption that strongly inaccessible cardinals exist.[5] So Wiles' original proof goes beyond the strength of ZFC. Nonetheless, Colin McLarty in [197] later proved that the proof of Fermat's Last Theorem required, in fact, much less than ZFC.

The large cardinal programme, i.e., adding higher axioms of infinity, to provide a solution for CH, in the beginning was encouraged by Gödel himself. Gödel says:

> First of all the axioms of set theory by no means form a system closed in itself, but, quite on the contrary, the very concept of set on which they are based suggests their extension by new axioms which assert the existence of still further iterations of the operation "set of". These axioms can be formulated also as propositions asserting the existence of very great cardinal numbers [...] The simplest of these strong "axioms of infinity" asserts the existence of inaccessible numbers.[6]

Short after Cohen introduced the forcing method, Levy and Solovay [177] showed that Cohen's method does not affect the existence of very large infinite sets. So we cannot directly solve CH by adding large cardinals. This is not to say we cannot decide the truth value of CH based on other axiom candidates which have a direct consequence on CH. Chris Freiling [103] introduced the *Axiom of Symmetry*, which states that if f is a function that assigns every real number x to a countable subset of real numbers A_x, then there exist two real numbers x and y for which $x \notin A_y$ and $y \notin A_x$. In fact, Freiling imagined a thought experiment of throwing darts at the real line, landing, respectively, at x and y. He concludes that the location y of the second dart is not in the countable set A_x which has Lebesgue measure zero. Since x is fixed, so is $f(x)$. Hence, the event that y being in A_x has probability zero. By the "symmetry" of the choice of order in which the darts are thrown, the location of x, by the same argument, is not in A_y. Freiling's result is stated as follows.

Theorem 17.1 (Freiling, 1986) Under ZFC, the Axiom of Symmetry is true iff CH is false.

---

[4] Informally, a *Grothendieck universe* is a transitive set which is closed under pairings, unions, and power set. Trivial examples of Grothendieck universes are the empty set and the set of all finite sets. But in this context, we are interested in universes that are uncountable.

[5] See Williams [295] (1969) and Kruse [174] (1965) for a detailed account on the relationship between Grothendieck universes and inaccessible cardinals.

[6] Gödel, p. 476, 1964.

Proof. First we prove that the Axiom of Symmetry is a sufficient condition for the negation of CH. Assume that the Axiom of Symmetry holds. Also suppose that CH is true. Let $<$ be a well-ordering of $\mathbb{R}$. By CH, the length of this well-ordering is ω_1 (that is, $\aleph_1$). Now let $f(x) = \{y : y < x\}$. Then, f is a function from the set of reals to the set of countable sets of reals. Since a well-order is always a total ordering, either $x < y$ or $y < x$. Then, either $x \in f(y)$ or $y \in f(x)$. Then, CH implies the negation of the Axiom of Symmetry. Therefore, by contrapositive, the Axiom of Symmetry implies the negation of CH. Conversely, if CH fails, then for any choice of ω_1 many distinct reals x_α, for $\alpha < \omega_1$, the set $\bigcup_{\alpha < \omega_1} f(x_\alpha)$ has size ω_1. Then, since we assumed $\neg$CH, there exists some real number y such that $y \notin f(x_\alpha)$. Since $f(y)$ has at most countably many x_α elements, there must exist some x_α satisfying that $x_\alpha \notin f(y)$. So then we have that $y \notin f(x_\alpha)$ and $x_\alpha \notin f(y)$. Therefore, the Axiom of Symmetry is equivalent to $\neg$CH. $\square$

If Freiling's argument is valid, then it provides what Hamkins calls the *dream solution*. However, Freiling's axiom is not a widely accepted axiom candidate. The first problem stems from the naive use of the notion of measure and probability, and the implicit assumption that there is a uniform way of associating a probability to any subset of the reals, hence the argument does not rely on the formal treatment of the notion of measure. Hamkins, in his work [131], says:

> Ultimately, rather than being accepted as the longed-for solution to the continuum hypothesis, Freiling's argument is instead most often described as providing an attractive equivalent formulation of $\neg$CH [...] Freiling's simple philosophical argument is turned on its head, used not as a justification of the axiom, but rather as a warning about the error that may arise from a naive treatment of measure concepts, a warning that what seems obviously true might still be wrong.[7]

The second problem is pointed out by Maddy, in [182], where she writes:

> A common objection to this line of thought is that various natural generalisations contradict the axiom of choice as well. For example, if Freiling's principle is modified to cover assignments of sets of any cardinality less than that of the reals, the result immediately implies that there is no well-ordering of the reals. Similarly if we are allowed to throw $\omega + 1$ darts. Members of the Cabal suggest that Freiling's hypothesis yields a picture more like that of the full AD-world [...] than of the choiceful universe V.[8]

---

[7]Hamkins, pp. 141–142, 2015.
[8]Maddy, p. 500, 1988. AD abbreviates the Axiom of Determinacy

Returning back to our discussion on determining the status of CH, one option is that we can give up on finding a definite answer and believe that status of CH depends on our perspective of the set-theoretical universe. The multiverse conception of sets, which was discussed in the previous chapter, is a position that challenges the foundational view. It actually alters the notion of truth. From the viewpoint of the multiverse conception, CH has no definite answer. Nevertheless, the set theorist still desires to be able to have access to higher infinities. Koellner [169] says that 'the motivation behind the generic multiverse is to grant the case for large cardinal axioms and definable determinacy but deny that statements such as CH have a determinate truth value.'[9] This was ensured by the "generic multiverse". Let M be a countable transitive model of set theory. The *generic multiverse generated by* M, denoted by $\mathbb{V}_M$, is the smallest class of countable transitive models such that for every pair (M_0, M_1) of countable transitive models, where M_1 is a forcing extension of M_0, if either $M_0 \in \mathbb{V}_M$ or $M_1 \in \mathbb{V}_M$, then both models are in $\mathbb{V}_M$. Taking M to be the cumulative hierarchy V, we get the *generic multiverse*. The corresponding notion of truth in the generic multiverse is defined as follows. A sentence φ is *true* in the generic multiverse (generated by V) if φ is true in each universe of the generic multiverse. These are the universal laws of set theory. Since the laws cannot change when you pass to another Cohen blueprint, the forcing method cannot be used to show that a statement is independent in the generic multiverse notion of truth. In summary, the generic multiverse conception of truth admits three types of sentences: universal laws, i.e., statements that are true in every model of the generic multiverse, their negation, and statements that are true in some models yet false in others. This leads to a mathematical conjecture called the Ω-*conjuecture*, but for this we shall refer the reader to Woodin [302] for further details.[10]

17.1.1 Axiom of Constructibility

Our next option is to refine the concept of set. Recall the definition of the cumulative hierarchy of sets (see section 13.3 of Chapter 13), where we begin with the empty set, and at each next stage, we take the power set of what we defined in the previous stage. At limit ordinal stages, we take the union of all sets we formed. This defines the cumulative hierarchy. The power set operation, that is, the operation of taking *all* subsets of a given set V_α at stage α, may be giving us too many sets in the end. Gödel, in his consistency proof of CH, considered instead the "definable" subsets of V_α.

Definition 17.2 A set X is *definable* in a model M, with parameters $a_1, \ldots, a_n$, if there exists a formula φ in the language of set theory and some

---

[9] Koellner, §4.2, 2013.

[10] See Woodin [302] (2011), §5–9, for a summary. See Woodin [298] (1999), §10.4, for a detailed account.

$a_1, \ldots a_n \in M$ such that

$$X = \{x \in M : M \models \varphi(x, a_1, \ldots, a_n)\}.$$

Let $\operatorname{def}(M) = \{X \subseteq M : X \text{ is definable in } M\}$ be the set of all definable subsets of M. Note that $M \in \operatorname{def}(M)$ and that $M \subseteq \operatorname{def}(M) \subseteq \mathcal{P}(M)$. Using transfinite induction, we define

$L_0 = \emptyset$,
$L_{\alpha+1} = \operatorname{def}(L_\alpha)$,
$L_\alpha = \bigcup_{\beta < \alpha} L_\beta$, if α is a limit ordinal.
For all ordinals α, $L = \bigcup_\alpha L_\alpha$.

We call L the *constructible universe*. The statement that $V = L$ is called the *Axiom of Constructibility*, which was introduced by Kurt Gödel in his work [116]. The Axiom of Constructibility says that every set is definable.

Theorem 17.2 (Gödel, 1939) If $V = L$, then the General Continuum Hypothesis holds.

Although it looks tempting to accept $V = L$, it appears to be quite restrictive due to the theorem given below by Scott [256]. Let us say that a cardinal κ is called *measurable* if it is the critical point of a non-trivial elementary embedding $j : V \to M$ for some transitive class M.[11]

Theorem 17.3 (Scott, 1961) If $V = L$, then there are no measurable cardinals.

Many set theorists are inclined to reject the Axiom of Constructibility for its restrictiveness. Woodin says, on a pragmatic basis, that 'the axiom $V = L$ limits the large cardinal axioms which can hold and so the axiom is *false*'.[12] At this point, we should examine Maddy's objection to $V = L$. As a matter of fact, Gödel, as the inventor of the Axiom of Constructibility, is not in favour of his invention, that is, the idea that all sets are constructible.[13] Gödel says that 'certain facts (not known at Cantor's time) [...] seem to indicate that Cantor's conjecture will turn out to be wrong'.[14] Gödel continues to present implausible consequences of CH so as to justify the falsity of Cantor's conjecture on extrinsic grounds.[15] On a footnote, Gödel writes that $V = L$ 'states a minimum property' and that 'only a maximum property would seem to harmonise with the concept of set'.[16] This view apparently received lots

---

[11]The *critical point* of j is the first ordinal which is not mapped to itself. That is, if $j(\alpha) = \alpha$ for every $\alpha < \kappa$ such that $j(\kappa) > \kappa$, then κ is the critical point of j.
[12]Woodin, p. 504, 2011.
[13]See also the discussion we gave in section 12.3 of Chapter 12.
[14]Gödel, p. 479, 1964.
[15]ibid, p. 479. See Martin [194] (1976), p. 87, for an opposing view.
[16]ibid, p. 479.

of attention. Essentially, Maddy's arguments rely on the idea that $V = L$ is too restrictive and that it does not maximize the concept of set. Moschovakis [201] puts this as follows:

> The key argument against accepting $V = L$ [...] is that the axiom of constructibility appears to restrict unduly the notion of *arbitrary* set.[17]

He further claims that 'there is no *a priori* reason why every subset of ω should be definable from ordinal parameters'.[18] On the other hand, Fraenkel, Bar-Hillel, and Levy [96] give some supporting arguments for the Axiom of Constructibility. They write:

> As an additional axiom for set theory the axiom of constructibility is somewhat attractive [...] It has also good number of additional mathematical interesting consequences. The most dramatical consequence [...] is the negation of the famous *Souslin hypothesis* [...] The axiom of constructibility also appeals to one's sense of economy — if we think of the ordinal numbers as a fixed given totality then the axiom of constructibility asserts that there are no sets other than those which can be proved to exist.[19]

John Steel claims that what we are really attempting to maximize is the 'interpretative power of the language of set theory', that a believer in $V = L$ can always translate his sentence into the language of set theory used by the believer in the existence of measurable cardinals but not vice versa.[20] But recall Hamkins' multiverse axioms we gave in section 16.2 of Chapter 16 where we stated the axiom of absorption into L. For that 'even if we have very strong large cardinal axioms in our current set-theoretic universe V, there is a much larger universe V^+ in which the former universe V is a countable transitive set and the axiom of constructibility holds'.[21]

Maddy's objection against $V = L$ rests on her criticism of *definabilism*, the view that every mathematical theory or entity must be definable in some uniform way.[22] Throughout the history of mathematics, we witnessed many man-

---

[17] Moschovakis, p. 610, 1980.

[18] ibid, p. 610.

[19] Fraenkel, Bar-Hillel, and Levy, pp. 108–109, 1973.

[20] Feferman *et al*, p. 423, 2000.

[21] Hamkins, p. 28, 2014.

[22] As it might be appreciated, the exact conditions of "uniformity" is arguable. One of the earliest definabilist views is Pythagoreanism. The fact that there are irrational numbers in geometric magnitudes, however, refutes the Pythagoreanistic doctrine. But this does not put definabilism in the disposal. Despite the fact that $\sqrt{2}$ lies beyond the Pythagoreanistic conception of the universe, it is still algebraically definable, as $\sqrt{2}$ is the length of the hypotenuse of a right-angled triangle whose legs are of unit length. Moreover, $\sqrt{2}$ is a computable number, that is, given any natural number n, we can algorithmically determine its n digit in the decimal expansion. So from the point of view of the Pythagorean specification of the mathematical and physical universe, $\sqrt{2}$ is not definable. However, the criterion for definability is not static notion and in the more general sense, $\sqrt{2}$ is still a definable object.

ifestations of definabilism in various forms. Examples include the Pythagorean view of the universe, Descartes's suggestion of considering curves that are definable by algebraic equations as the only legitimate type of objects in analytic geometry, representations of functions, Gödel's Axiom of Constructibility, and the Church-Turing Thesis that effective computability is bounded by Turing-computability. Zermelo's [309] choice function was a central issue in definabilism. The Axiom of Choice received some resistance from Baire, Borel, and Lebesgue [15]. In regards to finding a way to define non-analytically representable functions, Lebesgue writes:

> The question comes down to this [...] Can we prove the existence of a mathematical object without defining it? [...] I believe we can only build solidly by granting that it is impossible to demonstrate the existence of an object without defining it.[23]

On the other hand, unlike Lebesgue, Hadamard supported Zermelo's position.[24] Definabilism also appears in the form of predicativity. Poincaré [221], a well-known proponent of predicativism, argued against Zermelo's Well-Ordering Theorem for making use of impredicative definitions. He notes:

> Never consider any objects but those capable of being defined in a finite number of words [...] Avoid non-predicative classifications and definitions.[25]

Although Russell and Whitehead introduced a type theoretic mathematical system on a predicative basis, Gödel writes:

> (Type theory) makes impredicative definitions impossible and thereby destroys the derivation of mathematics from logic, effected by Dedekind and Frege, and a good deal of modern mathematics itself [...] I would consider this rather as a proof that the vicious circle principle is false than that classical mathematics is false.[26]

In Gödel's words, 'the *vicious circle principle* is that no totality can contain members definable only in terms of this totality, or members involving or presupposing this totality'.[27]

In a 1934 lecture, Paul Bernays [33] introduced a competing view called *quasi-combinatorialism*, which he defined in the following manner:

> [A]nalysis [. . .] abstracts from the possibility of giving definitions of sets, sequences, and functions. These notions are used in a

---

[23] Baire *et al* (1905); in Moore [200] (1982), p. 314.
[24] See ibid, pp. 317–318.
[25] Poincaré, p. 63, 1909. Also in Poincaré [222] (1913), p. 31.
[26] Gödel, p. 455, 1944. Paranthesis added.
[27] ibid, p. 454.

"quasi-combinatorial" sense, by which I mean: in the sense of an analogy of the infinite to the finite.

Consider, for example, the different functions which assign to each member of the finite series $1, 2, \ldots, n$ a number of the same series. There are n^n functions of this sort, and each of them is obtained by n independent determinations. Passing to the infinite case, we imagine functions engendered by an infinity of independent determinations which assign to each integer an integer, and we reason about the totality of these functions.

In the same way, one views a set of integers as the result of infinitely many independent acts deciding for each number whether it should be included or excluded. We add to this the idea of the totality of these sets. Sequences of real numbers and sets of real numbers are envisaged in an analogous manner. From this point of view, constructive definitions of specific functions, sequences, and sets are only ways to pick out an object which exists independently of, and prior to, the construction.[28]

He continues to elaborate that 'the axiom of choice is an immediate application of the quasi-combinatorial concepts'.[29] From this viewpoint then, *all* possible functions from a given domain to another exist already. Furthermore, it is not required whether each of these mappings can be described uniformly according to some rule, and this is in complete contrast to the idea of constructivism.

Notice that the construction of $L_{\alpha+1}$, in Gödel's constructible universe, is based on taking the definable subsets of L_α. Interestingly, Gödel views L inherently non-constructive for the *a priori* use of the ordinals as indices. In a letter to Hao Wang [290], Gödel writes:

However, as far as, in particular, the continuum hypothesis is concerned, there was a special obstacle which *really* made it *practically impossible* for constructivists to discover my consistency proof. It is the fact that the ramified hierarchy, which had been invented *expressly for constructivistic purposes*, has to be used in an *entirely nonconstructive way*.[30]

Maddy affirms this problem as she says in a footnote that 'from the predicativist's point of view, $V = L$ is an axiom divided against itself: the formulas used to define sets at each stage are predicative, but the ordinals that index those stages need not be'.[31] Regarding the objection to the idea that every ordinal needs to be definable, Gödel argues that such an objection is plausible.

---

[28]Bernays, pp. 259–260, 1934.
[29]ibid, p. 260.
[30]Wang, p. 10, 1974.
[31]Maddy, p. 129, 1997.

He remarks in [119] that 'there is certainly some justification in this objection. For it has some plausibility that all things conceivable by us are denumerable, even if you disregard the question of expressibility in some language'.[32]

Up to this point, it is understood that adding $V = L$ as an axiom restricts the set-theoretical realm. Maddy continues on to compare $V = L$ with another axiom candidate, namely $V = \mathrm{HOD}$. Let us define what HOD is. A set X is called *ordinal definable* if there exists a formula φ such that

$$X = \{y : \varphi(y, \alpha_1, \dots, \alpha_n)\}$$

for some ordinals $\alpha_1, \dots, \alpha_n$. The *transitive closure* of a set S, denoted by $\mathrm{TC}(S)$, is defined as

$$\mathrm{TC}(S) = \bigcap \{T : S \subseteq T \text{ and } T \text{ is transitive}\}.$$

A set X is called *hereditarily ordinal definable* if every element of $\mathrm{TC}(X)$ is ordinal definable. In other words, the members of X, the members of members of X, and so on, are ordinal definable in this case. Let HOD denote the class of all hereditarily ordinal definable sets. The axiom $V = \mathrm{HOD}$ is an alternative axiom candidate to $V = L$ with more beneficial properties. Maddy raises two points concerning why we can be less judgmental about $V = \mathrm{HOD}$. She says:

> The first is simply that the definitions in HOD need not be pred-
> icative, which severs $V = \mathrm{HOD}$'s connection with the strongest
> form of Definabilism (i.e., predicativity); from the point of view
> we've been considering, this difference alone would warrant a less
> suspicious attitude towards $V = \mathrm{HOD}$. The second is that [...]
> everything set theorists have come up with is consistent with
> $V = \mathrm{HOD}$.[33]

Recall Gödel's intention on finding an axiom with a "maximum" property to harmonise with the concept of set (see footnote 16). Maddy takes a similar approach for deflating $V = L$ on the basis of her naturalistic philosophy. She introduces two methodological maxims called UNIFY and MAXIMIZE. The independence results in set theory led to many interpretations of the concept of set. As MacLane [196] puts it, 'for these reasons [...] the purported foundation of all Mathematics upon set theory totters'.[34] Maddy's methodological maxim UNIFY is obtained by 'running this argument in reverse'.[35] In some sense, UNIFY is the preservation of foundationalism, as she continues to claim that 'if set theorists were not motivated by a maxim of this sort, there would be no pressure to settle CH'.[36] It follows that an anti-UNIFY maxim must be in favour of the pluralist conception. Her second methodological maxim is the MAXIMIZE, which she defines as follows:

---

[32]Gödel, p. 151, 1946.
[33]Maddy, p. 131, 1997. Parenthesis added.
[34]MacLane, p. 359, 1986
[35]Maddy, p. 209, 1997.
[36]ibid, p. 209.

[T]he set theoretic arena in which mathematics is to be modelled should be as generous as possible; the set theoretic axioms from which mathematical theorems are to be proved should be as powerful and fruitful as possible. Thus, the goal of founding mathematics without encumbering it generates the methodological admonition to MAXIMIZE.[37]

Hamkins [130] points out that Maddy's rejection of $V = L$, on the basis of her maxims, implicitly assumes a monist conception of background ontology. He says:

[T]he $V \neq L$ via maximize argument relies on a singularist as opposed to pluralist stand on the question whether there is an absolute background concept of ordinal, that is, whether the ordinals can be viewed as forming a unique completed totality. The argument, therefore, implicitly takes sides in the universe versus multiverse debate, and I shall argue that without that stand, the $V \neq L$ via maximize argument lacks force.[38]

We are not going to delve into details on what Maddy means by a "restrictive" theory.[39] But we shall mention that she claims $V = L$ should be rejected on the ground that it lacks proving non-constructible isomorphism types.[40] Instead, she proposes to adopt the existence of another large cardinal called $0^{\#}$ (zero sharp). Roughly defined, it is the set of true formulas about "indiscernibles" in the constructible universe. Let us give some definitions before giving the exact definition of $0^{\#}$. Let κ be an infinite cardinal and let $\mathcal{M}$ be a model which contains all ordinals $\alpha < \kappa$. A set $I \subseteq \kappa$ is called a *set of indiscernibles* for the model $\mathcal{M}$ if for every natural number n and every formula $\varphi(x_1, \ldots, x_n)$,

$$\mathcal{M} \models \varphi(\alpha_1, \ldots, \alpha_n) \text{ if and only if } \mathcal{M} \models \varphi(\beta_1, \ldots, \beta_n)$$

whenever $\alpha_1 < \cdots < \alpha_n$ and $\beta_1 < \cdots \beta_n$ are increasing sequences of elements of I. Let $\overline{\varphi}$ denote the Gödel number of the formula φ. Define *zero sharp* as

$$0^{\#} = \{\overline{\varphi} : L_{\aleph_\omega} \models \varphi(\aleph_1, \ldots, \aleph_n)\}.$$

Studying the mathematical developments of the theory of large cardinals here is beyond the scope of this book.[41] However, we shall state a very nice equivalency theorem by Kunen regarding the characterisation of $0^{\#}$.

Theorem 17.4 (Kunen) $0^{\#}$ exists if and only if there exists a non-trivial (i.e., non-identity) elementary embedding $j : L \to L$.

---

[37] ibid, p. 211.

[38] Hamkins, p. 26, 2014.

[39] See Maddy (1997) III. 6 (i) for details.

[40] See ibid, III. 5, footnotes 7 and 14.

[41] See Kanamori [158] (2003) for a complete treatment.

Another interesting theorem we shall state is called *Kunen's inconsistency theorem*, which can be given as below.

Theorem 17.5 (Kunen, 1971) There exists no non-trivial elementary embedding $j : V \to V$.

Kunen's inconsistency theorem draws a line concerning which kinds of large cardinal axioms we can add to set theory. Although many of the large cardinal axioms are not known to be inconsistent with ZFC, the existence of such an embedding, as stated in the theorem, leads to a contradiction. The critical point of the non-trivial elementary embedding $j : V \to V$ is called a *Reinhardt cardinal*, introduced by William Nelson Reinhardt in his works [240] and [241]. It follows then from Kunen's inconsistency theorem that Reinhardt cardinals do not exist if ZFC is consistent. But Kunen's proof uses the Axiom of Choice. It is still an open question whether such an embedding is consistent with ZF.[42] Also note that Theorem 17.4 does not contradict Kunen's inconsistency theorem. It is known that $0^{\#}$ is a non-constructible set. If $0^{\#}$ exists, then $V \neq L$.

Now that we know what $0^{\#}$ is, let us return to our discussion on Maddy's argument for preferring ZFC + "$0^{\#}$ exists" over $V = L$. Her preference of the large cardinal axiom "$0^{\#}$ exists" is simply due to its benefits. She writes:

> [T]here are things like $0^{\#}$ that are not in L. And not only is $0^{\#}$ not in L, its existence implies the existence of an isomorphism type that is not realised by anything in L [...] So it seems that ZFC + $V = L$ is restrictive because it rules out the extra isomorphism types available from ZFC + '$0^{\#}$ exists'.[43]

At this point, however, one may question the applicability of arguments against definabilism. It can be said that the arguments used against $V = L$, that it restricts the use of arbitrary sets, only applies when we assume *all* sorts of sets do actually exist in the absolute realist sense. The notion of arbitrary set is only used in the hypothesis of universal statements. What is wrong with saying "If A is a definable set, then ..." *if* we actually happen to accept solely definable sets? For if this was the case, we would not end up restricting the set-theoretic domain due to that definable sets would be all there is. This is similar to the argument given by Michael Dummett to support the Axiom of Choice in constructivism.[44] Secondly, the definabilist maxim does not appear to override its application to other mathematical concepts due to the fact that, for example, as it has turned out that for about 90 years or so, the notion of *computability* is well captured by Turing mechanical principles. The Church-Turing Thesis *is* a definabilist hypothesis which describes the intuitive concept of computability with Turing machines. If definabilism

---

[42]See Akira [3] (1999) for a partial solution to this problem.
[43]Maddy, p. 216, 1997.
[44]See our section 4.2 of Chapter 4, footnote 27.

is an undesirable methodological maxim, shouldn't it be abandoned for all mathematical concepts? There is no *a priori* reason to exempt the concept of set from the need of providing an exact specification, while confining the notion of computability to a mechanical model and leave no room for its conceptual improvement, e.g., maximising the concept of computability. Can one be a definabilist on what a "proof" or "computation" should be but endorse the opposite view and adopt maximising principles to openly discover what a "set" might be? Can't we say that the Church-Turing Thesis is "restrictive" for that it does not allow us to do supertasks, as long as we consider them as legitimate effective procedures? If this is the case, the maximising principles can be perfectly applied to the Church-Turing Thesis. If not, then there seems to be a double-standard in our "minimisation" treatment for the notion of computability, being one of the foundational concepts in mathematics.

17.2 Inner Model Programme

We see that adding $V = L$ as an axiom is not really a plausible option, although it provides a very nice, neat, and well-behaved universe. An *inner model* of ZFC is a transitive class that contains all ordinals and satisfies the axioms of ZFC.[45] Gödel's constructible universe L is an inner model of ZFC, and so is HOD. In fact, L is the smallest possible inner model.[46] Since adding $V = L$ does not work for various limitations, the next thing to consider is to "enlarge" L to the extent that it contains genuine large cardinals, i.e., those bigger than measurable cardinals. This is called the *inner model programme*. The goal of the inner model programme is to enlarge the constructible universe so as to make the universe compatible with as many large cardinal axioms as possible.

However, prima facie, there seems to be a small problem with this approach. Due to the incremental nature of cardinals, in order to know if we have an ultimate expansion of the constructible universe, we need to know all notions of higher infinity. But this is simply impossible. Furthermore, we need to make sure that the enlargement does not get affected by any generalisation of Scott's theorem (see Theorem 17.3). By works of Woodin [300], [301], [303], and Hamkins [128], a large cardinal of a certain type, namely "supercompact" cardinals, played a major role in the later developments of the inner model programme. We first give some definitions to formulate the axiom $V = $ Ultimate-L. Much of the definitions can be found in Woodin [304]. For completeness though, we begin with the basic notions of topology.

---

[45] See Steel [274] (2010), which can also be found in Foreman and Kanamori [159] (2010), for an introduction to inner model theory. Also see Jensen [157] (1995) for a summary of the same subject.

[46] See Jech [155] (2003), pp. 182–188, for a proof.

A *topological space* is a set X together with a collection $\mathcal{T}$ of subsets of X satisfying the following conditions:

(i) The empty set and X itself belong to $\mathcal{T}$.

(ii) Any arbitrary (finite or infinite) union of members of $\mathcal{T}$ belongs to $\mathcal{T}$.

(iii) The intersection of any finite number of members of $\mathcal{T}$ belongs to $\mathcal{T}$.

We call the elements of $\mathcal{T}$ *open sets* and the collection $\mathcal{T}$ is called a *topology* on X.

If X is a topological space and p is a point in X, a *neighbourhood* of p is a subset of X that includes an open set U containing p. Suppose S is a subset of a topological space X. A point $x \in X$ is called a *limit point* of S if every neighbourhood of x contains at least one point of S different from x itself. A subset A of a topological space X is called *dense* in X if every $x \in X$ either belongs to A or is a limit point of A. A *nowhere dense* set is a set that is not dense in any non-empty open set. If X is a topological space and $A \subseteq X$, then we say that A is *meager* if it can be written as the union of countably many nowhere dense subsets of X.

A subset A of a topological space X has the *property of Baire* if there exists an open set $U \subseteq X$ such that $A \Delta U$ is meager, where $A \Delta U = (A \cup U) - (A \cap U)$.

The following definition is due to Feng, Magidor, and Woodin [90].

Definition 17.3 A set $A \subseteq \mathbb{R}^n$ is *universally Baire* if for all topological spaces Ω and for all continuous functions $\pi : \Omega \to \mathbb{R}^n$, the preimage of A under π has the property of Baire in the space Ω.

Definition 17.4 Suppose that $A \subseteq \mathbb{R}$. We define $L_\alpha(A, \mathbb{R})$ by induction on α as follows:

(i) (Base case) $L_0(A, \mathbb{R}) = V_{\omega+1} \cup \{A\}$,

(ii) (Successor case) $L_{\alpha+1}(A, \mathbb{R}) = \mathrm{def}(L_\alpha(A, \mathbb{R}))$,

(iii) (Limit case) $L_\alpha(A, \mathbb{R}) = \bigcup\{L_\beta(A, \mathbb{R}) : \beta < \alpha\}$.

$L(A, \mathbb{R})$ is the class of all sets X such that $X \in L_\alpha(A, \mathbb{R})$ for some ordinal α.

A sentence φ is a Σ_2-*sentence* if, for some sentence ψ, it is of the form

"There exists an ordinal α such that $V_\alpha \models \psi$".

Definition 17.5 Suppose that $A \subseteq \mathbb{R}$. Then, $\mathrm{HOD}^{L(A,\mathbb{R})}$ is the class HOD as defined within $L(A, \mathbb{R})$.

We note that the Axiom of Choice holds in $\mathrm{HOD}^{L(A,\mathbb{R})}$ even if $L(A, \mathbb{R}) \models$ AD.

Definition 17.6 Suppose $A \subseteq \mathbb{R}$ is universally Baire. Then, $\Theta^{L(A,\mathbb{R})}$ is the supremum of the ordinals α such that there exists an onto function $\pi : \mathbb{R} \to \alpha$ such that $\pi \in L(A, \mathbb{R})$.

Definition 17.7 A cardinal δ is called a *Woodin cardinal* if for all $A \subseteq V_\delta$, there are arbitrarily large $\kappa < \delta$ such that for all $\lambda < \delta$ there exists an elementary embedding $j : V \to M$ with critical point κ such that $j(\kappa) > \lambda$, $V_\lambda \subseteq M$, and $A \cap V_\lambda = j(A) \cap V_\lambda$.

We are now ready to state the axiom for $V =$ Ultimate-L.

The Axiom for $V =$ Ultimate-L.

1. There is a proper class of Woodin cardinals.

2. For each Σ_2-sentence φ, if φ holds in V then there is a universally Baire set $A \subseteq \mathbb{R}$ such that $\mathrm{HOD}^{L(A,\mathbb{R})} \models \varphi$.

The axiom $V =$ Ultimate-L provides a well described conception of the set-theoretical universe in which many independent problems are settled. The axiom is intended to describe a single universe conception and it undermines what is known as the *forcing axioms* that allow us to construct set-theoretical universes in which many combinatorial problems of cardinal arithmetic, independent of ZFC, are resolved.[47] The following theorem provides some of the consequences of $V =$ Ultimate-L.

Theorem 17.6 If $V =$ Ultimate-L holds, then

(i) CH holds,

(ii) $V = \mathrm{HOD}$,

(iii) V is not a forcing extension of any transitive class $N \subseteq V$.

We need to define a few more concepts before we state the "$V =$ Ultimate-L conjecture" which Woodin hopes that it would be proved in the future. First is the notion of supercompact cardinal, which is due to Solovay *et al* [272].

Definition 17.8 If $\kappa \le \lambda$, then κ is *λ-supercompact* if there is an elementary embedding j from V into a transitive inner model M such that

(i) j has critical point κ and $j(\kappa) > \lambda$,

(ii) M contains of all of its λ-sequences (i.e., sequences of length λ).

We say that κ is *supercompact* if κ is λ-supercompact for all $\lambda \ge \kappa$.

Secondly, we need the notion of "weak extender model" for a supercompact cardinal to approximate or generate an elementary embedding from a system of measures.[48] The following definition is due to Reinhardt [241].

---

[47]See Schatz [253] (2019) for an investigation of the conflict between $V =$ Ultimate-L and the forcing axioms.

[48]For this we refer the reader to Woodin (2017), in particular, p. 8.

Definition 17.9 A cardinal δ is called an *extendible cardinal* if for each $\lambda > \delta$ there exists an elementary embedding

$$j : V_{\lambda+1} \to V_{j(\lambda)+1}$$

such that the critical point of j is equal to δ and $j(\delta) > \lambda$.

Surprisingly, the enlargement for *exactly one* supercompact cardinal turned out to be provably the ultimate enlargement of L. As posited by Woodin, one just then needs to prove that the following conjuecture is true.

The $V =$ Ultimate-L conjecture.

Suppose that δ is an extendible cardinal. Then there exists a weak extender model N for the supercompactness of δ such that

1. $N \subseteq \mathrm{HOD}$,

2. $N \models$ "$V =$ Ultimate-L".

If one manages to prove the $V =$ Ultimate-L conjecture, then it turns out that the axiom $V =$ Ultimate-L is consistent with *all* large cardinal axioms.[49] So Woodin's $V =$ Ultimate-L axiom is most likely to be the missing piece of the puzzle to understand the universe V.

Set theory has two possible futures. Ronald Jensen proved in his work [156] a dichotomy theorem for L, stating that either L is very "close" to V, or very "far" from it. More precisely, he proved the following.

Theorem 17.7 (Jensen, 1972) Exactly one of the following holds:

1. Every singular cardinal γ is singular in L, and $(\gamma^+)^L = \gamma^+$. (L is "close" to V)

2. Every uncountable cardinal is inaccessible in L. (L is "far" from V)

A similar dichotomy for HOD was proved by Woodin in [303] based on his works [300] and [301].

Theorem 17.8 (HOD dichotomy) Assume that there exists an extendible cardinal κ. Then exactly one of the following holds.

1. For every singular cardinal $\gamma > \kappa$, γ is singular in HOD, and $(\gamma^+)^{\mathrm{HOD}} = \gamma^+$.

2. Every regular cardinal $\gamma \geq \kappa$ is measurable in HOD.

In the first alternative, HOD is "close" to V, and in the second case, HOD is "far" from V. The difference between the HOD dichotomy and Jensen's L

---

[49]See Magidor [192] (2012), p. 14, for some concerns.

dichotomy is that the existence of even modest large cardinals like $0^{\#}$ determines which side of the dichotomy we are on. If $0^{\#}$ exists, then we know L is far from V. However, since all traditional large cardinal axioms are consistent with $V = \text{HOD}$, in the case of HOD dichotomy, no traditional large cardinal axiom results in the "far" side of HOD dichotomy. In fact, the inner model programme aims to establish the first case of this dichotomy.

The second alternative, on the other hand, is established through the programme of large cardinals beyond choice, i.e., large cardinal axioms consistent with ZF.[50] We know that Reinhardt cardinals are inconsistent with ZFC, but it is an open question whether the existence of Reinhardt cardinals is consistent with ZF. Then choiceless large cardinal hierarchy merely begins with the Reinhardt cardinal. Most of large cardinals beyond choice become inconsistent if the $V = \text{Ultimate-}L$ conjecture holds. In contrast, if these large cardinal axioms are consistent, then HOD is "far" from V and in this case the axiom $V = \text{Ultimate-}L$ is false.

17.3 Hyperuniverse Programme

Meanwhile, on the European continent, Tatiana Arrigoni and Sy-David Friedman [10] delivered a programme to settle the same problem of providing a description for V. As our final discussion in this chapter, we summarise what they call the *hyperuniverse programme*.[51] The programme is described as follows:

> The Hyperuniverse Programme is an attempt to clarify which first-order set-theoretic statements (beyond ZFC and its implications) are to be regarded as true in V, by creating context in which different pictures of the set-theoretic universe can be compared. This context is the *hyperuniverse*, defined as the collection of all countable transitive models of ZFC. The comparison of such models evoke *principles* (principles of *maximality* and *omniscience*, as we will name two of them) that suggest *criteria* for preferring, on justifiable grounds, certain universes of sets over others.[52]

They further claim that 'in formulating the Hyperuniverse Programme, Platonism is nowhere invoked, either with regard to V or to the hyperuniverse. To the contrary, some of its characteristic features clearly ex-

---

[50]See Bagaria, Koellner and Woodin [14] (2019) for a detailed study of the large cardinal hierarchy without the Axiom of Choice.

[51]See Antos *et al* [6] (2018) for a more complete account including recent mathematical results.

[52]Arrigoni and Friedman, p. 79, 2013.

press an anti-Platonistic attitude, which makes the program radically different from Gödel's'.[53] For them, two types of statements are regarded as "*true* in V". The first type is the axioms of ZFC and the consistency of ZFC + large cardinal axioms, which are called *de facto* statements. Secondly, there are statements 'that obey condition for truth explicitly established at the outset'.[54] The authors call them *de jure* set theoretic truths and these are sentences which are true in all preferred universes of the hyperuniverse.

In comparing the hyperuniverse programme with $V =$ Ultimate-L, it can be said that the hyperuniverse programme does not quite endorse the naturalistic position, as the authors write:

> When declaring the intention of extending ZFC independent questions, one also requires that one be as *unbiased* as possible to the way such questions should be settled and as to which principles and criteria for preferred universes one should formulate. In particular, the latter must not be chosen at the outset so as to be apt for settling questions independent of ZFC, or for meeting the needs of some particular area of existing set-theoretical practice.[55]

Friedman, in his [105], formulates a principle called the *inner model hypothesis*. A statement is *internally consistent* iff it holds in some inner model, under the assumption that there are inner models with large cardinals. He describes the *inner model hypothesis* as follows:

> The *inner model hypothesis (IMH)* asserts that the universe has been maximised with respect to internal consistency in the following sense: If a statement φ without parameters holds in an inner model of some outer model of V (i.e., in some model compatible with V), then it already holds in some inner model of V. Equivalently: If φ is internally consistent in some outer model of V then it is already internally consistent in V.[56]

Interestingly, Arrigoni and Friedman state that 'although the IMH is compatible with the internal consistency of very large cardinals (i.e., their existence in inner models), it contradicts their existence in the universe V as a whole'.[57] Since the hyperuniverse is the class of all transitive models of ZFC, it can be regarded as a multiverse of models. In fact, as explained earlier, Woodin's generic multiverse serve a similar purpose. The difference between the hyperuniverse and the generic multiverse relies on the notion of *class forcing*, introduced by Friedman [104]. Arrigoni and Friedman say:

---

[53]ibid, p. 80.

[54]ibid, p. 80.

[55]ibid, p. 81.

[56]Friedman, p. 596, 2006.

[57]Arrigoni and Friedman, p. 82, 2013. See Friedman, Welch and Woodin [106] (2008) for upper and lower bounds for the consistency strength of IMH in the large cardinal hierarchy.

Earlier work of the second author of this paper led to the introduction of the *class-generic multiverse around L*, obtained by closing L under class-forcing and class-generic ground models, as well as inner models of class-generic extensions that are not necessarily themselves class-generic [...] The *set-generic multiverse* and the *class-generic multiverse* are quite different: the former preserves large cardinals notions and does not lead beyond set forcing, whereas the latter can destroy large cardinals and leads to models that are not directly obtainable by class forcing.[58]

Arrigoni and Friedman announce three desiderata for a successful realisation of their hyperuniverse programme.[59]

Desideratum 1. The multiverse should be as rich as possible but it should not be an ill-defined or open-ended multiplicity.

Desideratum 2. The hyperuniverse is not an ultimate plurality. One can express preferences for certain members of it according to criteria based on justified principles.

Desideratum 3. Any first-order property of V is reflected into a countable transitive model of ZFC which is a preferred member of the hyperuniverse.

The question that which universes of the hyperuniverse are preferred is also considered by the authors. The factors are based on what they call *maximality* and *omniscience* criteria. The term maximality is used here in a similar sense of that Gödel's. Instead of structural maximality, they argue for logical maximality, 'for there is no tallest countable transitive model of ZFC'.[60] The logical maximality is divided into two subrequirements, namely *ordinal (vertical) maximality* and the *power set (horizontal) maximality*.[61]

Another factor for the preference of the universe in the hyperuniverse is the *omniscience* criterion. More precisely, it is defined in the following manner:

---

[58]ibid, pp. 83–84. See Woodin [299] (2009), p. 107, for his objection.

[59]See Arrigoni and Friedman (2013), pp. 85–87, for a detailed discussion of the desiderata.

[60]ibid, p. 88.

[61]The distinction between the two, as to why they are called *vertical* and *horizontal*, makes more sense if we look at the depiction of the cumulative hierarchy (see Figure 13.3). The height of the v-shaped hierarchy depicts the amount of "sizes of sets" in the universe, e.g., large cardinals. Whereas the width shows how many, say, real numbers exist in each level of the hierarchy. We cannot extend the height of V by the forcing method since V is all there is. So Cohen's method instead expands the width of the hierarchy by adding generic reals. This is why there are so many independent problems in set theory and that it is not hard to come up with statements independent from the axioms of ZFC. This is also the reason why set theorists desire to find a structure of the set-theoretical universe so as to be able to discover what there is within the structure, rather than arbitrarily defining what they are.

A universe is *omniscient* if it is able to describe what can be true in alternative universes. A precise criterion based upon this principle is the following.

Criterion of Omniscience: Let Φ be the set of sentences with arbitrary parameters from v which can hold in some outer model of v. Then Φ is first-order definable in v.[62]

The quest for new axioms has been a long standing ambitious project to understand what the set-theoretical universe V might be. The main goal of the hyperuniverse project, thus, is to provide a structure for set theory. It is through the understanding of the set-theoretical universe we begin to discover set theoretic truths. Set theory is overwhelmed by independence results. Consequently, many theories are being *formed* by the aid of pragmatic choices of set theorists, and they arbitrarily define the truth values for independent statements in their assumptions. The multiverse conception, therefore, emerged from the result of these choices made by set theorists. In the end, we can plot two possible futures for set theory, as usually put forward by Hugh Woodin. Either this enlargement programme succeeds and we finally have a plausible structure of set theory in which we can actually *do* mathematics, or else set theory remains as a mathematical discipline where we define the truth value of statements that we posit as we please, by using Cohen's method.

Discussion Questions

1. Might there be other motivations in enlarging our axiomatic system apart from solving independent problems?

2. Can Freiling's argument of the Axiom of Symmetry be seen as an informal proof similar to the proof of the fact that there are "computable" functions that are not primitive recursive?

3. Do you think there are intrinsic justifications for adding $V = L$ as an axiom? More specifically, what might be the *a priori* reasons for believing that all subsets of natural numbers are definable from ordinal parameters?

4. What are the trade-offs between definabilism and naturalism?

5. Discuss whether or not Maddy's MAXIMIZE restrains the multiverse conception. That is, if there are competing theories for our set conception, which one should be taken as our primary theory to preserve MAXIMIZE? Should they grow independently or is it best if MAXIMIZE is applied on a single theory?

6. In the large cardinal hierarchy, some large cardinal axioms have a higher

---

[62]ibid, p. 91. Here, the authors define "v" as a variable that ranges over the elements of the hyperuniverse (see ibid, p. 88).

consistency strength than others. This means that if an axiom φ has a higher consistency strength than that of ψ, then the consistency of ZFC+φ implies the consistency of ZFC + ψ. In some sense, consistency strength of a theory measures the hierarchical distance to inconsistency. Do you think that the results obtained from ZFC + ψ are more reliable than the same results entailed by ZFC + φ?

7. The structure of sets pretty much depends on what V might be. We still may not be able to find a satisfying answer no matter of the proposed axiom candidates. What might be the role of the fact that set theory is a foundational theory in the struggle of our quest for finding the structure of sets? In other words, would it be easier to find what V was if set theory was not a foundational theory for mathematics? Is the struggle merely due to the fact that set theory is a foundation of mathematics?

18

Mathematical Nominalism

We have saved the most dramatic philosophical view for the final chapter of this book. Despite the fact that, as a mathematician, I was at first reluctant to elaborate on such a position which might have prima facie come across as "nihilistic" to the mathematical community, it is one of the mainstream views about which many philosophers of mathematics contributed a large number of studies. Regardless of its criticisms—and I believe that nominalism still does maintain its importance in the philosophical literature—it most certainly deserves a careful study. We should note that the material given here is not intended to present a complete account on nominalism but rather give a summary of various nominalistic views, mainly Hartry Field's *fictionalism* [91] and [93], including its revisions by Mark Balaguer [17], Charles Chihara's *modal nominalism* [59], and Jody Azzouni's *deflating nominalism* [13]. We also look at a critical survey of nominalism by Burgess and Rosen [51].

In the philosophy of mathematics, there are two contrasting positions regarding the ontological status of mathematical objects to one of which we may in many cases "blindly" commit, so to speak. These positions are *realism* and *nominalism*, where the former being the view that mathematical objects exist as non-spatiotemporal, non-mental and causally inert abstract beings which are purportedly independent of the physical world, and the latter being the view that there are no abstract objects (i.e., metaphysically speaking, there are no universals but only particulars).[1] Mathematical nominalism is a special case of the latter metaphysical view which argues for that mathematical entities do not exist as abstract objects.[2] Nominalism is not the only anti-realist view however. An anti-realist in the general sense may accept the existence of abstract objects but argue that they are, say, mind-dependent. In most cases though, anti-realism refers to the rejection of the existence of universals.

---

[1] An *abstract* object is one that is non-spatiotemporal and causally inefficacious. Objects that are not abstract are called *concrete*.

[2] Some versions of nominalism argue that mathematical objects exist as linguistic constructs. Some argue that they are physical objects. Field's position holds the view that mathematical objects are fictional.

DOI: 10.1201/9781003223191-18

18.1   Problems of Realism

Mathematical nominalism stems from various ontological and epistemological problems that occur in realism. Apart from the long standing general problem of how we, as physical beings, obtain any knowledge about the mathematical realm, a more contemporary one is Benacerraf's identification problem (see section 14.2 of Chapter 14). Given a natural number n, we may represent n with different set-theoretical notations. But then which set notation corresponds to the "true" representation of n? As Field puts it, 'there is no uniquely natural set-theoretic explication of natural numbers'.[3] Nevertheless, Crispin Wright [305] does not find this problematic as he states the following:

> Benacerraf complains that nothing in our use of numerical singular terms is sufficient to determine which, if any, classes they stand for. But the same is evidently true of singular terms.[4]

A more serious challenge to realism is again seen in Benacerraf's 1973 paper [29] which relies on the *causal theory of knowledge*, a view that is largely abandoned today.[5] The problem that Benacerraf worries about realism is purely epistemological and it is very broadly described as follows. If mathematical objects exist as non-spatiotemporal abstract beings in the platonic sense, then platonists are required to explain how we come to know anything about mathematical objects. Benacerraf's concern is 'having a homogeneous semantical theory in which semantics for the propositions of mathematics parallel the semantics for the rest of the language, and the concern that the account of mathematical truth mesh with a reasonable epistemology'.[6] His general thesis is that 'almost all account of the concept of mathematical truth can be identified with serving one or another of these masters at the expense of the other'.[7] Benacerraf is particularly worried about the conflict of two requirements. The first requirement is concerned with the theory of truth. In fact, it is demanded that any theory of mathematical truth should conform to a general theory of truth. The second requirement concerns the relationship between the *knower* and what is *known*. Benacerraf says that 'it must be possible to link up what it is for p to be true with my belief that p'.[8] The nominalist philosopher exploits the missing epistemic relation between the abstract mathematical realm and the observable universe.

---

[3]Field, p. 20, 1989.

[4]Wright, p. 125, 1983.

[5]See Goldman [110] (1967).

[6]Benacerraf, p. 403, 1973. Page references to the reprint version in Putnam and Benacerraf (1983).

[7]ibid, p. 403.

[8]ibid, p. 409.

18.2 Field's Fictionalism

Studies in mathematical nominalism begin with Goodman and Quine [113] in which they present what they call *constructive nominalism*. But since they impose finitism in their nominalistic structures, the syntactic and proof-theoretic notions defined in their paper still fall short of the usual platonist counterparts. However, Quine later changed his views, rather in a reluctant way, in favour of existence of mathematical objects by arguing that mathematics is indispensable for science.[9] Hartry Field ingeniously provided in his *Science without Numbers* [91] a nominalistic account of the crucial parts of physics with no reference to, or quantification over, abstract mathematical entities. We investigate in this section Field's nominalisation programme of science by pointing out the essential parts and some criticisms raised in the literature.[10]

Field states that 'there is one and only one serious argument for the existence of mathematical entities', and that is the Quine-Putnam indispensability argument, that 'we need to postulate such entities in order to carry out ordinary inferences about the physical world and in order to do science'.[11] Field does not propose to reformulate classical mathematics nominalistically, but he rather demonstrates that the mathematics we need for our basic physical sciences does not have to refer to abstract entities, and so the scientific theories do not need to use objects that refer to, or quantify over, numbers, functions, sets, etc. Essentially, this is the rejection of the second premise of the indispensability argument. Field sees no reason to regard the part of mathematics containing references to abstract entities as true.[12] He admits that the axioms of the real number system is with no doubt important and that they are non-arbitrary. But he adds that this arbitrariness does not imply that the axioms are true.[13] It is also worth noting that Field does not completely abandon the mathematical practice, as he claims that using mathematics, it is possible 'to draw nominalistically statable conclusions from nominalistically statable premises'.[14] A nominalist can do this 'not because he thinks those intervening premises are true, but because he knows that they preserve truth among nominalistically stated claims'.[15] The argument he provides for this relies on the analogy between formal proofs in recursion theory using Turing machines and informal proofs using the Church-Turing Thesis. Field claims

---

[9]See our section 13.2 of Chapter 13 for the indispensability argument.

[10]All references of *Science without Numbers* in this chapter will be to the second (2016) edition.

[11]Field, p. 5, 2016. See Putnam [226] (1971), §5–8, for a detailed argument for the indispensability of mathematics.

[12]Field, pp. 1–2, 2016.

[13]ibid, pp. 4–5.

[14]ibid, p. 10.

[15]ibid, p. 14.

that for every platonist proof of a nominalistic statement, from nominalistical premises, there exists a nominalistic proof of the same thing.[16]

Field's first task is to reformulate science using a nominalistic language.[17] By doing this so, it is ensured that any reference to abstract entities is intended to be eliminated from scientific theories. But what about its strength compared to platonistic theories? Second task then is to show that the platonistic theory of science is "conservative" over the nominalistic theory. We will explain what is meant by this shortly.

The realisation of the first task can be thought of as something quite similar to what Hilbert did in his axiomatisation of the Euclidean geometry on a formalist basis. As Field puts it, 'we can say that Hilbert's theory is one in which the quantifiers range over regions of physical space, but do not range over numbers'.[18] Euclidean geometry can be treated on two separate grounds, namely *synthetically* and *analytically*.[19] The difference between the two treatments is the use of "mathematical" entities for basic terms like point, line, and other geometric concepts. The basic concepts in "synthetic" geometry, the way they were presented in ancient Greece, do not make use of abstract "points" as if they refer to real number values. The Cartesian development of "analytic" geometry, on the other hand, offers another approach to model n-dimensional spaces with reference to real numbers and treating every point as a coordinate of n-tuples of real numbers. The nominalist, therefore, prefers the synthetic approach over the analytic approach. In fact, in synthetic geometry, the points, straight lines, planes can be replaced with arbitrary objects as long as the axioms of geometry are satisfied.[20] Similarly, Field's approach is synthetic. He says:

> I believe that such 'synthetic' approaches to physical theory are advantageous not merely because they are nominalistic, but also because they are in some ways more illuminating than metric approaches: they explain what is going on without appeal to extraneous, causally irrelevant entities.[21]

Regarding the first task, Field begins with the nominalisation of the Newtonian gravitational theory. Field argues that points and regions in spacetime are concrete objects. His spacetime account is *substantivalist*, i.e., one that characterises space (or spacetime) with objects that exist on their own right. This is in opposition to *relationism*, where one characterises space in terms of the relations of objects, possibly described in a mathematical manner. Field's ulterior motive in defending substantivalism against the relationalist view is explained as follows.

---

[16] See ibid, p. 14, footnote 10 for details.

[17] A nominalistic language is one that does not refer to abstract objects whatsoever and that which does not quantify over such entities (cf. section 13.2 of Chapter 13).

[18] ibid, p. 27.

[19] Field refers to analytic geometry as the *metric approach*. See ibid, p. 42 for details.

[20] See section 6.2 of Chapter 6 footnote 11 for Hilbert's comment.

[21] Field, p. 43, 2016.

> A substantivalist viewpoint is important to my position on the
> applications of mathematics in physical theories. But my main
> point has been to illustrate my contention that the use of modal-
> ity to draw non-modal conclusions almost always turns on an
> equivocation in the modal concepts employed.[22]

The substantivalist position does not need to be restricted to just physically observable medium sized objects. In fact, the term mathematical "fictionalism" appears to be due to that the objects that constitute spacetime
regions can as well be merely theoretical physical entities like quarks, strings,
and so on, almost like fictional objects that are postulated to aid the entire
the physical theory. In some sense, points in a nominalistic spacetime are
quasi-concrete.

Field begins by supplying the conditions for "conservativeness" of platonic
mathematical theories over nominalistic theories so as to complete the second
task mentioned above. He first notes that it would be wrong to say that $N + S$
is a "conservative extension" of N, where N a is body of nominalistic assertions and S is a mathematical theory in the usual platonic sense. The reason
is that since N is a nominalistic theory, it may assert things that rule out the
existence of abstract entities, so in this case, $N + S$ would be inconsistent. To
overcome this problem, Field first introduces a unary predicate $M(x)$, meaning that "x is a mathematical entity". Secondly, given a nominalistic assertion
A, let A^* be the same assertion except that we restrict the quantification of
A to those entities x which satisfy $\neg M(x)$. Finally, if N is a body of nominalistic assertions, let N^* denote the collection of all A^* for A in N. Field calls
N^* the *agnostic version* of N. He gives the example that 'if N says that all
objects obey Newton's laws, then N^* says that all non-mathematical objects
obey Newton's laws, but it allows for the possibility that there are mathematical objects that don't'.[23] He states three related versions of his principle of
conservativeness. For simplicity, we shall state the first one.

Principle of Conservativeness. Let A be any nominalistically statable assertion, and N be any body of such assertions; and let S be any mathematical
theory. Then A^* is not a consequence of $N^* + S + $ '$\exists x \neg M(x)$' unless A is a
consequence of N.

In other words, the principle says that the mathematical theory S is conservative over a body N of nominalistic assertions if for any nominalistic
assertion A, A is not a consequence of $S + N$ unless A is a consequence of
N just alone.[24] The relationship between conservativeness and consistency is
immediate for pure mathematical theories. Concerning this relation, Field [92]
says:

---

[22]Field, p. 42, 1989.
[23]Field, p. 11, 2016.
[24]See ibid, ch. 1 for a detailed account on conservativeness.

When S is a pure mathematical theory, i.e., a theory whose variables range over mathematical entities alone, then the conservativeness of S is an obvious consequence of its consistency. But impure mathematical theories, such as set theories with urelements, have variables ranging over mathematical and nonmathematical entities alike; and for some such set theories conservativeness is a bit stronger than consistency.[25]

The purpose of conservativeness is to bypass mathematical truth in its application to our scientific theories. Mathematics does not need to be true so as to be useful for science. Field gives the general procedure of application of mathematics to a synthetically represented branch of science as follows.

Suppose N is a body of nominalistically stated premises; in the case that will be of primary interest, N will consist of the axioms of a nominalistic formulation of some scientific theory [...] the key to using a mathematical system S as an aid to drawing conclusions from a nominalistic system N lies in proving in $N^* + S$ the equivalence of a statement in N^* alone with some other statement (which I'll call an *abstract counterpart* of the N^*-statement) which quantifies over abstract entities. Then if we want to determine the validity of an inference in N^* (or equivalently, of an inference in N), it is unnecessary to proceed directly; instead we can if it is convenient 'ascend' from one or more statements in N^* to abstract counterparts of them, then use S to prove from these abstract counterparts an abstract counterpart of some other statement in N^*, and 'descend' back to that statement in N^*.[26]

The first chapter of Field's *Science without Numbers* is entirely devoted to the demonstration that mathematics is conservative.[27] The notion of conservativeness—since the concept of "consequence" is interpreted by either semantic or syntactic means—relies on some metamathematical requirements, particularly the completeness of first-order logic. If we ask whether mathematics is really conservative, then these metamathematical requirements must be used in the formulation of conservativeness in nominalistic terms, and in the demonstration that set theory is conservative. Now the nominalist is relied on the completeness of first-order predicate logic to produce an appropriate formulation of conservativeness without reference to abstract mathematical entities. Due to the Completeness Theorem of first-order logic (see Theorem 11.2 in section 11.1 of Chapter 11), the aforementioned distinction is not relevant for first-order sentences. The problem arises in second-order logic since there is no completeness theorem for second-order logic. Linnebo observes that

---

[25] Field, p. 240, 1985.
[26] Field, p. 22, 2016. Also see pp. 23–24 for an example of this general procedure.
[27] See ibid, Appendix to ch. 1.

if N is a second-order logic and P is an extant mathematical physics, then 'P is not conservative over N, since P proves many nonlogical truths'.[28]

Within the synthetic geometry that Field constructs, we may represent arithmetic and numbers. Shapiro [257] claims that this leads to a problem in Field's programme. If sufficient amount of arithmetic can be represented within the synthetic theory, then there must be a synthetic Gödelian statement. Shapiro argues that set theory with urelements, call is S, is not deductively conservative over nominalistic physics. In other words, there is a sentence θ which can be formulated in the language of a given nominalistic theory N such that θ is a deductive consequence of $S + N$ but not of N. His proof relies on the idea that one can capture the structure of arithmetic in the given nominalistic interpretation of spacetime, which isomorphic to $\mathbb{R}^4$.[29] In reply to Shapiro's objection, Field considers Shapiro's worry as a worry about whether the account of the application of mathematics to science requires a strong extended version of the representation theorem (see below). Field argues that if this extended-representation theorem holds, then no model of a certain nominalistic theory N can be formulated in first-order logic.[30] According to Field, such a nominalistic theory makes use of a strong form of logic called *mereology*.[31] Mereology is a theory of the part/whole relation. Mathematically, the axioms of mereology are the axioms for a complete Boolean algebra with the Axiom of Choice. It very much resembles the axioms of set-theoretic inclusion operator $\subseteq$ without the assumption of existence of atomicity. ZFC set theory is a bottom up construction (see the cumulative hierarchy), whereas mereology is a two-way up and down construction starting from a reference point and building parts and wholes, respectively, as we go down and up. Burgess and Rosen notes that mereology has existential implications. They claim that 'if it is accepted, then the acceptance of some initial entities involves the acceptance of many further entities, arbitrary wholes having the initial entities as parts'.[32]

To form the synthetic version of Newtonian gravitational theory, the abstract counterparts of sentences concerning the space/spacetime substitute real number valued coordinates with "points" and subtitute sets of coordinates with "regions", where the latter is conceived as a *Goodmanian sum* of points.[33] Such a substitution is ensured by the "representation theorem", originally proved by Hilbert, which we discuss in the next paragraph. But let us now continue with the first task of Field's nominalisation programme and discuss how he formulises the nominalistic version of scientific theories.

---

[28]Linnebo, p. 110, 2017.

[29]See Shapiro (1983), p. 526–527 for the complete argument. Also see Burgess and Rosen (1997), p. 119.

[30]Field, p. 245, 1985.

[31]Field, p. 131, 1989.

[32]Burgess and Rosen, p. 156, 1997.

[33]See Goodman [111] (1972), Part IV for Goodman's calculus of individuals.

Field begins his nominalisation with the Newtonian gravitational theory and the space/spacetime structure. For this he introduces various relational symbols so as to provide a nominalistic interpretation of geometrical concepts that are used in physics. For finding abstract counterparts, he states the representation theorem, that is, the existence of a homomorphism that bridge between concrete objects and their abstract counterparts. He defines a ternary predicate

$$y \text{ Bet } xz$$

meaning that y is between x and z. More precisely, y Bet xz means "y is a point on the line-segment whose endpoints are x and z". Then he defines a four place predicate

$$xy \text{ Cong } zw$$

meaning that "the distance from point x to point y is the same as the distance from point z to point w". He also denotes the distance between point x and y by $d(x, y)$. Such a distance function is ensured to exist by the *representation theorem* which Hilbert also proved in the axiomatisation of Euclidean geometry. Given an arbitrary model of the axiom system for space, there is at least one function d which maps pairs of points onto the non-negative real numbers, satisfying the following *homomorphism conditions*:

(a) for any points x, y, z and w, xy Cong zw if and only if $d(x, y) = d(z, w)$;

(b) for any points x, y and z, y Bet xz if and only if $d(x, y) + d(y, z) = d(x, z)$.

Field replies some objections regarding the Hilbertian axiomatisation of space that the space contains uncountably many points. The reason is that there is no difference between postulating a rich physical space and postulating real numbers. But recall from our earlier discussion, Field notes that the existence of uncountably many concrete objects does not violate nominalism, yet postulating the existence of 'even one real number' does.[34] Although Field's spacetime is isomorphic to $\mathbb{R}^4$, its elements, i.e., space-time points, are non-mathematical. So in Field's motivation there is no concern about the cardinality of objects as long as no abstract entity is posited to exist. It is more about how one makes sense of the cardinal quantification without using mathematical objects.[35]

The allowance of infinity in Field's programme is one major difference from Goodman and Quine's nominalistic treatment as the latter forbids the use of infinity. Field's nominalisation of spacetime is slightly different in that aspect. In a letter Quine wrote to Field, dated April 23, 1980, he says:

> In our little "Steps toward a constructive nominalism" of long ago, Goodman and I welcomed mereology of physical objects,

---

[34]See our section 14.1 of Chapter 14, footnote 4, for Field's reply to the cardinality objection and the note in the second edition.

[35]Thanks to Hartry Field for pointing this out in a personal correspondence.

but denied ourselves space-time regions and all postulation of infinity.[36]

Field continues with the nominalisation of quantity-related concepts that exist within spacetime, like temperature, mass, and so on. For example, he defines a ternary relation "y Temp-Bet xz" which means that "y is a spacetime point at which the temperature is (inclusively) between the temperatures of points x and z". Similarly, he defines a four place relation "xy Temp-Cong zw" which intuitively means that "the temperature difference between points x and y is equal in absolute value to the temperature difference between points z and w". He even gives a synthetic account of other complex notions like continuity, derivative, gradients, differentiation of vector fields, and so on.[37]

In fact, Field also formulates the Continuum Hypothesis in his synthetic version of physics. In his [93], he says:

> A number of people have observed that the continuum hypothesis has an analogue for physical space, statable with purely physical resources (i.e., without set theory): it will say that for every subregion of a line, there is either a graph portraying a 1-1 correspondence to another line or a graph portraying 1-1 correspondence to a discrete region of type ω.[38]

Field claims that the difficulty in deciding the Continuum Hypothesis presumably applies to its physical analogue as well.

The majority of Field's programme makes use of *intrinsic explanations* of nominalised structures rather than *extrinsic explanations*, 'e.g. in Bayesian psychology, using relations of comparative credence rather than numerical credence functions'.[39] For Field, the introduction of a frame of reference and metric units like "meter", "kilograms", or "celsius degree" is an arbitrary choice of convention. He states a non-nominalistic principle along the lines that 'underlying every good extrinsic explanation there is an intrinsic explanation'.[40] Field remarks that the nominalised space structure is best explained 'by the facts about physical space which are laid down without reference to numbers in Hilbert's axioms'.[41] An intrinsic explanation, thus, is one which does not depend on any convention, and from Field's viewpoint, the nominalist would be more interested in intrinsic explanations.

Field's programme received a number of criticisms. We already mentioned Shapiro's concern about the conservativeness. David Malament [193] wrote another critique regarding the applicability of Field's approach on quantum mechanics. Malament gives 'two examples where Field's strategy would not

---

[36]Field, P-55, 2016.
[37]See Field (1980/2016), §7–8.
[38]Field, p. 48, 1989.
[39]Field, P-5, 2016.
[40]ibid, p. 44.
[41]ibid, p. 29.

seem to have a chance: classical Hamiltonian mechanics, and ordinary (non-relativistic) quantum mechanics'.[42] Balaguer [17] presents a nominalisation of quantum mechanics along the lines of Field's fictionalist approach.[43] Balaguer's main claim is stated in the following manner.

> [T]he closed subspaces of our Hilbert spaces can be taken as representing *physically real properties* of quantum systems. In particular, they represent *propensity* properties, for example, the r-strengthed propensity of a state-Ψ system to yield a value in Δ for a measurement of A (or, to give a more concrete example, the 0.5-strengthed propensity of a z+ electron to be measured spin-up in the x direction).[44]

Otávio Bueno [47] argues against Balaguer's formulation for his 'use of propensity properties'.[45] Bueno claims that Balaguer's suggestion is incompatible with certain interpretations of quantum mechanics, in particular, van Fraassen's version of the modal interpretaion. Burgess, in his work [50], where he claims to present a general nominalisation method applicable to a large class of physical theories, notes that the main difference between Field's and his approach is 'in the strategy for showing that the alternative presentation captures the content of the usual presentation'.[46] Burgess also gives some advantages and disadvantages compared to Field's approach. One advantage of Field's strategy is his choice of background theory, and a disadvantage is that it encounters certain technical obstacles.

Before examining further, we shall look at Balaguer's defense of fictionalism as an anti-platonist theory. Although some of his ideas were already mentioned in section 16.1 of Chapter 16, we now look at his *Platonism and Anti-Platonism in Mathematics* [17] a bit more to have an idea about his fictionalist account on the philosophy of mathematics.[47] Balaguer's main thesis is that we can never discover whether mathematical objects exist or not. In other words, his metaphysical conclusion is that there is no fact of the matter as to whether there exist abstract objects. Epistemologically speaking, he claims that there is no good argument for or against the existence of mathematical objects.[48] This means that Platonism and anti-Platonism are both equally plausible views and that there is no good reason for endorsing one over another. For Balaguer, the only tenable version of Platonism, he claims, is what he calls *full-blooded Platonism* (FBP), which is described as the view that any logically consistent object exists.[49] It pretty much resembles the

---

[42]Malament, p. 533, 1982.

[43]See Balaguer (1998), §6, for details.

[44]Balaguer p. 120, 1998.

[45]Bueno, p. 1427, 2003.

[46]Burgess, p. 394, 1984.

[47]See also Balaguer [18] (2009).

[48]See Balaguer (1998), p. 152, for more details on the difference between the epistemic and metaphysical conclusion.

[49]He also calls this *plenitudinous platonism*. This is in contrast with another view he calls

Hilbertian criterion of existence.[50] On FBP, one does not need causal contact with the objects of a logically consistent theory. If Balaguer is correct on FBP, then any consistent theory is true by virtue of itself.[51]

Balaguer's anti-Platonistic view is a fictionalist position. In his work [18], he develops his fictionalist philosophy. He argues that fictionalism is the only viable version of anti-Platonism. He also notes in one of his earlier works that 'most versions of anti-Platonism are not *importantly different than fictionalism*'.[52] Balaguer lists many similarities between FBP and fictionalism.[53] He concludes that they agree on almost everything except that the former accepts the existence of mathematical objects but the latter does not. He points out that the reason we can't settle the dispute whether mathematical objects exist or not is due to the Benacerrafian epistemic worry that we cannot have access to the Platonic universe that exists (if it does) outside of spacetime for which there is no causal connection between us and the abstract objects. Balaguer's metaphysical conclusion shares essential similarities with the principles of logical positivism. According to Balaguer, the sentence "there exist abstract objects" does not have any truth conditions, for he says that 'the factual emptiness of the dispute over abstract objects is rooted in *semantic* problems'.[54] What about his personal view then? Balaguer neither endorses Platonism nor anti-Platonism to its fullest. Instead, he rather accepts everything FBP and fictionalism say about the mathematical theory and practice, but disagrees with their exclusive claims, namely FBP's claim that there are abstract objects and the fictionalist claim that there do not exist such objects. Nonetheless, in his work [18], he develops what he claims as the best version of fictionalism, called *intention-based fictionalism* (IBF), which is essentially the fictionalist counterpart of his intention-based platonism (IBP). For Balaguer, fictionalism is the view that

(a) our mathematical theories do purport to be about abstract objects; but

(b) there are no such things as abstract objects; and so

(c) our mathematical theories are not literally true.[55]

On IBF, mathematical truth is defined in a non-literal sense within "the story of mathematics".[56] For Balaguer, statements like "2+2=4" are not taken

---

sparse platonism according to which that of all the different kinds of mathematical objects that might exist, only some of them actually exist. According to Balaguer, FBP is superior to sparse platonism.

[50]See our section 6.2 of Chapter 6, footnote 14.

[51]cf. intention-based Platonism (IBP).

[52]Balaguer, p. 97, 1998.

[53]See Balaguer (1998), pp. 152–155, for a detailed account.

[54]ibid, p. 159.

[55]Balaguer, p. 132, 2009.

[56]Regarding fictionalism, a similar nominalist view called *mythological Platonism* was presented by Charles Chihara in [58]. This appears to be an early version of fictionalism

as literally true. On IBF, is it said that a sentence S of mathematics is *correct* iff, in the story of mathematics, S is true in all parts of the mathematical realm that count as intended in the given branch of mathematics; and S is *incorrect* iff, in the story of mathematics, S is false in all intended parts of the mathematical realm. Otherwise, there is no fact of the matter as to whether S is correct or incorrect.[57]

One might ask if the notion of "fictional truth" could be replaced with "vacuous truth". We know that any statement could be written in the form of a conditional statement, where the antecedent can always be rephrased as the ontological assumption "If the related object exists". For example, "4 is even" is translated as "If 4 exists, then 4 is even". Or the Continuum Hypothesis (CH) is translated as "If all sets used in the set-theoretical formula for the CH exist, then CH holds". But on fictionalism there are no abstract objects. So instead of regarding the statement "4 is even" as fictionally true (in the story of arithmetic), can we just take it as a vacuously true statement? Prima facie, it might seem that we really do not need the concept of "fictionally correct" here. That is, does a statement S being "fictionally correct" really mean "S is correct if it were the case that abstract objects existed"? Couldn't this be captured by the notion of vacuous truth? For Balaguer, two views are equivalent except that they disagree about what is the *right* way to interpret *actual* mathematical claims, and for this he gives the following example.[58]

(i) When non-philosophers say things like "3 is prime", what they *really* mean is that *if* numbers existed, then 3 would be prime.

(ii) When non-philosophers say things like "3 is prime", they are making claims about abstract objects.

Balaguer mentions that fictionalists think that (ii) is true, whereas (i) is endorsed by the "vacuous truth" view.

18.3   Is Mathematics a "Subject with No Object"?

The provoking question given in the section title is surely one that should be argued and reasoned from both sides by nominalists and anti-nominalists. Burgess and Rosen's *A Subject with No Object* [51] delivers a general comparative study and a critical survey of different versions of mathematical

---

where he argues about the difference between statements that are "objectively true" and those which are "true-to-a-story". See Chihara (1973), pp. 62–75, for details.

[57]See Balaguer (2009), pp. 149–151, for more details. See also Balaguer [21] (2018) for an extensive survey on fictionalism in the philosophy of mathematics.

[58]Personal communication dated June 26, 2020. Added with permission.

nominalism. In a passage, Burgess and Rosen describe a stereotypical nominalist in the following manner:

> We nominalists hold that reality is a *cosmos*, a system connected by causal relations and ordered by causal laws, containing entities ranging from the diverse inorganic creations and organic creatures [...] to the various unobservable causes of observable reactions that have been inferred by scientific theorists [...] Surely anti-nominalists owe us a detailed explanation of how anything we do here (i.e., the empirical world) can provide us with knowledge of what is going on over there (i.e., the platonic universe).[59]

Since abstract objects are causally inert, the nominalist concludes that she has no causal connection with abstract entities. A stereotypical antinominalist, on the other hand, is visualised by Burgess and Rosen as a naturalistic second philosopher in the Maddian sense. Burgess and Rosen say that anti-nominalists 'come to philosophy believers in a large variety of mathematical and scientific theories—not to mention many deliverances of everyday common sense—that are up to their ears in suppositions about entities nothing like concrete bodies we can see or touch [...]'.[60] Both nominalism and anti-nominalism have merits and defects. In a philosophical debate, each side assumes the burden of proof to be on the other. The nominalist is expected to provide a complete framework of scientific language which avoids reference to abstract entities, whereas the anti-nominalist is obliged to provide an acceptable epistemology for mathematical objects.

Burgess and Rosen also talk about the common strategy to any nominalisation process and that is *formalisation* of the theory in consideration. Given a platonistic theory, the nominalistic reconstruction usually begins with introducing a symbolic language for the synthetic counterpart of the platonistic theory. The first step of formalisation is what Quine calls *regimentation*. This is the process of translating a sentence in ordinary English to a language so as to make it look like a sentence in logic written in natural language. For example, the sentence "Either you stay home or, as long as you feel good, you go to the library" would be translated to something like "(you stay home) or (if you feel good, then you go to the library)". Next stage is *symbolisation*, where we denote each proposition by a propositional variable and denote each propositional connective by its corresponding symbol. Thus, the symbolisation of the given compound statement in our example would be written as $p \vee (q \rightarrow r)$. All mathematical reference to numbers, functions, relations, sequences, continuity, and so on, can be eliminated by reducing all referents to pure sets using set theory. Burgess and Rosen also give a formal presentation of the method of reducing abstract things to concrete things. For the most straightforward reduction of one type of entity to another type, they give *objectual reduction* as an example: 'each X of the former sort is assigned an x

---

[59]Burgess and Rosen, p. 29, 1997. Parentheses added.
[60]ibid, p. 34.

of the latter sort as a proxy or surrogate, an understudy taking on its role, an impersonator assuming its identity. Words referring to an X can then be reinterpreted as referring to the x representing it. Predicates applying to Xs can then be reinterpreted as predicated applying to the xs representing them. Quantification over Xs [...] would be reinterpreted as quantifications over the xs representing them'.[61] Upon defining the notion "x represents X", it is required that we assume (i) every x represents at most one X, (ii) every X is represented by at least one x, (iii) every X is represented by at most one x. For Burgess and Rosen, the success of the method depends on the satisfaction of condition (i). To ensure this, it is sufficient that no two distinct physical entities of one sort coincide with each other. Distinct abstracta should correspond to distinct concreta.[62] Another type of reduction Burgess and Rosen mention is the *contextual reduction*, where one drops the condition (iii) and is allowed to represent X by more than one x. A broader sense of reduction that the authors discuss later on is the *Tarskian reduction*, introduced first by Tarski, Mostowski, and Robinson [281], to which we shall refer the reader for further information.[63]

Burgess and Rosen discuss three major nominalistic reconstruction in their book. The first is called the geometric strategy. In fact, this is what Hartry Field followed, via the substantivalist treatment of spacetime as opposed to a relationalist approach. The synthetic approach of nominalising spacetime, as we saw, relies on the abandonment of numerical values in favour of quasi-concrete geometric entities. According to Burgess and Rosen, the main dilemma that geometrical nominalism faces is explained in the following manner.

> The case for accepting geometric entities as concrete draws on realistic, contemporary, twentieth-century physics; but the most elegant elimination of numerical entities in favour of such geometric entities can be carried out only for unrealistic, classical, nineteenth-century physics. It remains an open question how attractive a nominalistic alternative to up-to-date physics can be developed.[64]

Modality in Nominalism. Another version of nominalism regards mathematical entities as linguistic constructs in which a mathematical statement is taken as a *façon de parler*, rather to replace the existence of mathematical objects by invoking certain modal operators that avoid ontological commitments.[65] For example, instead of saying that "the number 5 *actually* exists", the modal constructivist says "the number 5 *possibly* exists". We may call the

---

[61]ibid, p. 80.

[62]See Quine [231] (1950), p. 627, for further elaboration and a visual illustration of the problem.

[63]See Tarski *et al* (1953), §1 for details.

[64]Burgess and Rosen, p. 98, 1997.

[65]For an overview of this position, see Burgess and Rosen (1997), II-B.

general view *modal nominalism*.[66] Prominent supporters of this position include Charles Chihara [59], Geoffrey Hellman [141], and partly Hilary Putnam [225]. The primary aim of modal constructivists is to reinterpret mathematics on the grounds of modality. Chihara's discourse begins with a presentation of a Chinese puzzle consisting of a square cut up into seven pieces of different kinds. Any object that is formed by combining any of these seven shapes is what he calls a tangram. Given this square, it is possible to construct different sorts of tangrams having different shapes. Chihara's motivation stems from the problem that what sorts of tangrams we can possibly construct. He says:

> The phrase 'it is possible to construct a tangram such that' functions very much like a quantifier. Indeed, it will be shown in this work how one can interpret the existential quantifier of a version of classical mathematics to be functioning very much like the constructibility.[67]

Similar to the existential quantifier $\exists x$, Chihara introduces a constructibility quantifier Cx that is understood as

"It is possible to construct x such that".

Prima facie, the reader might wonder if Cx would coincide with the modal operator $\Diamond \exists x$. Chihara notes that two operators are interpreted differently in infinite domains.[68] Chihara's mathematical system concerns the constructibility of open sentences. Instead of talking about "all" things that are made of wood, Chihara takes "x is made of wood" and applies the constructibility operator to open sentences of the latter kind so as to avoid any form of ontological commitment. His Fregean style treatment reveals a type-theoretic structure. Type 0 objects are those which are "to be found", type 1 objects are those which are properties of those objects (i.e., open sentences that are satisfied by type 0 objects), type 2 objects are those which are properties of those objects (i.e., open sentences that are satisfied by type 1 open sentences), etc. In this structure, the "falling under" relation is replaced by "membership", and a property of objects is replaced by the set of those objects that have the property. For this reason, Chihara's system can be regarded as a simple type theory. However, type theory can assert the *actual* existence of a mathematical object of, say, type n. Whereas in Chihara's system, it is only *possible* to construct a type n open sentence. Putnam holds a similar position. According to Putnam [227], there is another possible way of doing mathematics, i.e., the modal way. He writes:

> [T]he mathematician [...] makes no existence assertions at all. What he asserts is that certain things are *possible* and certain

---

[66]cf. Hellman (1989), which was discussed in our section 14.3 of Chapter 14.
[67]Chihara, p. 24, 1990.
[68]See Chihara (1990), pp. 37–38, for details.

things are *impossible* — in a strong and uniquely mathematical sense of 'possible' and 'impossible'.[69]

Burgess and Rosen come to notice an obstacle for the modal strategy, specifically that 'there aren't enough concrete tokens of numerals'.[70] They suggest considering not just what the numerals are, but also what numerals there could have been. According to Burgess and Rosen, defending modal nominalism needs two requirements. Firstly, a modal nominalist must favour *intentionalism* and argue against *extensionalism*. Secondly, the modal nominalist must accept modal logical distinctions as undefined concepts, hence go against *reductivism*.

For Burgess and Rosen, there are two types of conceptions that a reconstructive nominalist endorses, namely the *revolutionary conception* and the *hermeneutic conception*. The goal of the revolutionary conception is to provide 'a novel mathematical and scientific theories to replace current theories'.[71] The hermeneutic conception on the other hand is the hypothesis that the platonistic theory, when interpreted correctly, already makes no reference to abstract entities, and that the nominalistic reconstruction of the platonistic theory provides the deep meaning of the words and the structures appearing in the platonistic theory.

18.4   Deflating Nominalism

Deflating nominalism is a view predominantly supported by Jody Azzouni, established in his [13], though it traces back to his earlier studies.

In his earlier (1994) work [12], Azzouni calls *linguistic realism* 'the doctrine that takes mathematical talk at face value' where the truth of mathematical statements is defined in the same way as the truth of empirical statements.[72] The primary aim in his [12] is to defend linguistic realism. Most of the discussions in that work revolve around the nature of reference and mathematical truth. Azzouni states that ontological commitment is a mere product of grammar.[73] As a matter of fact, Azzouni's account on ontology and reference is similar to Quine's conception of *truth by convention* [229]. Azzouni draws an epistemic distinction between *thick* posits and *thin* posits.[74] A reference to an object is called *thick* if it can be explained by an epistemology through a causal connection between the knower and the referred object or an object of

---

[69]Putnam, p. 70, 1975.

[70]Burgess and Rosen, p. 124, 1997.

[71]ibid, p. 6.

[72]Azzouni, p. 4, 1994.

[73]ibid, p. 87.

[74]See Linnebo [179] (2018) for an extensive study on thin objects. See also Maddy's *thin realism* in Maddy (2011), Part III.

the same kind. A reference to an object is called *thin* if the reference appears by postulating a theory for that object.[75] But for Azzouni, mathematical statements operate with even thinner epistemic requirements in such a way that 'mathematical posits, even those in applied mathematics, are *ultrathin*'.[76]

Azzouni's ontological philosophy evolved into his 2004 work which shaped the deflating nominalist position. He summarises his view as follows:

> I take true mathematical statements as literally true; I forgo attempts to show that such literally true mathematical statements are not indispensable to empirical science, and yet, nonetheless, I can describe mathematical terms as referring to nothing at all.[77]

His "separation thesis" claims that there is a distinction between existential truth and ontology. In other words, Azzouni separates ontological commitment from quantifier commitment, in particular, those involving existential quantifiers. From the commitment of an existential quantification, it does not follow any ontological commitment. For Azzouni, '*ontological independence*, in the appropriate sense, should be the decisive consideration on which the existence of something should be decided'.[78] Azzouni writes:

> Our general acceptance that fictional objects exist in no sense at all isn't a brute intuitive fact *about fictional objects*. Rather, this intuition is an application of a more general intuition that if something is entirely "made up" or is ontologically dependent on our linguistic practices or psychological states, then it exists in no sense at all.[79]

For Azzouni, thus, mathematical objects are dependent on our linguistic practices or psychological states. Even so that they may be indispensable for science—on deflating nominalism it is perfectly acceptable that mathematics is indispensable and that mathematical statements are true—we do not need to commit to their existence.

We see some criticisms made by various philosophers againts deflating nominalism. According to Bueno [48], deflating nominalism offers an easy road to nominalism 'which does not require any form of reformulation of mathematical discourse, while granting the indispensability of mathematics'.[80] However, Azzouni's account is not the only one which follows this trend.[81] Mark Colyvan [66] criticises all these accounts by claiming that there is no easy road to nominalism and the nominalist is left with Field's hard road.

---

[75] See also Azzouni (2004), pp. 136–137.
[76] Azzouni, p. 119, 1994. See also p. 74 for a detailed explanation of ultrathin objects.
[77] Azzouni, pp. 4–5, 2004.
[78] ibid, p. 120.
[79] ibid, p. 98.
[80] Bueno, §5.1, 2013.
[81] See for example Melia [198] (1995), and Yablo [307] (2005).

Discussion Questions

1. Suppose that an object is *abstract* iff it is mental. How would this definition affect the nominalist philosophy of mathematics?

2. Field's nominalisation programme is built on the claim that mathematics is dispensable for science. But Field also uses the methods in mathematics. Hence, one might argue for that science is not entirely mathematics-free. Does Field's programme then refute only a weaker version of the second premise of the indispensability argument?

3. Consider the synthetic treatment of geometry that makes reference to non-abstract objects to denote points and lines. Such objects could be just theoretical physical entities. How are we supposed to nominalistically justify reference to unobservable objects of the kind which are merely theoretical posits? How could we explain within nominalism the usage of these non-abstract yet unobservable theoretical objects in our scientific theories?

4. Principle of Conservativeness states that a mathematical theory S is conservative over a body N of nominalistic assertions if for any nominalistic assertion A, A is not a consequence of $S + N$ unless A is a consequence of N. Might there be exceptions to this principle?

5. For a nominalist, postulating the existence of abstract objects—no matter how many—might run against nominalism, whereas postulating the existence of arbitrary number of concrete objects might not. Does this line of argument run into trouble when the size of the objects that are claimed to exist belongs to a higher place in the ordering of cardinals? For example, what about postulating the existence of κ many concrete objects where κ is a large cardinal or even a higher infinite that contradicts ZFC?

6. Articulate further on the difference between "being true in the story of mathematics" and "being vacuously true".

7. On Balaguer's IBF (intention-based fictionalism), if a sentence S is true in some parts of the intended mathematical realm and false in other parts that we count as intended, then it is said that there is no fact of the matter as to whether S is correct or incorrect. As far as fictionalism goes, can we not invent a story of mathematics in which independent statements are always determined? In other words, can epistemic independence be avoided in fictionalism?

8. Argue whether or not the Law of Excluded Middle might be a limitation for Chihara's constructibility operator in model nominalism. More specifically, is the sentence "It is *possible* to construct x" entailed by the sentence "It is not the case that it is not possible to construct x"? Symbolically, do you think $\neg\neg Cx \rightarrow Cx$ is a legitimate inference in modal nominalism?

Bibliography

[1] Aczel, P., **Non-Well-Founded Sets**, CSLI Lecture Notes, Vol. 14 (1988).

[2] Adleman, L. M., *Molecular computation of solutions to combinatorial problems*, Science, **266**(5187), pp. 1021–1024 (1994).

[3] Akira, S., *No elementary embedding from V into V is definable from parameters*, Journal of Symbolic Logic, **64**(4), pp. 1591—1594 (1999).

[4] Al-Dhalimy, H., Geyer, C. J., *Surreal time and ultratasks*, The Review of Symbolic Logic, **9**(4), pp. 836–847 (2016).

[5] Antos, C., Friedman, S. D., Honzik, R., Ternullo, C., *Multiverse conceptions in set theory*, Synthese, **192**(8), pp. 2463–2488 (2015).

[6] Antos, C., Friedman, S. D., Honzik, R., Ternullo, C. (eds.), **The Hyperuniverse Project and Maximality**, Birkhäuser, Basel (2018).

[7] Apostle, H. G., **Aristotle's Philosophy of Mathematics**, University of Chicago Press, Chicago (1952).

[8] Aristotle, **Physics**. Trans. by W. D. Ross, Oxford University Press, Oxford (1936).

[9] Aristotle, **Metaphysics**. Trans. by W. D. Ross, Oxford University Press, Oxford (1924).

[10] Arrigoni, T., Friedman, S. D., *The hyperuniverse programme*, Bull. Symbolic Logic, **19**(1), pp. 77–96 (2013).

[11] Ayer, Alfred. J., **Language, Truth and Logic**, Victor Gollancz, London, 1936; reprinted by Penguin Books, London (2001).

[12] Azzouni, J., **Metaphysical Myths, Mathematical Practice**, Cambridge University Press, Cambridge (1994).

[13] Azzouni, J., **Deflating Existential Consequence**, Oxford University Press, New York (2004).

[14] Bagaria, J., Koellner, P., Woodin, W. H., *Large cardinals beyond choice*, Bulletin of Symbolic Logic, **25**(3), pp. 283–318 (2019).

[15] Baire, R., Borel, E., Hadamard, J., Lebesgue, H., *Five letters on set theory*, 1905; translated in Moore (1982), pp. 311–320.

[16] Balaguer, M., *Non-uniqueness as a non-problem*, Philosophia Mathematica, **6**, pp. 63–84 (1998).

[17] Balaguer, M., **Platonism and Anti-platonism in Mathematics**, Oxford University Press, New York (1998).

[18] Balaguer, M., *Fictionalism, theft, and the story of mathematics*, Philosophia Mathematica, **17**(2), pp. 131–162 (2009).

[19] Balaguer, M., *Full-blooded platonism*; in R. Marcus and M. McEvoy (eds.), **An Historical Introduction to the Philosophy of Mathematics**, pp. 719–732, Bloomsbury Publishing, New York (2016).

[20] Balaguer, M., *Mathematical pluralism and platonism*, Journal of Indian Council of Philosophical Research, **34**(2), pp. 379–398, Springer (2017).

[21] Balaguer, M., *Fictionalism in the philosophy of mathematics*; in Zalta, E. N. (ed.), **The Stanford Encyclopedia of Philosophy** (2018).

[22] Banach, S., Tarski, A., *Sur la décomposition des ensembles de points en parties respectivement congruentes*, Fundamenta Mathematicae, **6**, pp. 244–277 (1924).

[23] Barton, N., *Forcing and the universe of sets: Must we lose insight?*, Journal of Philosophical Logic, **49**, pp. 575–612 (2020).

[24] Beall, J. C., *Is Yablo's paradox non-circular?*, Analysis, **61**, pp. 176–187 (2001).

[25] Beeson, M. J., **Foundations of Constructive Mathematics: Meta-mathematical Studies**, Springer-Verlag, Berlin (1985).

[26] Bell, J. L., **Intuitionistic Set Theory, Studies in Logic**, **50**, College Publications (2014).

[27] Benacerraf, P., *Tasks, super-tasks, and modern eleatics*, The Journal of Philosophy, **59**(24), pp. 765–784 (1962).

[28] Benacerraf, P., *What numbers could not be*, The Philosophical Review, **74**(1), pp. 47-73 (1965).

[29] Benacerraf, P., *Mathematical truth*, The Journal of Philosophy, **70**(19), pp. 661–679, 1973; reprinted in Benacerraf and Putnam (1983), pp. 403–420.

[30] Benacerraf, P., Putnam, H. (ed.), **Philosophy of Mathematics: Selected readings**, Cambridge University Press, New York (1983).

[31] Beringer, T., Schindler, T, *A graph-theoretic analysis of the semantic paradoxes*, Bulletin of Symbolic Logic, **23**, pp. 442–492 (2017).

[32] Berkeley, G., **The Analyst; or, a Discourse addressed to an Infidel Mathematician**, J. Tonson in the Strand London (1734).

[33] Bernays, P., *On Platonism in Mathematics*, 1934; reprinted in Benacerraf and Putnam (1983), pp. 258–271.

[34] Bernays, P., *Hilbert, David*, in **Encyclopedia of Philosophy**, Vol. 3, pp. 496–504, P. Edwards (ed.), MacMillan Publishing Co. and The Free Press, New York (1967).

[35] Bishop, E., **Foundations of Constructive Analysis**, McGraw-Hill, New York (1967).

[36] Black, M., **The Nature of Mathematics: A critical survey**, Paterson, New Jersey (1959).

[37] Blum, L., Shub, M., Smale, S., *On a theory of computation and complexity over the real numbers*, Notices of the American Mathematics Society, **21**(1), pp. 1–46 (1989).

[38] Boole, G., **An Investigation of the Laws of Thought**, 1854; reprinted by Watchmaker Publishing, Gearhart, OR (2010).

[39] Boolos, G., *The iterative conception of set*, The Journal of Philosophy, 68(8), pp. 215–231, 1971; reprinted in his **Logic, Logic and Logic**, pp. 13–29, Harvard University Press, Cambridge (1998).

[40] Brooks, B., *Black hole quantum computers*, Technical Report, University of Colorado at Boulder (2008).

[41] Brouwer, L. E. J., *Over de grondslagen der wiskunde*, Doctoral dissertation, University of Amsterdam (1907).

[42] Brouwer, L. E. J., *Intuitionism and Formalism*, 1912; in Benacerraf and Putnam (1983), pp. 77–89.

[43] Brouwer, L. E. J., *Begründung der Funktionenlehre unabhängig vom logischen Satz vom ausgeschlossenen Dritten*, Erster Teil, Stetigkeit, Messbarkeit, Derivierbarkeit, KNAW Verh. 1e sectie, deel XIII, **2**, pp. 1–24 (1923).

[44] Brouwer, L. E. J., *Intuitionistische Zerlegung mathematischer Grundbegriffe*, Jahresbericht der Deutschen Mathematiker-Vereinigung, **33**, pp. 251–256, 1923; reprinted in L. E. J. Brouwer, **Collected Works: Philosophy and Foundations of Mathematics vol. 1**, A. Heyting (ed.), pp. 275–280, North-Holland, Amsterdam (1975).

[45] Brouwer, L. E. J., *Consciousness, philosophy, and mathematics*, 1948; in Benacerraf and Putnam (1983), pp. 90–96.

[46] Brown, J. R., **Philosophy of Mathematics: A Contemporary Introduction to the World of Proofs and Pictures**, Routledge, London (2008).

[47] Bueno, O., *Is it possible to nominalize quantum mechanics?*, Philosophy of Science, **70**, pp. 1424–1436 (2003).

[48] Bueno, O., *Nominalism in Philosophy of Mathematics*; in Zalta, E. N. (ed.), The Stanford Encyclopedia of Philosophy (2013).

[49] Burali-Forti, C., *Una questione sui numeri transfiniti*, Rendiconti del Circolo Matematico di Palermo, **11**, pp. 154–164 (1987).

[50] Burgess, J. P., *Synthetic mechanics*, Journal of Philosophical Logic, **13**(4), pp. 379–395 (1984).

[51] Burgess, J. P., Rosen, G., **A Subject with No Object**, Oxford University Press, Oxford (1997).

[52] Burgess, J. P., **Mathematical Rigor and Structure**, Oxford University Press, New York (2015).

[53] Button, T., Walsh, S., **Philosophy and Model Theory**, Oxford University Press, New York (2018).

[54] Carnap, R., **Logische Syntax der Sprache**, Springer, Wien (1934). Trans. by A. Smeaton as **The Logical Syntax of Language**, London: Kegan Paul (1937).

[55] Carnap, R., *Empiricism, semantics, and ontology*, Revue Internationale de Philosophie, **4**, pp. 20–40, 1950; in Benacerraf and Putnam (1983), pp. 241–257.

[56] Carroll, L., *What the Tortoise Said to Achilles*, Mind, **4**(14), pp. 278–280 (1895).

[57] Chaitin, G., **Algorithmic Information Theory**, Cambridge University Press, Cambridge (2004).

[58] Chihara, C., **Ontology and the Vicious-Circle Principle**, Cornell University Press, London (1973).

[59] Chihara, C., **Constructibility and Mathematical Existence**, Oxford University Press, Oxford (1990).

[60] Church, A., *An unsolvable problem in elementary number theory*, American Journal of Mathematics, **58**, pp. 345–363 (1936).

[61] Clark, P., Read, S., *Hypertask*, Synthese, **61**(3), pp. 387–390 (1984).

[62] Cohen, P. J., *The independence of the continuum hypothesis, I, II*, Proceedings of the National Academy of Sciences of the United States of America, **50**, pp. 1143–1148 (1963).

[63] Cohen, P. J., **Set Theory and the Continuum Hypothesis**, Dover Publications, New York (2008).

[64] Colyvan, M., **The Indispensability of Mathematics**, Oxford University Press, New York (2001).

[65] Colyvan, M., *The ontological commitments of inconsistent theories*, Philosophical Studies, **141**, pp. 115–123 (2008).

[66] Colyvan, M. *There is no easy road to nominalism*, Mind, **119**(474), pp. 285–306 (2010).

[67] Cook, R. T., *There are non-circular paradoxes (But Yablo's isn't one of them!)*, The Monist, **89**, pp. 118–149 (2006).

[68] Cooper, S. B., **Computability Theory**, Chapman & Hall, CRC Press, Boca Raton, FL, New York, London (2004).

[69] Curry, H. B., **Outlines of a Formalist Philosophy of Mathematics**, North-Holland, Amsterdam (1951).

[70] Curry, H. B., *Remarks on the definition and nature of mathematics*, Dialectica, **8**, pp. 228–233, 1954; reprinted in Benacerraf and Putnam (1983), pp. 202–206.

[71] Cusa, Nicolas, **Nicolas of Cusa: Selected Spiritual Writings**, trans. by H. L. Bond, Paulist Press, New York (1997).

[72] Çevik, A., *ω-circularity of Yablo's paradox*, Logic and Logical Philosophy, **20**(3), pp. 325–333 (2020).

[73] Çevik, A., **Matematik Felsefesi ve Matematiksel Mantık**, Nesin Yayıncılık, İstanbul (2019).

[74] Dalen, D. van, *From Brouwerian counter examples to the creating subject*, Studia Logica, **62**, pp. 305–314 (1999).

[75] Davies, E. B., *Pluralism in Mathematics*, Philosophical Transactions of the Royal Society A: Mathematical, Physical and Engineering Sciences **363**(1835), pp. 2449–2460 (2005).

[76] Davis, M., **Computability and Unsolvability**, Dover Publications, New York (1982).

[77] Deolalikar, V., Hamkins, J. D., Schindler, R., $P \neq NP \cap co\text{-}NP$ *for infinite time turing machines*, Journal of Logic and Computation, **15**(5), pp. 577–592 (2005).

[78] Deutsch, D., *Quantum Theory, the Church-Turing principle and the universal quantum computer*, Proceedings of the Royal Society of London, A 400, pp. 97–117 (1985).

[79] Diaconescu, R., *Axiom of choice and complementation*, Proceedings of the American Mathematical Society, **51**, pp. 176–178 (1975).

[80] Downey, R., Hirshfeldt, D., **Algorithmic Randomness and Complexity**, Springer-Verlag, Berlin (2010).

[81] Dummett, M., **Elements of Intuitionism, Oxford Logic Guides**, 2nd edition, Clarendon Press, Oxford (1977)

[82] Dummett, M., *Frege and Kant on geometry*, Inquiry: An Interdisciplinary Journal of Philosophy, **25**(2), pp. 233–254 (1982).

[83] Earman, J., *Bangs, Chrunches*, **Whimpers, and Shrieks**, Oxford University Press, New York (1995).

[84] Enderton, H. B., **A Mathematical Introduction to Logic**, 2nd edition, Academic Press, Cambridge (2001).

[85] Etesi, G. and Nemeti, I., *Non-Turing computations via Malament-Hogarth spacetimes*, International Journal of Theoretical Physics, **41**, pp. 341–370 (2002).

[86] Feferman, S., *Predicativity*; in **The Oxford Handbook of Philosophy of Mathematics and Logic**, Shapiro S. (ed.), pp. 590–624, Oxford University Press, Oxford (2005).

[87] Feferman, S., *Are there absolutely undecidable problems? Gödel's dichotomy*, Philosophia Mathematica, **14**(2), pp. 134–152 (2006).

[88] Feferman, S., *Is the Continuum Hypothesis a definite mathematical problem?*, Exploring the Frontiers of Independence, Harvard lecture series (2011).

[89] Feferman, S., Friedman, H. M., Maddy P., Steel, J. R., *Does mathematics need new axioms?*, The Bulletin of Symbolic Logic, **6**(4), pp. 401–446 (2000).

[90] Feng, Q., Magidor, M., Woodin, W. H., *Universally Baire sets of reals*; in **Set theory of the continuum**, pp. 203–242, Springer, Berlin (1992).

[91] Field, H., **Science without Numbers**, Princeton University Press, New Jersey, 1980; 2nd Edition by Oxford University Press, Oxford (2016).

[92] Field, H., *On conservativeness and incompleteness*, Journal of Philosophy, **82**, pp. 239–260 (1985).

[93] Field, H., **Realism, Mathematics and Modality**, Basil Blackwell Ltd., Oxford (1989).

[94] Fletcher, P., *Brouwer's weak counterexamples and the creative subject: A critical survey*, Journal of Philosophical Logic, **49**, pp. 1111–1157 (2020).

[95] Forster, T., *The axiom of choice and inference to the best explanation*, Logique et Analyse, **49**(194), pp. 191–197 (2006).

[96] Fraenkel, A., Bar-Hillel, Y., Levy, A., **Foundations of Set Theory**, 2nd rev. edn., North Holland, Amsterdam (1973).

[97] Frege, G., **Begriffsschrift, eine der arithmetischen nachgebildete Formelsprache des reinen Denkens**, Halle a. S.: Louis Nebert, 1879; **Concept Script, a formal language of pure thought modelled upon that of arithmetic**, trans. by S. Bauer-Mengelberg, J. van Heijenoort (ed.), in **From Frege to Gödel: A Source Book in Mathematical Logic**, pp. 1879–1931, Harvard University Press, Cambridge (1967).

[98] Frege, G., **Foundations of Arithmetic**, 1884. Trans. by J. Austin, 2nd edition, Harper, New York (1960).

[99] Frege, G., **Grundgesetze der Arithmetik I**, Olms Hildesheim, Hildesheim (1893).

[100] Frege, G., **Grundgesetze der Arithmetik II**, Olms Hildesheim, Hildesheim (1903).

[101] Frege, G., **Conceptual Notation and Related Articles**, Terrell Ward Bynum (ed.), Clarendon Press, Oxford (1972).

[102] Frege, G., **Philosophical and Mathematical Correspondence**, Gottfried Gabriel, Hans Hermes, Friedrich Kambartel, Christian Thiel, Albert Veraart, Brian McGuinness, and Hans Kaal (eds.), Blackwell Publishers, Oxford (1980).

[103] Freiling, C., *Axioms of symmetry: Throwing darts at the real number line*, The Journal of Symbolic Logic, **51**(1), pp. 190–200 (1986).

[104] Friedman, S. D., *Fine structure and class forcing*, Bulletin of Symbolic Logic, **7**(4), pp. 522–525 (2001).

[105] Friedman, S. D., *Internal consistency and the inner model hypothesis*, The Bulletin of Symbolic Logic, **12**(4), pp. 591–600 (2006).

[106] Friedman, S. D., Welch, P., Woodin, W. H., *On the consistency strength of the inner model hypothesis*, The Journal of Symbolic Logic, **73**(2), pp. 391–400 (2008).

[107] Friend, M., **Pluralism in Mathematics: A New Position in Philosophy of Mathematics**, Springer, Berlin (2014).

[108] Galilei, G., **Discorsi e Dimostrazioni Matematiche Intorno a Due Nuove Scienze** Elsevier, Leiden (1638).

[109] Goldblatt, R., **Topoi: The Categorial Analysis of Logic**, Dover Books, New York (1979).

[110] Goldman, A., *A causal theory of knowing*, Journal of Philosophy, **64**, pp. 355–372 (1967).

[111] Goodman, N., **Problems and Projects**, Bobbs-Merrill, Indianapolis (1972).

[112] Goodman, N., Myhill, J., *Choice implies excluded middle*, mathematical Logic Quarterly, **24**, p. 461 (1978).

[113] Goodman, N., Quine, W. V. O., *Steps toward a constructive nominalism*, Journal of Symbolic Logic, **12**, pp. 105–122 (1947).

[114] Gödel, K., *Über die Vollständigkeit des Logikkalküls*, PhD Thesis, University of Vienna (1929).

[115] Gödel, K., *Über formal unentscheidbare Sätze der Principia Mathematica und verwandter Systeme I*, Monatsch Math. und Phys., **38**, pp. 173–198 (1931).

[116] Gödel, K., *Consistency-proof for the Generalized Continuum Hypothesis*, Proceedings of the National Academy of Sciences of the United States, **25**, pp. 220–224 (1939).

[117] Gödel, K., *Undecidable Diophantine Propositions*, 193?; in Gödel, K. **Collected Works**, Volume III, Unpublished Essays and Lectures, S. Feferman (ed.), pp. 164–175, Oxford University Press (1995).

[118] Gödel, K., *Russell's Mathematical Logic*, in **The philosophy of Bertrand Russell**, Paul Arthur Schilpp (ed.), Northwestern University, Evanston and Chicago, pp. 123–153 (1944). Reprined in Benacerraf and Putnam (1983), pp. 447–469.

[119] Gödel, K., *Remarks before the Princeton bicentinnial conference on problems in mathematics*, 1946; in Gödel, K., **Collected Works**, Volume II, S. Feferman (ed.), pp. 150–153, Oxford University Press (1990).

[120] Gödel, K., *What is cantor's continuum problem?*, The American Mathematical Monthly, **54**(9), pp. 515–525 (1947). Reprinted in Gödel, K., **Collected Works**, Volume II, S. Feferman (ed.), pp. 176–188, Oxford University Press (1990).

[121] Gödel, K., *Some basic theorems on the foundations of mathematics and their implications*, 1951; in Gödel, K., **Collected Works**, Volume III, Unpublished Essays and Lectures, S. Feferman (ed.), pp. 304–323, Oxford University Press, Oxford (1995).

[122] Gödel, K., *The modern development of the foundations of mathematics in the light of philosophy*, 1961; in Gödel, K., **Collected Works**, Volume III, Unpublished Essays and Lectures, S. Feferman (ed.), pp. 375–387, Oxford University Press, Oxford (1995).

[123] Gödel, K., *What is Cantor's Continuum Problem?*, Revised version of the 1947 paper, 1964; in K. Gödel, **Collected Works**, Volume II, Solomon Feferman (ed.), pp. 254–270, Oxford University Press, New York (1990). Also reprinted in Benacerraf and Putnam (1983), pp. 470–485.

[124] Grover, L. K., *A fast quantum mechanical algorithm for database search*, STOC '96 Proceedings, 28th Annual ACM Symposium on the Theory of Computing, pp. 212–216 (1996).

[125] Hagar, A., Korolev, A., *Quantum Hypercomputation—Hype or Computation?*, Philosophy of Science, **74**(3), pp. 347–363 (2007).

[126] Halbeisen, L., Shelah, S., *Relations between some cardinals in the absence of the Axiom of Choice*, Bulletin of Symbolic Logic, **7**(2), pp. 237–261 (2001).

[127] Hamilton, A. G., **Logic for Mathematicians**, Cambridge University Press, Cambridge (1988).

[128] Hamkins, J. D., *Extensions with the approximation and cover properties have no new large cardinals*, Fundamenta Mathematicae, **180**(3), pp. 257–277 (2003).

[129] Hamkins, J. D., *The set-theoretic multiverse*, Review of Symbolic Logic, **5**, pp. 416–449 (2012).

[130] Hamkins, J.D., *A multiverse perspective on the axiom of constructibility*; in **Infinity and Truth**, Qi Feng, C. T. Chong, W. H. Woodin and T. A. Slaman (eds.), vol. 25, pp. 25–45, World Scitific Publishing, Hackensack, NJ (2014).

[131] Hamkins, J. D., *Is the dream solution of the continuum hypothesis attainable?*, Notre Dame J. Formal Logic, **56**(1), pp. 135–145 (2015).

[132] Hamkins, J. D., **Lectures on the Philosophy of Mathematics**, MIT Press, Cambridge (2021).

[133] Hamkins, J. D., Lewis, A., *Infinite time Turing machines*, Journal of Symbolic Logic, **65**(2), pp. 567–604 (2000).

[134] Hamkins, J. D., Welch, P. D., $P^f \neq NP^f$ *for allmost all* f, Mathematical Logic Quarterly, **49**(5), pp. 536–540 (2003).

[135] Hardy, G. H., **A Mathematician's Apology**, Cambridge University Press, Cambridge (1940).

[136] Hardy, J., *Is Yablo's paradox Liar-like?*, Analysis, **55**(3), pp. 197–198 (1995).

[137] Hawking, S. W., *Particle creation by black holes*, Communications in Mathematical Physics, **43**, pp. 199–220 (1975).

[138] Hedman, S., **A First Course in Logic: An Introduction to Model Theory, Proof Theory, Computability, and Complexity**, Oxford University Press, New York (2004).

[139] Heijenoort, J van, **From Frege to Gödel: A Source Book in Mathematical Logic**, Harvard University Press, Cambridge (1967).

[140] Heine, E., *Die elemente der funktionslehre*, Crelle Journal für die reine und angewandte Mathematik, **74**, pp. 172–188 (1872).

[141] Hellman, G., **Mathematics Without Numbers**, Oxford University Press, Oxford (1989).

[142] Hellman, G., Bell, J. L., *Pluralism and the foundations of mathematics*; in J. L. Bell (ed.), **Minnesota Studies in the Philosophy of Science Series**, Vol 19, pp. 64–79, University of Minnesota Press, Minneapolis (2006).

[143] Henkin, L., *The completeness of the first-order functional calculus*, Journal of Symbolic Logic, **14**, pp. 159–166 (1949).

[144] Heyting, A., **Intuitionism An Introduction**, North Holland, Amsterdam (1956).

[145] Heyting, A., *The Intuitionistic Foundations of Mathematics*, 1931; in Benacerraf and Putnam (1983), pp. 52–61.

[146] Hilbert D., **Grundlagen der Geometrie**, Leipzig, Teubner, 1898; **The Foundations of Geometry**, trans. by E. J. Townsend, La Salle, III, Open Court, Illinois (1950).

[147] Hilbert, D., *Über das Unendliche*, Mathematische Annalen, 95, pp. 161–190, 1925; *On the infinite*, in Benacerraf and Putnam (1983), pp. 183–201.

[148] Hilbert, D., Ackermann, W., **Grundzüge der theoretischen Logik**, Berlin, Springer (1928).

[149] Hilbert D., **Gesammelte Abhandlungen**, 1935; Springer Verlag (1970).

[150] Hogarth, M., *Does general relativity allow an observer to view an eternity in a finite time?*, Foundations of Physics Letters, 5, pp. 173–181 (1992).

[151] Horsten, L., Welch, P., **Gödel's Disjunction: The scope and limits of mathematical knowledge**, Oxford University Press, Oxford (2016).

[152] Hume, D., **A Treatise of Human Nature**, 1739; reprinted by D. F. Norton, M. J. Norton (eds.), Oxford University Press, Oxford (2000).

[153] Hogarth, M., *Non-Turing computers and non-Turing computability*, PSA: Proceedings of the Biennial Meeting of the Philosophy of Science Association 1, pp. 126–138 (1994).

[154] Irvine, A.D., *Epistemic logicism and Russell's regressive method*, Philosophical Studies, **55**, pp. 303–327 (1989).

[155] Jech, T.: **Set Theory**, Springer, Berlin (2003).

[156] Jensen, R., *The fine structure of the constructible hierarchy*, Annals of Mathematical Logic, **4**, pp. 229–308 (1972).

[157] Jensen, R., *Inner models and large cardinals*, The Bulletin of Symbolic Logic, **1**(4), pp. 393–407 (1995).

[158] Kanamori, A., **The Higher Infinite**, 2nd edition, Springer-Verlag, Berlin (2003).

[159] Kanamori, A., Foreman, M. (eds.), **Handbook of Set Theory**, Springer, Berlin (2010).

[160] Kant, I., *Concerning the ultimate ground of the differentiation of directions in space*, 1768; in **Theoretical Philosophy**, pp. 1755–1770, trans. by D. Walford and R. Meerbote, Cambridge University Press, New York (1992).

[161] Kant, I., **Critique of Pure Reason**, 1781; trans. by Paul Guyer and Allen W. Wood, Cambridge University Press, New York (1998).

[162] Kant, I., **Prolegomena To Any Future Metaphysics That Can Qualify as a Science**, 1783; trans. by Paul Carus, Open Court Publishing Company, Chicago (1997).

[163] Ketland, J., *Yablo's paradox and ω-inconsistency*, Synthese, **145**, pp. 295–302 (2005).

[164] Kleene, S. C., **Introduction to Meta-Mathematics**, Ishi Press International, New York (1950).

[165] Kleene, S. C., **Mathematical Logic**, Dover Books on Mathematics, New York (1967).

[166] Kline, M., **Mathematical Thought from Ancient to Modern Times**, Oxford University Press (1972).

[167] Koellner P., *On the question of absolute undecidability*, Philosophia Mathematica, **14**(2), pp. 153–188 (2006).

[168] Koellner, P., *Truth In Mathematics: The Question of Pluralism*; in **New Waves in the Philosophy of Mathematics**, O. Bueno and Ø., Linnebo (ed.), pp. 80–116, Palgrave Macmillan, London/New York (2009).

[169] Koellner, P., *The Continuum Hypothesis*, in Zalta, E. N. (ed.), **The Stanford Encyclopedia of Philosophy** (2013).

[170] Koellner, P., *Hamkins on the Multiverse*, preprint, available at `http://logic.harvard.edu/EFI_Hamkins_Comments.pdf` (2013).

[171] Koepke, P., *Turing computation on ordinals*, Bulletin of Symbolic Logic, **11**, pp. 377–397 (2005).

[172] König, J., *Zum Kontinuum-Problem*, in Adolf Krazer (ed.), **Verhandlungen des dritten Internationalen Mathematiker-Kongresses in Heidelberg**, Vom 8, Bis 13, pp. 144–147 Teubner: Leipzig (1904).

[173] Kripke, S., *The Church-Turing 'Thesis' as a special corollary of Gödel's Completeness Theorem*"; in B. J. Copeland, C. Posy, and O. Shagrir (eds), **Computability: Turing, Gödel, Church and Beyond**, pp. 77–104, MIT Press, Cambridge, MA (2013).

[174] Kruse, A. H., *Grothendieck universes and the super-complete models of Sheperdson*, Composito Mathematica, **17**, pp. 96–101 (1965).

[175] Kunen, K., **Set Theory An Introduction to Independence Proofs**, Elsevier, Amsterdam (1983).

[176] Lear, J., *Aristotle's philosophy of mathematics*, The Philosophical Review, **91**(2), pp. 161–192 (1982).

[177] Lévy, A., Solovay, R. M., *Measurable cardinals and the continuum Hypothesis*, Israel Journal of Mathematics, **5**(4), pp. 234–248 (1967).

[178] Linnebo, Ø., **Philosophy of Mathematics**, Princeton University Press, New Jersey (2017).

[179] Linnebo, Ø., **Thin Objects: An Abstractionist Account**, Oxford University Press, Oxford (2018).

[180] Linnebo, Ø., Shapiro, S., *Actual and potential infinity*, Noûs, **53**(1), pp. 160–191 (2019).

[181] Lucas, J. R., *Minds, machines and Gödel*, Philosophy, **36**, pp. 112–127 (1961).

[182] Maddy, P., *Believing the axioms. I*, Journal of Symbolic Logic, **53**(2), pp. 481–511 (1988)

[183] Maddy, P., **Realism in Mathematics**, Oxford University Press, Oxford (1990).

[184] Maddy, P., *Physicalistic Platonism*, in **A. D. Irvine (ed.), Physicalism in Mathematics**, pp. 259–289, Kluwer Academic Publishers, Dordrecht (1990).

[185] Maddy, P., *Indispensability and practice*, The Journal of Philosophy, **89**(6), pp. 275–289 (1992).

[186] Maddy, P., *Does V equal L?*, Journal of Symbolic Logic, **58**(1), pp. 15–41 (1993).

[187] Maddy, P., *Naturalism and ontology*, Philosophia Mathematica, **3**(3), pp. 248–270 (1995).

[188] Maddy, P., **Naturalism in Mathematics**, Oxford University Press, Oxford (1997).

[189] Maddy, P., **Second Philosophy**, Oxford University Press, Oxford (2007).

[190] Maddy, P., **Defending the Axioms: On the Philosophical Foundations of Set Theory**, Oxford University Press, New York (2011).

[191] Maddy, P., *Set-theoretic foundations*, Contemporary Mathematics, **690**, pp. 289–322 (2017).

[192] Magidor, M., *Some Set Theories Are More Equal*, unpublished paper, preprint available at http://logic.harvard.edu/EFI_Magidor.pdf (2012).

[193] Malament, D., *Review of Hartry Field, Science without Numbers*, Journal of Philosophy, **19**, pp. 523–534 (1982).

[194] Martin, D. A., *Hilbert's first problem: The Continuum Hypothesis*, in F. Browder (ed.), **Mathematical Developments Arising from Hilbert's Problems**, Vol. 28 of Proceedings of Symposia in Pure Mathematics, American Mathematical Society, Providence, pp. 81–92 (1976).

[195] Matiyasevich, Y. V., **Hilbert's Tenth Problem**, MIT Press, Cambridge, Massachusetts (1993).

[196] MacLane, S., **Mathematics: Form and Function**, Springer, New York (1986).

[197] McLarty, C., *What does it take to prove Fermat's Last Theorem? Grothendieck and the logic of number theory*, Bulletin of Symbolic Logic, **16**(3), pp. 359–377 (2010).

[198] Melia, J., *On what there's not*, Analysis, **55**(4), pp. 223–229 (1995).

[199] Mill, J. S., **A System of Logic, Ratiocinative and Inductive**, 1843; reprinted by University Press of the Pacific, Honolulu (2002).

[200] Moore, G. H., **Zermelo's Axiom of Choice**, New York, Springer (1982).

[201] Moschovakis, Y., **Descriptive Set Theory**, North-Holland, Amsterdam (1980).

[202] Mycielski, J., Steinhaus, H., *A mathematical axiom contradicting the axiom of choice*, Bulletin de l'Académie Polonaise des Sciences. Série des Sciences Mathématiques, Astronomiques et Physiques, **10**, pp. 1–3 (1962).

[203] Nagel, E., Newman, J. R., (ed.) Hofstadler, D. R., **Gödel's Proof**, NYU Press (2008).

[204] Neumann, J. von, *Eine axiomatiserung der Mengenlehre*, Journal für die reine und angewandte Mathematik, **154**, pp. 219–240, 1925; in van Heijenoort (1967), pp. 393–413.

[205] Neumann, J. von, *Die formalistische Grundlegung der Mathematik*, Erkentniss, **2**, pp. 116–121, 1931; trans. as *The formalist foundations of mathematics*, in Benacerraf and Putnam (1983), pp. 62–65.

[206] NG, Y. J. and Dam, H. van, *Spacetime foam, holographic principle, and black hole quantum computers*, International Journal of Modern Phys., A20, pp. 1328–1335 (2005).

[207] Nies, A., **Computability and Randomness**, Oxford University Press, Oxford (2009).

[208] Parsons, C., *The structuralist view of mathematical objects*, Synthese, **85**, pp. 303–346 (1990).

[209] Pedeferri, A., Friend, M., *Are mathematicians better described as formalists or pluralists?*, Logic and Philosophy of Science Vol. IX, **1**, pp. 173–180 (2011).

[210] Penrose, R., **The Emperor's New Mind: Concerning Computers, Minds and The Laws of Physics**, Oxford University Press, Oxford (1989).

[211] Piccinini, G., *Computationalism, the Church-Turing Thesis and the Church-Turing fallacy*, Synthese, **154**, pp. 97–120 (2007).

[212] Plato, *Euthypro*, in W.K.J. Guthrie, **A History of Greek Philosophy: Plato The Man and His Dialogues (Earlier Period)**, Cambridge University Press, Cambridge (1995).

[213] Plato, *Meno*, 2nd edition, trans. by G. M. A. Grube, Heckett Publishing Company, Inc., Indianapolis (1980).

[214] Plato, *Phaedo*, 2nd edition, trans. by G. M. A. Grube, Heckett Publishing Company, Inc., Indianapolis (1977).

[215] Plato, *Philebus*, 2nd edition, trans. by D. Frede, Heckett Publishing Company, Inc., Indianapolis (1993).

[216] Plato, *Republic*, 2nd edition, trans. by G. M. A. Grube, revised by C. D. C. Reeve, Heckett Publishing Company, Inc., Indianapolis (1992).

[217] Plato, *Sophist*, trans. by N. P. White, Heckett Publishing Company, Inc., Indianapolis (1993).

[218] Plato, *Statesman*, trans. by C. J. Rowe, Heckett Publishing Company, Inc., Indianapolis (1999).

[219] Playfair, J., **Elements of Geometry**, W. E. Dean (1846) W. E. Dean, New York.

[220] Poincaré, H., **La Science et l'Hypothèse**, 1902; trans. by William John Greenstreet, The Walter Scott Publishing, Co., Ltd, New York (1905).

[221] Poincaré, H., *The Logic of Infinity*, 1909; in Poincaré (1913), pp. 45–64.

[222] Poincaré, H., **Mathematics and Science**: *Last Essays* (1913), trans. John Bolduc, Dover Publications, New York (1963).

[223] Priest, G., *Yablo's paradox*, Analysis, **57**(4), pp. 236–242 (1997).

[224] Priest, G., *Mathematical pluralism*, Logic Journal of the IGPL, **21**(1), pp. 4–13 (2013).

[225] Putnam, H., *Mathematics without foundations*, Journal of Philosophy, **64**, pp. 5–22, 1967; reprinted in Benacerraf and Putnam (1983), pp. 295–311.

[226] Putnam, H., **Philosophy of Logic**, Harper Torchbooks, New York (1971).

[227] Putnam, H., *What is mathematical truth?*; in his **Mathematics, Matter and Method: Philosophical papers**, Vol. 1, pp. 60–78, Cambridge University Press, Cambridge (1975).

[228] Putnam, H., *Models and reality*, The Journal of Philosophy, **45**(3), pp. 464–482 (1980).

[229] Quine, W. V. O., *Truth by convention*, in **Philosophical Essays for Alfred North Whitehead**, O. H. Lee (ed.), pp. 90–124, Longmans, New York, 1936; reprinted in Benacerraf and Putnam (1983), pp. 329–354.

[230] Quine, W. V. O., *New foundations for mathematical logic*, American Mathematical Monthly, **44**, pp. 70–80 (1937).

[231] Quine, W. V. O., *Identity, ostension, and hypostasis*, The Journal of Philosophy, **47**(22), pp. 621–633 (1950).

[232] Quine, W. V. O., *Two dogmas of empiricism*, The Philosophical Review, **60**, pp. 20–43 (1951).

[233] Quine, W. V. O., *On Carnap's views on ontology*, Philosophical Studies: An International Journal for Philosophy in the Analytic Tradition, **2**(5), pp. 65–72 (1951).

[234] Quine, W. V. O., **Word and Object**, MIT Press, Cambridge (1960).

[235] Quine, W. V. O., **Ontological Relativity and Other Essays**, Columbia University Press, New York (1969).

[236] Quine, W. V. O., *Five milestones of empiricism* 1975; in Quine, pp. 67–72, The Belknap Press of Harvard University Press, Cambridge (1981).

[237] Quine, W. V. O., **Theories and Things**, Harvard University Press, Cambridge (1981).

[238] Quine, W. V. O., **Methods of Logic**, 4th edition, Harvard University Press, Cambridge (1982).

[239] Rahman, F., **Philosophy of Mullā Ṣadrā**, State University of New York Press, New York (1975).

[240] Reinhardt, W. N., *Topics in the metamathematics of set theory*, PhD thesis, University of California, Berkeley (1967).

[241] Reinhardt, W. N., *Remarks on reflection principles, large cardinals, and elementary embeddings*, Axiomatic set theory, Proc. Sympos. Pure Math., XIII, Part II, Providence, R. I., Amer. Math. Soc., pp. 189–205 (1974).

[242] Resnik, M. D., *On Skolem's paradox*, The Journal of Philosophy, **63**(15), pp. 425–438 (1966).

[243] Resnik, M. D., **Frege and the Philosophy of Mathematics**, Cornell University Press, Ithaca, NY (1980).

[244] Resnik, M. D., *Mathematics as a science of patterns: Ontology and reference*, Nous, **15**, pp. 529–550 (1981).

[245] Roland, J. W., *Naturalism and mathematics: Some problems*, in K. J. Clark (ed.), **The Blackwell Companion to Naturalism**, pp. 289–304, John Wiley & Sons, Inc. West Sussex (2016).

[246] Rosser, J. B., *Extensions of some theorems of Gödel and Church*, Journal of Symbolic Logic, **1**, pp. 87–91 (1936).

[247] Russell, B., *Recent work on the principles of mathematics*, The International Monthly, **4**, pp. 83–101 (1901).

[248] Russell, B., **The Principles of Mathematics**, 1903; Reprinted by W. W. Norton & Company, New York (1996).

[249] Russell, B., *The regressive method of discovering the premises of mathematics*, Cambridge Mathematics Club, 1907; in Russell, B., **Collected Works of Bertrand Russell, Vol 5: Toward Principia Mathematica, 1905–08**, G. H. Moore (ed.), pp. 571–580, Routledge, New York (2014).

[250] Russell, B., **The Problems of Philosophy**, 1912; reprinted by Oxford University Press, Oxford (1967).

[251] Russell, B., **Introduction to Mathematical Philosophy**, Allen and Unwin, London (1919).

[252] Russell, B., **My Philosophical Development**, Simon and Schuster, Inc., New York (1959).

[253] Schatz, J. R., *Axiom selection by maximization: V = Ultimate L vs forcing axioms*, PhD dissertation, University of California at Irvine (2019).

[254] Schechter, E., **Handbook of Analysis and Its Foundations**, Academic Press, Cambridge (1997).

[255] Schindler, R., *P ≠ NP for infinite time Turing machines*, Monatshefte für Mathematik, **139**, pp. 335–340 (2003).

[256] Scott, D. S., *Measurable cardinals and constructible sets*, Bulletin de l'Académie Polonaise des Sciences, **9**, pp. 521–524 (1961).

[257] Shapiro, S., *Conservativeness and incompleteness*, Journal of Philosophy, **80**, pp. 521–531 (1983).

[258] Shapiro, S., **Foundations without Foundationalism: A Case for Second–Order logic**, Oxford University Press, Oxford (1991).

[259] Shapiro, S., **Philosophy of Mathematics: Structure and Ontology**, Oxford University Press, New York (1997).

[260] Shapiro, S., **Thinking about mathematics: Philosophy of mathematics**, Oxford University Press, New York (2000).

[261] Shoenfield, J. R., *Axioms of set theory*, in J. Barwise (ed.), **Handbook of Mathematical Logic**, pp. 321–344, North Holland, Amsterdam (1977).

[262] Shor, P., *Algorithms for quantum computation: Discrete log and factoring*, Proceedings of the 35th annual IEEE symposium on Foundations of Computer Science - FOCS, pp. 20–22 (1994).

[263] Sieg, W., *Calculations by man and machine: Conceptual analysis*, in W. Sieg, R. Sommer, C. Talcott (ed.), **Reflections on the Foundations of Mathematics: Essays in Honor of Solomon Feferman**, Lecture Notes in Logic, **15**, pp. 390–410, Cambridge University Press, Cambridge (2002).

[264] Sieg, W., *Church without dogma: Axioms for computability*, in B. Cooper, B. Lowe, and A. Sorbi (eds), **New Computational Paradigms: Changing Conceptions of What is Computable**, pp. 139–152, Springer Verlag, New York (2008).

[265] Sierpiński, W. *L'hypothèse généralisée du continu et l'axiome du choix*, Fundamenta Mathematicae **34**, pp. 1–5 (1947).

[266] Simpson, S. G., **Subsystems of Second Order Arithmetic: Second Edition** (Perspectives in Logic), Cambridge University Press, New York (2009).

[267] Skolem, T., *Some Remarks on Axiomatized Set Theory*, 1922; in J. van Heijenoort, **From Frege to Gödel: A Source Book in Mathematical Logic**, pp. 291–301, Harvard University Press, Cambridge (1967).

[268] Smith, P., **An Introduction to Gödel's Theorems**, Cambridge University Press, Cambridge (2007).

[269] Soare, R. I., **Recursively Enumerable Sets and Degrees**, Perspectives in Mathematical Logic, Springer-Verlag, Berlin (1987).

[270] Soare, R. I., *Computability and recursion*, Bulletin of Symbolic Logic, **2**, pp. 284–321 (1996).

[271] Soare, R. I., **Turing Computability: Theory and Applications**, Springer-Verlag, Berlin Heidelberg (2016).

[272] Solovay, R. M., Reinhardt, W. N., Kanamori, A., *Strong axioms of infinite and elementary embeddings*, Annals of Mathematical Logic, **13**, pp. 73–116 (1978).

[273] Sorensen, R., *Yablo's paradox and kindered infinite liars*, Mind, **107**, pp. 137–156 (1998).

[274] Steel, J., *An outline of inner model theory*; in Kanamori, A., Foreman M. (eds.), **Handbook of Set Theory**, vol. 3, pp. 1595–1684, Springer, Berlin (2010).

[275] Stillwell, J., **Reverse Mathematics: Proofs from the Inside Out**, Princeton University Press, New Jersey (2018).

[276] Suppes, P., **Introduction to Logic**, Litton Educational Publishing, Inc., New York (1957).

[277] Suppes, P., **Axiomatic Set Theory**, Dover Publications, Inc., New York (1972).

[278] Syropoulos, A., **Hypercomputation Computing Beyond the Church–Turing Barrier**, Springer, New York (2008).

[279] Takeuti, G., **Proof Theory**, Dover Publications, New York (2013).

[280] Tarski, A., *Der Wahrheitsbegriff in den Formaliserten Sprachen*, Studia Philosophica, **1**, pp. 261–405, 1936; trans as *The concept of truth in formalized languages*, in **Logic, Semantics and Metamathematics**, J. Corcoran (ed.), pp. 152–278, Hackett Publishing Company, Inc., Indiana (1983).

[281] Tarski, A., Mostowski, A., Robinson, R. M., **Undecidable Theories**, North-Holland Publishing Company, Amsterdam (1953).

[282] Ternullo, C., *Maddy on the multiverse*, in Deniz Sarikaya, Deborah Kant & Stefania Centrone (eds.), **Reflections on the Foundations of Mathematics**, pp. 43–78, Springer Verlag (2019).

[283] Thomae, J., **Elementare Theorie der analytischen Functionen einer complexen Veränderlichen**, 2nd Edition, Halle, Jena, Germany (1898).

[284] Thomson, J. F., *Tasks and supertasks*, Analysis, **15**(1), pp. 1–13 (1954).

[285] Tieszen, R. L., **After Gödel: Platonism and Rationalism in Mathematics and Logic**, Oxford University Press, New York (2011).

[286] Tritton, D. J., **Physical Fluid Dynamics**, 2nd Edition, Oxford University Press, Oxford (1988).

[287] Troelstra, A. S. and Schwichtenberg, H., **Basic Proof Theory** 2nd edition, **Cambridge Tracts in Theoretical Computer Science**, **43**, Cambridge University Press (2000).

[288] Turing, A. M., *On computable numbers with an application to the entscheidungsproblem*, Proceedings of the London Mathematical Society, Ser. 2, **42**, pp. 230–265 (1937).

[289] Visser, A., *Semantics and the liar paradox*, Handbook of Philosophical Logic, **4**, pp. 617–706 (1989).

[290] Wang, H., **From Mathematics to Philosophy**, Routledge & Kegan Paul, London (1974).

[291] Welch, P. D., *The length of infinite time Turing machine computations*, Bulletin of the London Mathematical Society, **32**(2), pp. 129–136 (2000).

[292] Welch, P. D., *Eventually infinite time Turing degrees: infinite time decidable reals*, Journal of Symbolic Logic, **65**(3), pp. 1193–1203 (2000).

[293] Whitehead, A. N., Russell, B., *Principia Mathematica*, Vol. 1-2-3, Cambridge University Press, Cambridge (1910–1913).

[294] Wiles, A., *Modular elliptic curves and Fermat's Last Theorem*, Annals of Mathematics, **141**(3), pp. 443–551 (1995).

[295] Williams, N. H., *On Grothendieck universes*, Composito Mathematica, **21**(1), pp. 1–3 (1969).

[296] Wittgenstein, L., **Tractatus Logico-Philosophicus** 1922; Translated by D. F. Pears and B. F. McGuinness, Routledge & Kegan Paul, London (1974).

[297] Wittgenstein, L., **Philosophical Investigations**, 1953; Translated by G. E. M. Anscombe, Basil Blackwell Ltd, Oxford (1958).

[298] Woodin, W. H., **The Axiom of Determinacy, Forcing Axioms, and the Nonstationary Ideal**, De Gruyter, Berlin (1999).

[299] Woodin, W. H., *The realm of the infinite*, in **Infinity: New Research Frontiers**, Michael Heller and W. Hugh Woodin (eds.), pp. 89–118, Cambridge University Press, Cambridge (2009).

[300] Woodin, W. H., *Suitable extender models I*, Journal of Mathematical Logic, 10, (01n02), pp. 101–339 (2010).

[301] Woodin, W. H., *Suitable extender models II: Beyond ω-huge*, Journal of Mathematical Logic, **11**(2), pp. 115–436 (2011).

[302] Woodin, W. H., *Strong Axioms of Infinity and the Search for V*, Proceedings of the International Congress of Mathematicians 2010 (ICM 2010), pp. 504–528 (2011).

[303] Woodin, W. H., Davis, J., Rodríguez, *The HOD Dichotomy*; in **Appalachian Set Theory 2006–2012**, Cummings, J., Schimmerling, E. (eds.), London Mathematical Society, Lecture Note Series 406, pp. 397–419, Cambridge University Press, Cambridge (2012).

[304] Woodin, W. H., *In search of Ultimate-L: The 19th Midrasha Mathematicae Lectures*, The Bulletin of Symbolic Logic, **23**(1), pp. 1–109 (2017).

[305] Wright, C., **Frege's Conception of Numbers as Objects**, Aberdeen University Press, Aberdeen (1983).

[306] Yablo, S., *Paradox without self-reference*, Analysis, **53**(4), pp. 251–252 (1993).

[307] Yablo, S., *The myth of the seven*; in Kalderon, M. (ed.), **Fictionalism in Metaphysics**, pp. 88–115, Oxford University Press, Oxford (2005).

[308] Young, L. C., **Lectures on the Calculus of Variations and Optimal Control Theory**, AMS Chelsea Publishing, Rhode Island (2000).

[309] Zermelo, E., *Beweis, dass jede Menge wohlgeordnet werden kann*, Mathematische Annalen, **59**(4), pp. 514–516 (1904). English trans. available in van Heijenoort (1967), pp. 139–141.

[310] Zermelo, E. *Über Grenzzahlen und Mengenbereiche: neue Untersuchungen über die Grundlagen der Mengenlehre*, Fundamenta Mathematicae, **16**, pp. 29–47 (1930).

[311] Zukav, G., **The Dancing Wu Li Masters: An Overview of the New Physics**, Bantam Books, New York (1979).

Index

Printed in the United States
by Baker & Taylor Publisher Services